ENABLING

AMERICAN

INNOVATION

History of Technology Series

Jane Morley, Series Editor

ENABLING AMERICAN INNOVATION

ENGINEERING AND THE
NATIONAL SCIENCE FOUNDATION

DIAN OLSON BELANGER

Purdue University Press
West Lafayette, Indiana

02   01   00   99   98   5   4   3   2   1

∞ The paper used in this book meets the minimum requirements of American
National Standard for Information Sciences—Permanence of Paper for Printed
Library Materials, ANSI Z39.48-1992.

Printed in the United States of America

Library of Congress Cataloging-in-Publication Data

Belanger, Dian Olson, 1941–
        Enabling American innovation : engineering and the National Science
    Foundation / Dian Olson Belanger.
                p.   cm. — (Purdue University Press history of technology series)
        Includes bibliographical references and index.
        ISBN 1-55753-111-0 (cloth : alk. paper)
        1. Engineering—United States—History. 2. National Science Founda
    tion (U.S.)—History. 3. Technology—United States—History. I. Title. II.
    Series: History of technology series (West Lafayette, Ind.)
    TA23.B45 1997
    620'.007'2073—dc21   97-28311
                                                                            CIP

Contents

This publication is based on work supported by the National Science Foundation under contract number LPA 90-02432. Any opinions, findings, and conclusions expressed herein are those of the author and do not necessarily reflect the views of the National Science Foundation.

FOREWORD

History provides us a unique perspective from which to see broad themes that recur in the present and can be projected into the future. While history does not repeat itself exactly, events and issues of the past do have a tendency to reappear, albeit in slightly different form. That is why the aphorisms, "Study the past" and "What is past is prologue"—which are carved into the entrance to the National Archives—are cited so often. History can and should be a valuable component of sound policy, particularly for public policy makers. Policy makers risk reinventing the wheel when they address contemporary problems unless they are well informed about the context surrounding earlier decisions of a similar nature. What alternatives were considered? Why were certain ones chosen? What personal and institutional forces shaped a particular policy?

This book is one of several planned volumes encompassing selected areas supported by the National Science Foundation. The goal of these books is to provide a historical analysis of Foundation policies and its support of particular disciplines. They will focus on the Foundation's management of the disciplines within the larger context of agency policy making, showing both the success and failure of particular funded programs and projects.

Dian Belanger recounts the efforts of engineers to gain standing in the National Science Foundation, a federal agency founded to promote fundamental research in all fields of science and engineering as well as in science and engineering education. Given wide-ranging agency access, Dian has mined Foundation records as well as other primary and secondary sources. She describes and interprets the ebb and flow of the fortunes of the engineering leadership in its attempt to share in the management and policy making activities of the Foundation. Running through the book is a pervasive theme: How we must recognize the differences and unique strengths of both science and engineering and how engineering has not been as apparent in the process. This part of the larger drama is significant to historians, policy analysts, and others interested in understanding the place of both science and technology in society. But the subject is not meant for these people alone. Dian's unfolding story is particularly instructive for today's policy makers, for it provides the necessary historical perspective to craft decisions for the future. She has accomplished the goal of the project with a story line that is engaging and worthy of study.

—George T. Mazuzan
National Science Foundation Historian

ACKNOWLEDGMENTS

My colleague, historian Joan M. Zenzen, made possible the timely completion of this study. She pursued a great deal of the research, organized and assembled the document files, prepared a comprehensive working bibliography and synopses of oral history interviews, and produced summaries and analyses of various topics. She generated charts and graphs, formatted the endnotes, and oversaw the final preparation of the manuscript, including careful proofreading. She did all this with meticulousness, solid scholarship, good judgment, and a willing spirit.

NSF historian George T. Mazuzan, who directed this project for the National Science Foundation, was ever helpful and supportive. He provided intellectual guidance, leads to sources, introductions to key people and institutions, and invaluable commentary on the manuscript as it progressed. He seemed to know just when to give advice, when to leave us alone, when to offer a reassuring word. He made our professional association a pleasure. His continuing faith and helpfulness through publication has been appreciated beyond measure.

Indeed, everyone at NSF cooperated with us fully, providing access to document collections, information, and professional perspective. At the risk of an unintentional omission, I thank the following individuals, hoping I have not misused or misinterpreted their words. Several key players, having varying relationships with the Foundation, agreed to taped interviews, which provided uniquely useful insight: former Connecticut representative Emilio Q. Daddario, who led the legislative effort to expand NSF's original charter to include applied research; NSF director Erich Bloch, who was then about to retire; former National Science Board chair and engineering champion Eric Walker; former earthquake engineering program manager Michael Gaus; former RANN chief and assistant director for Engineering Jack Sanderson; former National Science Board chair Lewis Branscomb; former assistant director for Engineering Nam Suh; and NSF Senior Engineering advisor William Butcher.

Many other former and current NSF engineering leaders also participated in untaped interviews detailing their respective roles: Oscar Dillon, "rotating" mechanics program director; Dale Draper, budget officer and staff associate for engineering infrastructure development; Carl Hall, former acting assistant director for engineering, who also authorized the initial funding for this project; Paul Herer, planning and resources officer; John Lehmann, former engineering officer, now computer sciences deputy division director; Lynn Preston, Engineering Research Centers deputy director; John Scalzi, structures and building systems program director; Elias Schutzman, former RANN official; and Joel Widder, congressional affairs section head. Hall, Herer, Preston, and Widder also provided valuable documentary materials.

In addition, the staff of the National Science Board office extended us courtesies and generous access to files, especially Vicki DeHullu, Catherine Flynn, and Andrea McIntyre, as did NSF librarian Florence Heckman. Elizabeth Vanderputten designed and ran computer programs to retrieve engineering grant files, although we all learned that the awesome power of the computer could not overcome the problem of limited input. Thomas Cooley and Joyce Latham helped with budgetary and audio-visual materials, respectively. The budget office's John Perhonis, while on temporary assignment, read and critiqued early chapters, while former NSF historian J. Merton England, Mazuzan, Butcher, and Hall all reviewed the entire work, to its certain benefit. Joseph Bordogna, NSF acting deputy director and former assistant director for engineering, provided both funding and crucial approvals for moving the publication forward.

David Hoeppner, David Pershing, and others of the University of Utah provided published material and observations by telephone on the Advanced Combustion Engineering Research Center; Bruce Woodson gave me an informative tour by telephone of the MIT Biotechnology Process Engineering Center. Finite element analysis expert Ray Clough of the University of California, Berkeley, shared recollections and papers on the technique he helped pioneer. Engineering consultant Emory Kemp contributed to my understanding of a number of technical areas and perceptively commented on the manuscript. Engineer Gordon Millar also reviewed early chapters.

Research at History Associates Incorporated, where this book was written, was invariably a team effort, and this project enjoyed the research support of Andrew Kirkendall and Gregory Wright, who also prepared a preliminary chronology, as well as that of Joan Zenzen.

Zenzen's spouse Stuart Weinstein took her researched figures and produced the original computer-generated graphs. Senior vice president Richard Hewlett offered ongoing support and kind but penetrating analyses of the chapters as they were completed and again when the manuscript was revised. Support staff who helped all along the way and perfected the final manuscript included Gail Mathews, Judy Bressler, Maryanne Glover, and Renee Sabin. HAI president Philip Cantelon assigned this significant and engaging history to me, and Ruth Dudgeon provided continuing administrative guidance, for which I remain appreciative.

I must thank series editor Jane Morley, who recommended that Purdue University Press publish this book; Martin Reuss, whose insightful commentary helped me improve its truth and presentation; and Bruce Seely, who offered important suggestions in critiques of earlier drafts. It has been a pleasure working with Margaret Hunt, Purdue's managing editor, whose cheerful prodding and observant, sensitive questioning firmly but gently brought the work to its final form.

Being able to include technical subjects of particular interest to my son Marcel Belanger, a mechanical engineer, and my husband Brian Belanger, an electrical engineer and my most frequently consulted adviser, added to my personal satisfaction. My daughter Lia Belanger Book, then a student, now a veterinarian, was always there with respect and encouragement for a woman's work. Their love and support nurtured me and therefore also this history.

PROLOGUE

ENGINEERING

An American Tradition

> [Engineers' and technologists'] strategic position between
> the endless frontiers of new knowledge on the one hand, and
> the equipment proved by success on the other, enables them
> to act as the most effective revolutionists of our time.
> —W. H. G. Armytage, *A Social History of Engineering*, 324

Engineering and science are often spoken of in the same breath, as almost interchangeable parts of a great national striving for technical excellence, achievement, and prestige. But science and engineering have also lived in a sort of antithetical intellectual rivalry, often argued—though too narrowly—in value-laden expressions of pure science versus applied, with pure (or basic) science enjoying the higher social rank. Today engineers, associated with applied science—although again too narrowly—assert that their unique and vital contribution must be defined and judged on its own terms. This study traces engineers' struggle to win intellectual, financial, and organizational recognition within the National Science Foundation (NSF), a federal agency formed in 1950 to support basic

1

scientific research. It reflects a dynamic, often tense relationship that not only pervades NSF's history but is as old as the New World.

Engineering—if we define it as the application of knowledge to produce things and processes to solve practical problems for human betterment—has ever been with us. European pioneers, who found in America a huge, untamed, largely uninhabited continent where natural resources were abundant but human resources scarce, soon learned the value of technology. With necessity their imperative and the skills of a centuries-old craft tradition their inheritance, frontier technologists, some of them armed with a little mathematics, often self-taught, adapted or devised tools and mechanical contrivances to ease manual labor and understand their environment. They designed lighter, better-balanced felling axes to clear vast forests, carved hardwood clock movements in the absence of brass. They fashioned instruments to observe, measure, and record everyday phenomena such as land area, topography, and relative position.

From the ranks of these hands-on practitioners came the nation's first "engineers," interested in *doing,* in designing functional—not perfect—solutions to ubiquitous practical problems. By contrast, scientists, who pursued knowledge for the sake of knowing, to understand nature as it is, seeking "generality and exactitude," were necessarily gentlemen of means and leisure—rare indeed in youthful America. That the term "Yankee ingenuity," which fittingly describes early America's zesty, improvisational enterprise, derives from the same root as "engineer," speaks to the importance of engineering in the new land, however informally the latter term might have been applied at the time.[1] Thus, in 1787, ingenious Oliver Evans designed and patented an automatic flour mill that moved grain vertically by an elevator consisting of a series of buckets on an endless belt and horizontally with an Archimedean screw device, later powered by steam engines of his own design. It worked. It saved labor and, therefore, cost. It was good (enough). No one expected it to last forever, or thought it should. Materials—especially wood—to create a new one seemed limitless. Building for permanence and beauty was an old-country attitude discarded in the new, where serviceable, economical, and fast carried a higher premium.[2]

While a recognizable profession of engineering did not emerge in the United States until the nineteenth century, these early, small-scale "engineering" accomplishments bore the mark of British attitude and practice. As Britain's spectacular industrial and commercial growth in the eighteenth century offered employment opportunities for workers with technical expertise, "civilian" (not military—later called "civil") engineers developed an identity. They were largely apprentice-trained and relied on practical, empirical problem-solving techniques. Meanwhile, as France consolidated its monarchical power and pursued its military ambition in the seventeenth and eighteenth centuries, it created two corps of engineers, which the government trained to design artillery and fortifications for its growing, increasingly mechanized permanent army and to build a network of roads, bridges, and other improvements. By the mid-eighteenth century, special schools provided this training, with emphasis on scientific and mathematical principles. By 1794, with the new Ecole polytechnique institutionalizing the shift to an academic approach, engineering was becoming a well-established occupation in France, its focus primarily military. Nascent American engineering borrowed from and "blended" these European engineering traditions with "indigenous elements" to develop a unique New-World engineering style.[3]

When the Revolutionary War created a great demand for engineers to plan and build military works, General Washington found the finest in politically sympathetic France, and "perhaps two dozen" trained French engineers helped lead his inexperienced corps. In 1802, Congress finally acted on Washington's 1778 general order pleading for a school of engineering and established the United States Military Academy at West Point, modeled on the Ecole polytechnique, which early on trained its engineers for work "of a public as well as military nature." The need was clear. Daniel Hovey Calhoun, applying a standard excluding the self-taught tinkers, estimated that, "all allowances made, the number of engineers or quasi-engineers probably never averaged more than two per state in any year up to 1816," about thirty in all. The first West Point graduate working in civilian engineering appeared in 1818; ten years later there were 10; by 1838, 107. These alumni, having received formal instruction in cartography, surveying, and instrument making and rigorous tutoring in mathematics and the sciences, especially physics, thus became the nation's first domestically trained professional engineers. They conducted government surveys and built roads, canals, bridges, and railroads—all projects critical for the growing, westward-moving nation.[4]

A considerable portion of U.S. scientific and technical knowledge continued to be imported, however. The French-descended, British-born, German-trained Benjamin Henry Latrobe, for example, who arrived in 1790, became an eminent American engineer and architect during the first two decades of the nineteenth century. While his classical revival designs for the Capitol and White House reconstructions have secured his immortality, his engineering feats were arguably more significant. Latrobe's Philadelphia Waterworks project was the first comprehensive urban water system in the United States. Using two steam engines, he pumped water from the Schuylkill River to an elevated reservoir, distributing it thence via wooden pipes (though he would have preferred a more lasting material). He also contributed significantly to road, canal, railroad, and bridge projects. The cosmopolitan, learned Latrobe admired French engineers, who (over the British) had "generally, I may say *always,* a better education in the *Science* of their profession."[5]

By contrast, most nineteenth-century American engineers were trained by apprenticeship or in practice, in the manner of the British craft tradition. Native-born John B. Jervis, who took his formal education in "occasional sessions at the common school," was typical of the American engineer of the time, although his prodigious achievements—astonishing in their number, variety, and quality—placed him among the giants. He learned engineering by helping to build the Erie Canal, which Elting Morison called the "first—and quite possibly the best—school of general engineering in this country." Jervis, who began in 1817 as a tree-cutting laborer, mastered every aspect of canal planning and construction; he observed intently, practiced techniques, read the available literature, and welcomed the mentorship of "solicitous superiors,"— although they were students of the school of experience, too. In a fifty-year career, he built the Delaware and Hudson Canal, with an inclined railroad at one end. He ordered the country's first locomotive from England; when the rigid machine bounced and took curves unsatisfactorily on the crude American track, he designed a free-swinging, four-wheel pilot truck that quickly became the American paradigm. He designed and built the forty-mile Croton Aqueduct, including a daring dam, reservoir, and bridge, to bring water into New York City. Despite the disastrous failure of an earthen embankment during construction, the system opened and endured as a "Herculean" triumph. Jervis built other canals, other railroads, always seeking more information and simpler, less costly, more effective solutions to problems.[6]

Thus, with that "passion for practicality" that Alexis de Tocqueville found remarkable in 1831, Americans pursued and embraced a technological revolution in the nineteenth century that put engineers, the "stewards of technology," in the forefront. It produced improvements in the production of iron, the design and utilization of steam engines, the mechanization of textile manufacture, and the fabrication of precision machine tools. Americans learned to manufacture in quantity using uniform, interchangeable parts, creating a production format that soon became known as the American System.[7]

Fast, efficient, economic production and simplicity of design became American trademarks and, by midcentury, brought the United States respectful recognition, even grudging admiration. At London's Crystal Palace Exhibition in 1851, for example, English and Continental observers were taken aback by the audacious Americans' ingenuity, workmanship, and utility (but not beauty) of design. Proportional to its population, the United States took more prizes at this first world's trade fair than any other country.[8] Similarly, engineering, personified by the great Corliss engine that powered all of Machinery Hall, was the object of attention at the Philadelphia Centennial Exhibition in 1876. American engineers made an even greater impression in 1878 at the Paris Exhibition.[9] Raymond Merritt called these American engineers "functional intellectuals"; they critically, pragmatically, used the methods and discipline of scholars but considered ideas as tools for change, not aesthetic experiences. To them Merritt credited the "functional and professional leadership" for the "great transformation" that technological development had brought to American society by the mid-nineteenth century, both materially and culturally.[10]

As their products transformed society, engineers turned their vocation into a true profession, becoming more specialized and institutionalized in the process. Flourishing machine technologies, whether used in stationary steam engines or textile mills, for reaping wheat or sending instant messages by wire, encouraged engineers to rely on scientific principles as well as the inventions of mechanics. As Edwin Layton put it, America's early engineering practice outgrew its craft tradition and was "grafted onto science," borrowing its methods.[11] In the new profession of mechanical engineering, a protracted internal struggle over whether the proper venue for training practitioners was the traditional machine shop or the engineering school exemplified the tensions of change.[12]

Coming from the opposite direction, scientific toward technical, electrical engineering emerged from academic physics departments following Alexander Graham Bell's invention of the telephone in 1876 and Thomas Edison's 1879 demonstration of the potential of electricity. Establishing laboratories to study electrical phenomena gave contemporary physics a welcome claim to utility, but these roots also demanded that the new field be scientifically and mathematically based. The first formal course in electrical engineering appeared in 1882. Chemical engineering followed in the early years of the twentieth century, peeling off from chemistry when its practitioners had difficulty defining their relationship with the parent science. Later, industrial engineers further systematized production processes and addressed problems relating to burgeoning industries, such as automobile manufacturing, which in turn stimulated work on materials such as steel, petroleum, rubber, and plate glass. In the twentieth century, still more engineering disciplines proliferated—aeronautical, radio, electronics, nuclear, computer.[13]

Evidence of their growing professionalism throughout this period was engineers' eagerness to associate with one another for the promotion of their work. The oldest engineering profession had organized the American Society of Civil Engineers in 1852, effective nationally after reorganizing in 1867; the American Institute of Mining and Metallurgical Engineers split off from the ASCE in 1871. The American Society of Mechanical Engineers followed in 1880. Those in the young but growing field of electricity created the American Institute of Electrical Engineers in 1884, while it took until 1908 to form the American Institute of Chemical Engineers after a struggle within the chemistry profession for engineering's legitimacy—and so on.[14]

Between 1880 and 1919, the number of engineers in the United States increased seventeen-fold, from about 8,000 to 136,000. By 1930 there were 226,000 in a wide variety of specialties. With steadily improving educational and professional credentials, the various engineering disciplines grew in stature and importance. Some individual engineers, such as Edison and bridge builder and entrepreneur James Eads, enjoyed fame of heroic proportions. But no encompassing professional society gave engineers the "unified and authoritative voice," not to mention prestige, that the National Academy of Sciences, established in 1863, offered its members. The early NAS elected a few engineers, such as Eads (who had no formal engineering education), but their numbers shrank in subsequent years, as the scientists presumably deemed engineers' culture and contributions less worthy than their own. Although efforts to form a national engineering society began as early as 1886, the National Academy of Engineering (NAE) was not established until 1964—a century later than the sciences' fraternity—and, then, under the NAS charter, without congressional sanction of its own.[15] So while the engineering profession increased its standing, it remained overshadowed by science.

Tensions over engineering's status relative to science, explicit and implied, thread the entire American engineering story and have been an enduring concern for engineers. After 1950, such questions became a perennial issue for the National Science Foundation. Roots of the problem appeared as early as the seventeenth century, when scientists began to "assume a position of conscious superiority to the 'rude mechanicals,'" although gentleman-scientists had little contact with the baser-born engineers who worked in a baser world.[16]

An illuminating example of engineering's struggle for professional recognition in the United States is provided by a nineteenth-century interplay between Harvard University and the fledgling Massachusetts Institute of Technology (MIT). In 1847, textile magnate Abbott Lawrence endowed what came to be known as the Lawrence Scientific School at Harvard in order to train students as "engineers or chemists or, in general as men of science applying their attainments to practical purposes." But, says Bruce Sinclair, the school "almost immediately subverted" Lawrence's instructions and "became instead a preserve of pure science ideology while its engineering enrollments languished." Meanwhile, in 1863, William Barton Rogers founded MIT to provide intensive instruction leading to practical technical careers. Some said MIT would never have been born if Harvard had done the job that Lawrence had intended.

Harvard tried in 1870, 1878, and 1897 to absorb MIT, but its attitude of intellectual superiority so irritated the technical school that MIT maintained its independence, even though merger would have brought economic benefits to the financially strapped institution. MIT went on to develop its own vision of an alternative but equally rigorous and valid education, rejecting Harvard's attitude that "the more explicitly training led to employment, the more intellectually restricted it was." But MIT also broadened its own cultural and intellectual offerings. As graduate John Ripley Freeman wrote, "If MIT stayed as it was while Harvard created extensive programs of pure and applied science, the Institute would end up educating the corporals of industry instead of its captains." In 1916 MIT president Richard Maclaurin declared basic science indistinguishable from the applied work his engineers did. But he also said that applied science was dependent upon the pure—thoughts that would reappear.[17]

In an 1892 address before the American Association for the Advancement of Science, Ohio State University president B. F. Thomas denounced the "contempt felt by the devotees of pure science for those who seek to put her truths to some use." He credited the success of American engineering graduates to their avowedly practical education. Yet, he also foreshadowed an emerging emphasis: Although "pocket book engineers" could get by with relatively simple mathematics, the "superior" engineer understood that the "truths of mechanics and physics" were so "broad and far-reaching," that it was "absolutely impossible to express them fully or to show their bearing and relations without the use of more or less advanced mathematical methods."[18]

Thus, education was increasingly the key to the advancement of engineering as engineers' work became more complex and demanding. And with the profession's increased recognition and remuneration, engineering education began to change profoundly, apprenticeships and shop learning giving way to college training. The pace was slow, however; more engineers continued to enter their work through "experiential, nonacademic channels" for many decades.[19] Early scholastic choices were few beyond the military academy. In 1819, West Point graduate Alden Partridge founded the country's first civilian school of engineering, the American Literary, Scientific and Military Academy in Norwich, Vermont, its program deliberately less theoretical and more practical and flexible than that of West Point. As Norwich University (since 1834), it first conferred degrees in 1837. Opened in 1824, in Troy, New York, Stephen Van Rensselaer's Rensselaer School, which became Rensselaer Polytechnic Institute (RPI) in 1861, was the United States' first endowed independent technical school for training in specialized industrial careers. Gradually introducing engineering courses, RPI granted its first four degrees in civil engineering in 1835.[20] Washington Roebling, who built the monumental Brooklyn Bridge from his father's design, earned his degree in civil engineering at RPI in the mid-1850s. The Franklin Institute of the State of Pennsylvania for the Promotion of the Mechanic Arts, established in Philadelphia in 1824, offered courses in mineralogy, chemistry, natural philosophy, mechanics, architecture, machine design, and mathematics. By midcentury, it was an important center for the American engineering profession.[21]

A trickle of other schools of varying rigor and technical curricula followed, then a small stream. By midcentury, some seventy colleges offered engineering training of some sort; in 1862, about a dozen were recognizable engineering schools. By 1872, there were seventy engineering colleges, an astonishing expansion, credited largely to the landmark Morrill Act of 1862. This legislative response to the growing need for advanced engineering education granted federal land to each state—30,000 acres for each senator and representative—for the establishment and maintenance of mechanical and agricultural schools. By 1865, engineer George Morison could note with some accuracy that the age of "Yankee ingenuity," which was based on knowing what had been done, had been replaced by "persistent study and observation."[22] The land-grant colleges, in fact, became the primary trainers of the nation's engineers. The colleges with the largest undergraduate engineering enrollments in 1900 were almost all land-grant institutions. For their part, engineers dominated these schools, accounting for nearly 60 percent of the enrollment at America's sixty-five land-grant colleges in 1901.[23]

Engineering curricula grew increasingly rigorous by the end of the nineteenth century, with more emphasis on the physical sciences, advanced mathematics, and laboratory training, although leading educators complained that engineering colleges still offered too much "shop work" and too little "head work" (meaning science). Increasingly complex problems and engineers' desire to "rationalize" their techniques by applying mathematical analysis to their empirical results helped to push engineering education away from its historical reliance on rules of thumb based on painstaking observation, trial, and testing. Robert Thurston, who thought of engineering as "a union of science [the knowledge base] and art [its useful application]" and advocated scientific education to augment shop training for mechanical engineers, introduced laboratory instruction at the Stevens Institute of Technology in Hoboken, New

Jersey, in 1871. His engineering laboratory exemplifies an area of engineering education that Americans pioneered, offering "conspicuous leadership," according to Lawrence Grayson. The emerging fields of electrical and chemical engineering also furthered the scientific approach. Since the professors of these subjects tended to be physicists and chemists, they brought with them the attitudes and practices of academic science.[24]

Another boost came from new industries, especially those with a science-based chemical and electrical focus, and the great industrial laboratories, which were emerging in response to "the steady convergence of scientific knowledge and industrial practice." These endeavors demanded mathematically literate graduates, who would receive their task-specific training on the job. At General Electric (GE) in Schenectady, New York, Charles Steinmetz pressed for a research laboratory promoting two missions, the "commercial application of new principles" and the "discovery of those principles." As industrial laboratories proliferated, research became more organized, purposeful, and scientific. Some of it was very basic, although the overriding goal was to develop new products and processes that the company could exploit, which commonly called for engineering.[25]

In general, engineering by the turn of the century had become integral to American industry as the latter grew rapidly and required more and more complex technological systems in order to function. Engineers hired to design and maintain electric power grids or assembly lines frequently rose to executive levels, managing people and money as well as corporate technologies. Such engineers needed a broad-based education, grounded in theoretical science, mathematics, and laboratory training but also economics and management (another American contribution, in Grayson's view). College degrees, and even graduate degrees, were becoming essential. A total of 226 students earned engineering degrees in 1880; that number leapt to 11,000 degrees by 1930. In 1938–39, 119 American engineering schools enrolled well over 82,000 undergraduates and more than 5,000 graduate students. Cooperative engineering education programs between universities and industry, such as the early twentieth-century course for electrical engineering students that Dugald C. Jackson developed at MIT with General Electric, strengthened academic-corporate ties and served the research needs of both institutions, though not without tension over their respective values. Engineers found themselves "caught between the business culture of their corporate employers and the scientific culture of their professional colleagues."[26]

An enormous influence on the evolution of engineering training was the emergence of a new way of thinking in the mid-1850s. Scottish engineering professor W. J. M. Rankine developed the notion of "engineering science," which he saw as an "intermediate mode of knowledge" between studies promoting knowledge and understanding (theory) and studies creating objects of practical convenience and profit (practice). For example, he applied to considerations of stresses on structures his observation that the laws of nature created a state or condition in a substance while the material's properties modified those laws: "Science could determine the stability of an ideal structure, and the engineer could determine the experimental strengths of the materials used in the structure, but neither approach could accurately predict how the material might act in a real structure." Rankine's engineering science approach "established a way to connect the static [idealized] forces that scientists dealt with and the experimental data of the [materialist, real-world] engineers."[27] David Channell, describing engineering science as a "translator between the languages of science and technology," noted that in linking theory with practice it emphasized the *how,* not the *what.*[28] Or, as Layton put it in discussing the hydraulic turbine, "engineers borrowed and adapted the methods and spirit of basic science in order to generate a body of science tailored to the needs of technology."[29]

In the United States, engineering educators like Robert Thurston, at Cornell University after 1885, actively promoted Rankine's engineering science ideas (though not frequently using that term, according to Ronald Kline). In Bruce Seely's words, they "created a synthesis between science and engineering by linking scientific methods to engineering values." That is, engineers using scientific (mathematical) information and methods served their own central value, which was to *do*—to solve a practical problem, to design a useful device. Science's highest value, in contrast, was simply to *know*. The result was a push for more scientific fundamentals in the engineering curriculum, general agreement about which was reached by about 1900 among leading educators, although widespread curricular implementation of engineering science took another forty years at least. World War II was the catalyst for massive change, but even in 1952, New York University engineering dean Thorndike Saville, noting increasing attention to mathematics and physical sciences in engineering schools, called the trend "discernible but not marked during the past fifty years."[30]

Engineering education reached a new plateau in 1893 at the World's Columbian Exposition, when a number of congresses on important branches of knowledge and industry met, including one on engineering. The members of the division on engineering education, overwhelmingly academic and led by Illinois engineering professor Ira Osborn Baker, formed the Society for the Promotion of Engineering Education (SPEE), the first such focus by a major profession. The society hoped to promote college as the appropriate training ground for engineers and a curriculum based on scientific and mathematical principles. In 1907 the SPEE asked other professional engineering societies to join in examining all aspects of engineering education, including research. University of Chicago professor Charles R. Mann, who headed the effort, issued a final report on this first detailed study in 1918. The report's call for emphasis on fundamentals and a unified curriculum went largely unheeded in the busy postwar period. But World War I did demonstrate engineers' contributions in industrial management and production techniques, and that resulted in greater curricular emphasis on the administrative and economic aspects of engineering.[31]

SPEE sponsored another, even more comprehensive education study between 1923 and 1929, chaired by William E. Wickenden, AT&T vice president and former MIT professor. His two-volume report on sixteen separate studies, which generally viewed engineering education as a "compromise between academic and professional study" and required courses in economics and the humanities, had a profound effect.

The 1932 creation of the Engineers' Council for Professional Development (ECPD) to accredit engineering curricula implemented one of the report's chief recommendations, although not under the auspices of SPEE. Another was greater emphasis on graduate degrees, which began to be more common. By the 1920s, graduate engineering enrollments were multiplying manyfold.[32]

Wickenden's associate Harry P. Hammond, dean of engineering at Pennsylvania State College, led a new investigation on graduate study in 1935. He wrote that graduate education for engineers had become a "fixture"—which was an optimistic exaggeration—but disagreement continued over whether graduate work should emphasize research or specialized course work. Hammond later directed two more education studies for SPEE, in 1940 and 1944.[33]

Engineering research, like engineering education—of which it was a key part—was also experiencing a transformation. Research in engineering colleges was rare before 1900 and usually targeted toward solving specific, practical, often local technical problems. Only a handful of professors were involved in research, usually as consultants with industry. They seldom pursued original inquiries for the sake of curiosity. Indeed, SPEE continued to consider teaching more important than research right up to World War II.[34] Laboratory investigation increased significantly in volume with the establishment of state-run engineering experiment stations (akin to the prospering agricultural experiment stations) at the land-grant colleges, beginning with the University of Illinois in 1902. This, too, was practical, applied research—some of it for industry and some as public service, such as testing road-building materials.[35]

Wickenden wrote that it was difficult to classify academic engineering research "under such heads as actual research for fundamental knowledge, first-aid problems for industry, routine engineering testing, and what may be called quasi-scientific puttering." Further revealing his sense of its quality, he continued, "Confessedly the amount of real 24-carat research would be relatively small, for the reason that few teachers of engineering are capable of doing it or directing it." Blaming the "backwardness" of research in engineering colleges less on a lack of money than on a shortage of competent researchers—in turn a result of academic preparation—Wickenden concluded that it was critical to improve engineering education in order to generate first-class engineering research.[36] Sixty years later, Seely put this relationship in reverse order, finding that only after engineering research became more basic (with studies of structural mechanics replacing investigations of bridges, for example), did engineering curricula become more generalized and theoretical.[37] In any case, adequately sized and financed engineering research programs were still unusual in American colleges in 1930, and the Depression effectively halted the growth of what few there were. At the close of that decade, however, Jackson observed an "increasing recognition" of the importance of research in engineering education, especially for advanced undergraduates, which he called "a notably fruitful trend." That trend was greatly accelerated in the war years.[38]

The involvement of the federal government, which in 1990 funded nearly half of all academic research, is a persistent theme in the history of American science and engineering and essential to an understanding of the future National Science Foundation. The U.S. Constitution itself (article 1, section 8) directed Congress to "promote the Progress of Science and useful Arts," which it did with patents—the overwhelming number of which in the nineteenth century rewarded practical mechanical technologies—and "continuous, if at times modest," support for research.[39] Eclipsed during the robber baron period, the federal role nonetheless grew overall, especially after the Morrill Act and the creation of the Department of Agriculture in 1862. Various agencies established scientific research branches for purposes as diverse as the mapping activities of the Geological Survey and the statistical analyses of the Census Bureau. The Army Corps of Engineers continued its research, supporting public-works projects. Responding belatedly in 1901 to Thomas Jefferson's urging a century earlier, Congress created the National Bureau of Standards, which became a sort of national physical laboratory, pursuing basic-science and industry-related investigation as well as the "abstruse research involved in defining and maintaining standards," work that industry wanted but could not or would not do for itself. The Bureau of Mines and the Public Health Service each carved out their own research niches. The National Advisory Committee for Aeronautics (NACA), formed in the unsettled peacetime of 1915, soon became the nation's first war-research agency. Its organization and operation also served as a model for the originators of the National Science Foundation concept.[40]

Advocates attempted in 1896 and again in 1916 to get Congress to establish engineering experiment stations through the land-grant college system. Political wrangling killed these efforts (non-land-grant technical colleges like the Georgia Institute of Technology protested being left out, while some schools wanted such largess distributed by merit only), but several states created their own engineering experiment stations, following Illinois's 1902 example. By 1916, there were around twenty; by 1931, forty.[41]

Hunter Dupree, in his study of science and the federal government, saw contrasting types of research emerging. By the beginning of the twentieth century, as universities broadened their curricula, raised their standards, and organized graduate schools, they embraced basic research and widely assumed that it "belonged" in the academy, "leaving only applied research to the government." He found exceptions in schools like MIT and the land-grant colleges, which emphasized applied science, and in the government's Smithsonian Institution, which delved into a wide variety of basic research. The burgeoning private foundations, too, engaged primarily in fundamental science. Dupree also noted that the "cloistered seeker for pure knowledge" enjoyed superior status to the "grubby civil servant chained to mundane, grinding routine investigation."[42]

As the twentieth century advanced, the federal government sought broader control over scientific and technological policy than could be obtained from individual agencies following their own agendas. In 1918 Woodrow Wilson signed an executive order making permanent the National Academy of Sciences' wartime National Research Council, which he charged with stimulating scientific research and advising the government. The NRC, first funded by the Engineering

Foundation, prepared legislation in the 1930s to support meritorious research projects, regardless of institutional or geographical distribution. But opposition to the elitism of granting most awards to the few institutions already conducting the preponderance of research defeated the measure, a scenario to be revisited during NSF's legislative birthing. The contemporaneous Naval Consulting Board, formed to obtain engineering advice for the war effort, was not continued, but from its roots sprang the Naval Research Laboratory in 1923, another model for NSF and an important source of its early personnel.[43]

In the mid-1920s, when industry dominated research funding (most of it in-house and product directed), Commerce Secretary Herbert Hoover called for government support for basic research. To him the national scientific program was an interrelated whole: basic scientific pursuit should be continuously encouraged lest applied science not be able to maintain itself over the long haul, another sentiment to enjoy high favor a generation later. Hoover was an exception among Republicans; most science leaders of the period, elite and conservative, "tended to fear government interference with the autonomy of science more than they welcomed its succor."[44] In the 1930s, Franklin Delano Roosevelt's agriculture secretary, Henry Wallace, emphatically supported government research, but he insisted that the social sciences be given as much attention as the natural sciences in order to promote his goals of economic security and a better life. This dichotomy—between Republicans favoring privately funded, independent basic research and Democrats promoting a significant government role but toward practical results— would lead to a protracted struggle to establish a federal agency devoted to basic science.[45]

In 1933 Roosevelt created the Science Advisory Board—to be headed by MIT's president, Karl Compton, under the National Research Council—to study and make recommendations on government science policy, personnel, and coordination of scientific work. But conservative opposition within the SAB and natural scientists' unwillingness to include the social sciences, which they generally considered intellectually inferior, put off new federal funding of science research in universities during the

1930s.[46] Legislation to support more basic scientific research in nonprofit institutions by earmarking for them half of a substantially increased National Bureau of Standards research budget, and to fund engineering research stations in every state also failed after a five-year effort during the 1930s. The Bureau of the Budget thought it too costly when judged against priorities arising from the Depression and then the push for war preparedness. Some academic institutions resisted federal intrusions, which they felt threatened their autonomy, although others were happy to take what federal money there was, such as that of various military laboratories.[47]

Roosevelt showed his growing interest by asking the National Resources Committee—an enlargement of a board he formed in 1935— to address the issue of federal science research policy. Its comprehensive report, *Research—A National Resource,* appeared in 1938. The report concluded that a national research effort could help the nation emerge from the Depression and emphasized the importance of universities as a source of research personnel and their training, repositories of expertise, and centers of pure research. It suggested that government stimulation of academic research would yield a high return on investment and recommended funding the highest-quality science through existing mechanisms, coordinated with the private foundations' system of grants. At that time, sixteen research universities employed almost half of the nation's distinguished scientists. In an egalitarian gesture, the report urged support of business and engineering research stations in each state, as well. But no action followed. The United States was between crises in 1938.[48]

By the late 1930s, federal funding for research was estimated to have exceeded $100 million annually. But unlike the situation in other countries, the abundance of private funding—through the Rockefeller, Carnegie, and other foundations (which paid little attention to engineering) and industrial laboratories, such as those of General Electric, DuPont, and Standard Oil—"effectively precluded government involvement in much of basic science." The National Research Council and other science leaders maintained an "ideological commitment to the private stewardship" of scientific research, and no federal administrative mechanism existed.[49] Despite their lack of success, though, all these efforts helped mold postwar legislation that provided for federal patronage of academic science.

With all these caveats, federal interest in science and technology had grown in both size and significance by the time the United States became enmeshed in the unprecedented challenges of World War II. To direct research toward the specific technologies needed for victory, the White House established an Office of Scientific Research and Development under the leadership of Vannevar Bush, former MIT vice president and dean of engineering who had become a political insider as president of the Carnegie Institution of Washington and chair of the National Advisory Committee for Aeronautics. Federal money flowed in a gush that would never again recede to prewar levels. Much of it went to universities, strengthening and cementing existing government-academy links and creating new forms of interaction, such as the contract and grant systems that would later be adapted for postwar research. Results of this mobilization of civilian scientists and engineers as well as parallel military research included radar, the proximity fuse, and solid-fuel rockets, all of which made a difference, and the atomic bomb, which climaxed the nation's research effort and ended the war.[50]

In 1944 Roosevelt asked Bush to help him consider the role of government science in the postwar world. The president was dead before Bush was ready with his formal response in 1945, but his seminal *Science—The Endless Frontier* won instant acclaim. While many of Bush's ideas had appeared earlier, the nation now seemed ready for them. This thin book dominated the thinking that led to the birth of the National Science Foundation in 1950.

Thus, between the end of the nineteenth century and World War II, engineering in the United States underwent dramatic changes. Specialized professional engineering became centered in larger, more complex organizations, especially corporations and industrial laboratories. Government also came to employ growing numbers of engineers, especially in public works—more so during the Depression, in economy-stimulating efforts such as the Tennessee Valley Authority and WPA construction projects.[51] To train engineers for these needs and their own, universities (where most NSF funding would go) developed markedly different attitudes, approaches, curricula, and research goals for academic engineering. Engineering colleges increasingly offered a curriculum couched in engineering science—in Layton's words, a body of knowledge and a pattern for action considered "intrinsically objective but operating within a value-laden context."[52] Engineering research became more broadly directed toward an understanding of principles, not solutions to specific problems. The pace of change was rapid and accelerating. Outside forces for change, always significant, included abundant theoretical groundwork laid by European engineers over the span of two centuries, increasingly sophisticated technologies that required more fundamental knowledge, the embarrassing wartime example of scientists performing some of the best engineering, and now, the promise of continued massive federal funding.[53]

As the history of engineering in NSF unfolds, much of it will be told in terms of organizational change, driven by both external and internal imperatives. For example, early administrative patterns reflected the institutionalization of engineering's increasing specialization, by department in the academic environment and in separate professional societies. But NSF's engineering leaders quickly saw the need for engineers with different but complementary training and interests to work together to solve complex problems that knew no artificial disciplinary boundaries. So interdisciplinary research was encouraged early on, and from it came a growing interest in an integrated approach (later called systems engineering)—looking at each part in terms of the broadly defined whole. In the mid-1980s NSF engineering leaders attempted to break down classical disciplinary designations for organizational units in favor of more functional labels that would better describe and serve the nature of the work.

The tense and dynamic engineering-science relationship will be a key theme of this study, since engineering has continually fought for its place in the historically science-dominated National Science Foundation. The tools and arguments of the struggle altered over time, but engineers felt it ever necessary to assert the value of their contribution, which they usually measured in comparative budgetary terms. Internal politics will appear in several guises. One sore point, or at least an illustration of one, was that while engineers often called themselves scientists, the latter did not reciprocate. Layton ascribes some of this emotional difficulty to evolving word usage, but the issue is ideological as well.[54]

Despite the fact that postwar engineers stressed their scientific-ness and training in engineering science, engineers and scientists are inherently different. Eugene Ferguson, unhappy with the trend toward making engineering entirely mathematical and analytical, argues that engineers do much of their thinking intuitively, nonverbally; they "see" in the "mind's eye." Engineers ultimately focus on the needs of prac-tice, which means design, and that necessarily involves "loss of gener-ality and acceptance of approximate solutions" or "relative degrees of perfection." While engineers use scientific theory, the "complexities of practice transcend" it. Such an approach would necessarily clash with that of pure scientists devoted to the pursuit of generalizable, timeless truth. Unlike scientists, who analyze and explain things as they are, engineers concern themselves with how things ought to be, says Fer-guson. They synthesize, aiming toward a functional solution; when they fail, they fail at judgment, not calculation.[55] Thus, engineering, which had established its own culture and record of accomplishment in Ameri-can history, had to scrabble for a place within the science establish-ment—first on science-determined terms, eventually (if not finally) on its own.

Engineering is never separate from its environment. Society's pressures—political, economic, social—are always present in an engineer's choices. It is not surprising that bursts of engineering activity and advancement, more than pure science, have historically accompanied cataclysmic societal dislocations. For example, during World War I— the first to see combat in the air—General Electric developed the first turbine-driven supercharger, which compressed thin air, enabling airplane engines to sustain full power at high altitudes. This technology later played a key role in the next war.[56] Even the Great Depression, which caused economic activity, including scientific research, to stagnate, encouraged engineering. While the WPA supported some university research by paying the salaries of otherwise unemployed research assistants, it is remembered more for the bridges, dams, and park and civic facilities that still stand as monuments to their builders' engineering skill if not their new ideas.[57] Similarly, during the late 1960s and 1970s, when American society seemed to be coming apart at the seams, NSF answered a call to solve identified problems; today economic competitiveness is the spur to action, and the focus is on technology.

NSF's engineering story will begin and remain bound up in the larger milieu of government-supported science, which emerged as a major force after World War II. Because political entities—Congress and the executive branch through the Office of Management and Budget—hold the purse strings, federal funding introduces the role of national politics—sometimes partisan, always reflecting politicians' desire to respond to the pressures they themselves encounter, especially from constituents. A federal agency like NSF, remarkably independent until its budget grew large enough for politicians to notice, has only limited choices in its policies and actions.

Engineering's organizational development within the Foundation proceeded in step with its increasingly sophisticated and crucial contributions. These advances solved some societal problems and created others. They perfected old methods and materials and introduced new ones. They found application in every physical environment and responded to the gamut of political and social demands. They added new knowledge. Sometimes engineers created artifacts that were "frozen music," to use Johann Wolfgang von Goethe's admiring description.[58] Sometimes they made monsters. Only a few such stories can be told here; they are selected to indicate the breadth and variety of engineering and engineering education that NSF has supported, with varying results and impact.

The National Science Foundation as an institution, like engineering within it, both influenced, and was molded by, these continuing themes.

Chapter 1

Engineering under a Mandate for Basic Science

> Under the pressure for immediate results, and unless deliberate policies are set up to guard against this, applied research invariably drives out pure.
>
> —*Science—the Endless Frontier,* 83

Engineers enthusiastically endorsed the formation of the National Science Foundation. Although engineering played a minor role in the prolonged public debate, it was included organizationally, in the Division of Mathematical, Physical and Engineering Sciences, from the beginning. Philosophically, perhaps even emotionally, however, its position seemed less than certain in this new government agency, which was established to support basic scientific research in the name of postwar national strength and security. Could research in engineering, by definition aimed at ultimate application toward the solution of practical problems, be considered basic in scope and significance? Despite affirmative answers to the direct question and their own hard work over half a century to anchor their profession in scientific and mathematical fundamentals, engineers operated from a vague sense of insecurity and defensiveness in the Foundation's early years. They insisted their work was as scientific as any other science. They took pains to portray and conduct their research in broad, basic terms within traditional disciplinary categories in order to satisfy NSF's pure science requirement. They enjoyed modest success but began to grow restive, feeling underrecognized and out of the circle.

CREATING A NATIONAL SCIENCE FOUNDATION

Vannevar Bush, presenting his landmark report *Science—The Endless Frontier* to President Harry Truman in July 1945, argued that scientific progress would be crucial in the modern postwar world to ensure the nation's health, welfare, and, especially, security.[1] The United States, he said, should establish a federally supported "National Research Foundation" to promote university research and education in basic science under the overall policy supervision of a board of eminent scientists. Not surprisingly, Bush is generally credited with fathering the National Science Foundation. But his words were not the first on the subject—nor the only.

Starting in 1942, Democratic Senator Harley Kilgore, a New Dealer from West Virginia, annually sponsored legislation to establish a national science foundation (the name originated with him), first for the purpose of acquiring and engaging technical knowledge more effectively for the war effort and later for solving national peacetime problems. Kilgore's background as a one-time student of engineering and a politician accountable for constituents' tax dollars dictated that his proposed agency direct science and technology toward tangible, practical public goals, although he did eventually include support for theoretical science as well—not "just to build it up," but to "do something for the betterment of humanity." The Republican Bush, electrical engineer and physicist as well as brilliant, sometimes overbearing administrator—at MIT, the Carnegie Institution, and the wartime Office of Scientific Research and Development (OSRD)—respected Kilgore's intelligence and intent but did not care for the senator's practical, politically based legislation any more than his liberal politics. Kilgore responded in kind.[2]

Here, between Bush and Kilgore, began an interplay of philosophical differences that forms a running theme throughout the history of the National Science Foundation. Bush's view, that the agency be devoted to basic research—that is, research pursued "without thought of practical ends"—ultimately prevailed. Scientists held the view that "basic" was the domain of scientists. So, central to an understanding of engineering's fit, and welcome, within the early NSF was the issue of whether engineering research could be considered basic science. At least some leading engineers of the time had no doubt. As early as 1935, Earle B. Norris, dean of engineering at the Virginia Polytechnic Institute, wrote that engineers had often "extended their activities into the realm of pure science" because they had "caught up with the scientists and have been forced themselves to become scouts in the vanguard of knowledge." He found it difficult to draw a "line of demarcation" between science and engineering. While engineers generally sought the "best route to some definite goal," they would often "come on attractive trails leading elsewhere and return with tales of new and unexplored territory." Engineers often *were* basic scientists, he seemed to be saying, and many engineers would have agreed.[3]

Kilgore's third national science agency bill in three years moved toward Bush's insistence upon maximum political independence for fundamental scientists and their work when the latter got his own opportunity to alter history. On 17 November 1944 Franklin Roosevelt wrote to Bush asking him to consider how the federal government could continue to encourage scientific inquiry in the coming peace. If Bush had not actually drafted the president's letter, he undoubtedly had significantly contributed to it, and as head of the OSRD, he had definite ideas on the subject. He was not reluctant to share them.[4]

Bush's *Science—The Endless Frontier* was in many ways a conservative's alternative to Kilgore's scenario. Bush agreed that the federal government should expand its role as a "patron of science" but focus on promoting fundamental science. He warned, "Our national preeminence in the fields of applied research and technology should not blind us to the truth that, with respect to pure research—the discovery of fundamental new knowledge and basic scientific principles—America has occupied a secondary place." American airplanes and radio were "spectacular developments," to be sure, but were based on basic scientific discoveries made in nineteenth-century Europe. Such borrowing stifled American innovation; worse, that source would soon dry up as the war-ravaged continent's economic prostration was fully felt. During the war Americans had also been drawing too deeply on their own limited reservoir of basic scientific knowledge, and now only pure research could replenish the well, wrote Bush. Practical Yankee ingenuity would no longer suffice in a future dominated by complex new fields, such as electronics and aerodynamics, in which technological advance would be "inseparable" from, and directly dependent on, new fundamental knowledge.[5]

Bush's own background and inclinations would not necessarily have led to such views. Based on his own engineering education at MIT, which had been rigorously grounded in mathematical and scientific principles—not the empirical experimentation that still lingered in many schools—Bush could comfortably dedicate the new agency to basic science without excluding engineering. At the same time, his own achievements had been largely problem-driven and practical, including the invention of a differential analyzer, an electronic calculating machine. Interestingly, his early 1945 correspondence shows him advocating a broad approach, arguing, for example, that "a diversified junk pile would have been of great value" to the Wright brothers, who were not even legitimate researchers by current academic standards. Lifelong engineering advocate Eric Walker remembered that Bush, whom he enormously admired, had envisioned a "national science and engineering foundation." When he chided Bush for abandoning engineering, Bush told him not to rock the boat and endanger passage of the bill. "And anyway, to most people, science includes engineering." For once, he was wrong, thought Walker. Bush quickly dropped that early view and never publicly diverged from the basic-science script, to which his subsequent commitment seemed entirely genuine. There is no documentation of a change of heart on Bush's part, but the politically astute leader was careful to adopt key phrases and strong ideas from his four working committees as they strove for a unified, focused report that would sway Congress. (This was a man who had referred to engineers as scientists during the war in order to gain them the reluctant respect of military leaders—and later regretted it when the "marvelously skillful engineering job" of landing astronauts on the moon was popularly described as a feat of science).[6]

The rotund, often rumpled Kilgore, on the other hand, preferred a broader interpretation but worried less about the type of research promoted than the importance of some centralized direction. Without it, he thought, critical technologies like synthetic rubber would have been unavailable to help wage the most mechanized war in history, and American postwar science might come to be dominated by monopolists, as Germany's I. G. Farben had in prewar Europe. He won important concessions in the management of the new foundation even as he promised "specific steps to make sure that it would function in a nonpolitical way and that scientific freedom would be safeguarded."[7]

The eight-year legislative history of the birth of the National Science Foundation has been often told, though not from the viewpoint of engineers or its impact on engineering.[8] Over the years more than a dozen bills for some kind of national scientific and/or technological research effort ran their course through Congress; numerous others surfaced briefly, only to die quietly in committee. For example, a recommendation by the Senate Subcommittee on War Mobilization for a forty-member Research Board for National Security was the subject of one ultimately unsuccessful bill in 1945. This board would have consisted of "civilian scientists, engineers, or industrialists" in approximately equal numbers with army and navy officers to plan and conduct scientific military research.[9]

In a 6 September 1945 message to Congress, Truman agreed in principle that "progress in scientific research and development is an indispensable condition to the future welfare and security of the Nation." Citing recent history as both "proof and prophecy of what science can do," he noted how organized effort had pushed "forward the frontiers of knowledge [and] forged the new weapons that shortened the war." Clearly referring to the atomic bomb, Truman was speaking of an accomplishment that was widely acknowledged to have been primarily engineering (though performed in considerable part by physicists). The president concluded that maintaining world leadership would demand development of the nation's full scientific and technological potential, building on the wartime example of cooperation among industry, academia, and the federal government.[10]

In early 1946 Kilgore's subcommittee submitted a report on the then pending S. 1850, a compromise between Washington Democrat Warren G. Magnuson's S. 1285, which embodied Bush's main ideas, and Kilgore's S. 1297 of the previous year. Having overcome turf battles and animosities born of differing ideas on the new agency's governance, the subcommittee urged the establishment of a national science foundation with the argument that "today no nation is stronger than its scientific resources." Echoing Bush, the report noted further that while the United States was "strong on applications," it was "weak in fundamental science," since during the war practically no basic research had been pursued. The Magnuson-Kilgore compromise would have created an agency of eight divisions, including a Division of Engineering and Technology "to be the support of research in the fundamental engineering sciences and other studies basic to the broad development of technology" but, pointedly, "not the engineering development of machines or processes."[11] This legislative effort also failed, and many more bills, bruised egos, and compromises were to pile up before scientists and politicians could agree among themselves as well as with each other.

Meanwhile, late in 1945, the Association of Land-Grant Colleges and Universities and the National Association of State Universities, which together oversaw a great deal of America's engineering education and academic research, joined forces to try one last time to secure federal funding for engineering experiment stations. They recommended that 30 percent of the new science agency's research monies (later reduced to 25 percent) be allocated to the states in proportion to their population and assigned to land-grant and tax-supported institutions "qualified to do scientific research." Engineering, prominent in such schools, would clearly benefit from such a provision. S. 1850 and other bills incorporated the stipulation, but the Bureau of the Budget objected. And Bush had long favored a "best science" approach, which meant in reality focusing on the few "elite" institutions where research was already strongest. Other attempts to specify a democratic (geographic) distribution of funds failed to survive the legislative process, but the final NSF law did note without elaboration that an objective of the foundation would be to "avoid undue concentration" of research and education support—a compromise to close a major point of political dispute.[12]

The engineering community actively promoted legislation for a science agency. The Engineering College Research Association (ECRA), which represented about seventy officially recognized engineering colleges and scientific organizations, testified during hearings in May 1946 on H.R. 6448, a companion bill to S. 1850. Charles E. MacQuigg, ECRA legislative committee chair and dean of engineering at Ohio State University, told the House committee that ECRA had been formed during the war for the very "purposes broadly envisaged" in the bill. Applauding the legislation's concept, ECRA requested that the president appoint members of the agency's nonpartisan board from names suggested by recognized scientific bodies, like itself.[13] But no foundation bill passed in 1946.

In 1947 more national science legislation appeared, with hearings in early March. Thorndike Saville, dean of engineering at New York University and chair of the Engineering College Research Council (ECRC, successor to the ECRA) of the American Society for Engineering Education (ASEE, formerly the SPEE), testified that the society "heartily favored" the creation of a national science foundation. The ASEE requested that the legislation name science *and* engineering specifically in its title and text to avoid ambiguity about either field, but this opportunity to shorten a long chapter of NSF's engineering history was not taken.

"We believe, sir, that engineering is as scientific as any other type of science," Saville declared to the House Committee on Interstate and

Foreign Commerce, defending engineering's inclusion in language. Asked if engineering was the "practical application of scientific basic research along physical lines," the conservation engineer replied, "Well, sir, I would say it was that and very much more." What more? The engineer, he explained, not only made use of the fundamental research of physicists, chemists, and mathematicians, he had to be "almost on a parity with them as to his knowledge before he [could] take those fundamental things and build a Hanford atomic pile, for example." Saville, a longtime government consultant on coastal engineering, argued that the government should support engineering "on a parity" with other disciplines, whether basic or applied. Building a bridge was an applied science, he acknowledged, but it depended on knowledge of elements as basic as soil mechanics, research on which the government "could well support."[14]

The Engineers Joint Council (EJC), which represented 80,000 "qualified American engineers" in five major engineering associations, established a national research foundation study panel whose chair, Boris Bakhmeteff, declared in 1946 that "in no branch of fundamental research is the void more acute and pressing than in the realm of basic engineering science." Bakhmeteff, a hydraulic engineer with government consulting experience, decried the widespread lack of appreciation of that work among the public, legislators, and even engineers. The EJC was irritated to have not received an invitation to testify at the 1947 hearings, having "actively cooperated" in revising earlier bills. It sent a letter anyway, which was included in the record. The council "wholeheartedly endorsed" the establishment of a federal science agency whose main function would be to support fundamental research, the "foundation of modern engineering." The engineer was the "active link" between pure research and technology, using basic scientific knowledge to master the resources and powers of nature in the interests of humanity, it said, in words that illustrate Ronald Kline's thesis that postwar engineers promoted engineering science "to gain a place at the federal trough," deferring to the hierarchy of the "pure science ideal" in a probable "effort to present a united front with the scientists." The EJC favored an institution governed by a board that, in turn, employed a director with broad powers over daily operations. The board should "necessarily include members representative of engineering," and engineering science should be recognized in the formation of the divisional structure.[15]

It soon became clear from the testimony of more than 150 witnesses—outstanding scientists, educators, public officials, and representatives of labor, industry, and the clergy—that virtually everyone with an interest in the subject liked the idea of a federally supported, independent national science foundation. Consensus came early that the foundation should work through the university system to support scientific research, grant scholarships and fellowships to train the next generation, and disseminate research results. Although engineering research was not mentioned in the earliest bills, it appeared specifically in all subsequent ones. A conspicuous exception to the general enthusiasm was National Academy of Sciences president Frank Jewett, who, fearing government control, preferred private financing of research. He specifically opposed including engineering in the new agency; it "certainly is not a science," he testified. While he called engineering "a technology based on fundamental science," it was "not of the same character" as basic scientific research.[16]

Differing views on various legislative details were slowly ground into agreement. Fairly simple provisions on patent rights and unspecified but permissive language on including the social sciences were two more important compromises. Early emphases on military and medical research faded as Congress established the Atomic Energy Commission (AEC) and the Office of Naval Research in 1946 and began providing larger appropriations to the National Institutes of Health, which took over the wartime research contracts of the OSRD's Committee on Medical Research. Indeed, by the time these new agencies and others—including the national laboratories that the AEC created from Manhattan Project facilities—claimed their research niches, NSF's potential scope was significantly diminished. Moreover, by the early 1950s, the army and air force also had their own research laboratories.[17]

Meanwhile, the most difficult legislative hurdle for NSF was the mechanism for governing the new agency. When, after two years of ardent debate, Congress in 1947 passed an NSF bill that, following the earlier National Advisory Committee for Aeronautics (NACA) model, lodged ultimate power (including the appointment of the director) in a board of distinguished scientists—as the influential and politician-distrusting Bush wanted—Truman regretfully vetoed it. He called that approach administratively unsound and an impediment to accountability; "divorced from control by the people," it showed a "lack of faith in the democratic process."[18] In the 1948 presidential campaign, Truman castigated the Republican-controlled 80th Congress for handing him an NSF bill that "eliminated the President from the Government of the United States." He preferred the approach of the so-called Steelman Report—a 1947 study, *Science and Public Policy,* by the President's Scientific Research Board—which recommended that the president appoint the director and the board, the latter as advisors only. He urged Congress to try again.[19]

Finally, after nearly three more years of complex maneuvering, supportive scientists and politicians worked through deep divisions on issues and soothed difficult personalities to craft a bill acceptable to both the Republican Congress and the Democratic president. An essential provision of S. 247, introduced in March 1948 and accepted as amended by conference committee in April 1949, was that the science agency would be governed by a twenty-four-member National Science Board and a director, both appointed by, and accountable to, the president. Truman signed the National Science Foundation into law on 10 May 1950, years after substantive agreement on its mission had been reached. NSF historian J. Merton England suggested that the most important reason for the long delay was NSF's "restriction to basic research," as cost-conscious lawmakers preferred to invest in applied science and technology to solve recognized problems; but the control issue loomed largest in the record. For now, engineers would cut their coat to fit the cloth; for most postwar engineering leaders, the fit felt comfortable.[20]

Founding NSF

The first National Science Board (NSB), which met in the White House Cabinet Room on 12 December 1950, was a distinguished lot. It was also a diverse one, by intention of the president if not the nation's science leadership. Among its two dozen members was one engineer, Russian-born Andrey A. Potter, dean of engineering at Purdue University. Educated at MIT and Kansas State College (D.Eng., 1925), Potter held a fistful of honorary degrees and national awards and was the author of numerous books on mechanical engineering. He built at Purdue one of the largest U.S. cooperative education programs; under his leadership Purdue's engineering experiment station grew to the nation's largest in the late 1920s. Not well known beyond the technical community, he was superbly qualified both to speak for engineering and to assist in the formation of a new research and education institution, although his voice was but one.[21] When Potter addressed the American Society for Engineering Education in June 1951 on engineers' role in the developing Foundation, he began his talk by calling NSF's focus on basic research crucial in the postwar period. He ended with the certainty that NSF would "need the active cooperation and advice of engineers in insuring that basic research in engineering is carried out on a scale commensurate with its importance to American industry and security."[22]

To the National Science Board fell the multiple tasks of determining overall policy direction for the infant Foundation, deciding on the details of organizational structure, and recommending the best possible personnel. Though he had topped no one's list of preferred candidates, the NSB and science leaders were pleased when on 6 April 1951 Truman appointed the modest but firm and politically knowledgeable Alan T. Waterman as the NSF's first director. The former deputy director and chief scientist of the Office of Naval Research came with solid credentials: a Princeton-educated (Ph.D., 1916) Yale physicist, he was also an accomplished musician and avid outdoorsman. Waterman served longer than two six-year terms and decisively shaped the character and direction of the new agency. He steadfastly maintained that NSF's business was and should be promoting fundamental science. Contemporary engineers generally accepted that emphasis, insisting they belonged within it.[23]

The National Science Board's charter members set to work, interpreting their charge within the context of federal precedent and expectation. For example, the NSF act called for promoting the education of scientists and engineers through fellowships and scholarships, which amounted to direct federal assistance. The government had traditionally left most educational matters to the states, but now critical technical needs and shortages of competent scientists and engineers suggested a reassessment of that hands-off policy, and the land-grant colleges served as a model for federal involvement. In another area, although NSF was assigned to develop a national science policy and coordinate and evaluate government research programs, the Foundation under Waterman resisted pursuing that avenue—either missing an opportunity to dominate the federal voice for science (unlikely given its small, late start) or saving the fledgling organization the embarrassment of failing to impose its will on mightier powers or incurring their displeasure with its judgments.[24]

A driving force behind all deliberations and decisions for the establishment of NSF was the nation's preoccupation with national security. The cold war took on a new dimension when the Soviet Union detonated a nuclear device in September 1949. World tensions exploded in June 1950 when Communist-led North Koreans attacked the Republic of South Korea across the barbed-wire border at the 38th Parallel. A *Chemical and Engineering News* editorial spoke for many when it worried that "The fast approaching bottleneck of too few scientists and technologists can well be the most efficient weapon possessed by Stalin and the Politburo." Yet, the NSB voted in particular not to establish an earlier-proposed division devoted entirely to military research—which would have been rich in engineering relevance—arguing that all the Foundation's progress in fundamental knowledge would "contribute in

the long run to the nation's military and technological strength." Besides, the Atomic Energy Commission and the Office of Naval Research
(ONR) had already staked out that territory.[25]

But little research of any kind was possible in the early years.
NSF's first, puny budget of $225,000, of which it spent only $153,000,
allowed for no more than getting organized. The Foundation began by
law with four programmatic units: the Division of Mathematical, Physical, and Engineering Sciences (MPE); Medical Research; Biological
Sciences (combined with the former in 1952); and Scientific Personnel
and Education. Paul Klopsteg, a professor of applied science and director of research at the Northwestern Institute of Technology, became
assistant director for MPE in November 1951. An engineer, physicist,
and published archery expert, Klopsteg had directed the development
of high-voltage X-ray machines during the war, for which he had been
decorated by the president. He had also chaired the board of governors
of Argonne National Laboratory for the Atomic Energy Commission
and the Advisory Committee on Artificial Limbs of the National Research Council. In May 1952 NSF tapped Ralph A. Morgen, a chemical and electrical engineer who had directed the Engineering and
Industrial Experiment Station of the University of Florida, to oversee
MPE's programs in engineering sciences.[26]

Conscious of its significance as a first, Harvard president James
B. Conant, chair of the National Science Board, proudly presented to
Truman NSF's painstakingly prepared annual report for 1950–51. Citing basic research as the heart of NSF, the report justified it at length and
expressed confidence that NSF's approach and planning were on target.
As Alfred P. Sloan of General Motors had said, fundamental knowledge,
gained through pure research, was like "ore in the ground," Conant
wrote. It had to be discovered before it could be extracted, refined, and
put to constructive use. While pure research could be considered a
gamble, and would in any case take time to pay off, prudent support of
a wide range of original ideas would surely yield a good return, with
useful application to follow. "Our real genius as a Nation has been the

power to convert scientific knowledge into practical utility."[27] NSF was, of course, less concerned with the second part.

On 23 May 1951 the president submitted to Congress NSF's first budget for research for fiscal year (FY) 1952—a mere $8.1 million. Half of these monies were slated for MPE, though Potter protested that "engineering was practically left out." Further cuts accompanied the authorization and appropriation processes; in the end NSF got a total of $3.5 million for everything, $1.1 million for research. Of that, MPE eventually saw 311,300 actual dollars for grants from a promise of $500,000. Engineering's share was not quite $42,000. NSF's research budget for FY 1953 started out even worse, at $6.6 million. Neither year's funds approached adequacy but were all that could be wrested from tight-fisted legislators; NSF's original statutory limit was $15 million. For context, by 1952 the United States was spending approximately $3 billion on all scientific research (up from $8 million in 1941), of which two-thirds, or $2 billion, was federal support, dominated by the Department of Defense. Industry contributed about one-third of the total; universities, about 3 percent. NSF support of basic research—the "stockpile" on which applied research for military security and economic growth depended—was about one-tenth of 1 percent of the U.S. total. Still, Waterman insisted that this modest beginning would promote NSF's goal of a "comprehensive program in really fundamental research on which is superimposed a gap-filling program," the latter assisted by the basic research of the mission agencies.[28]

When the House Appropriations Committee thought it appropriate to reduce the Foundation's budget request for FY 1952, arguing that NSF's "early aid in the present emergency" (Korea) was "not very tangible," the National Science Board countered that current military applications of science—in such areas as jet engine metallurgy and liquid fuels production—were slowed by gaps in basic knowledge. More research here could well provide "the vital margin of technical supremacy which will insure our survival." With the country engaged in a new war of uncertain duration and potential escalation in the terrifying nuclear shadow, such concerns bore weight. Lest Congress forget, the NSB also quoted Bush's belief that the outcome of World War II was "affected decisively by weapons unknown at the outbreak of hostilities." NSF's budget justifications to Congress throughout the 1950s hammered on the inescapable need for scientific superiority in an atmosphere of "intensified" cold war.[29]

Congress lifted NSF's constraining budgetary ceiling in 1953, and in 1954 a more confident Foundation asked Congress for a modestly increased appropriation to raise its support for university research and the training of new scientists. As Waterman argued, of the $115 million that the federal government was estimated to be spending on basic research, only about half of it supported research in universities, traditionally the "primary source for the discovery and dissemination of new knowledge." NSF also needed additional funding to support basic research transferred from mission agencies (relatively little in the end) and to correct the current "imbalance" between basic and applied research. President Dwight Eisenhower's Executive Order 10521 of 17 March 1954 reemphasized the importance of scientific investigation to national security and welfare, calling for expanded research in basic science in order to support "practical scientific progress." It also delineated a stronger role for NSF in funding "general-purpose basic research" and coordinating that of other agencies.[30] The linear progression from basic research to applied research to application was virtually unquestioned at that time, although agreement on the boundaries of these terms remained elusive. Waterman had decided by then that the intent of the researcher made the difference; basic scientists sought new knowledge, while applied scientists tried to make practical use of it. Two investigators working on a project with the same title would, with "difference in intent," produce "quite different results." That was a popular view, voiced as early as World War I, but there were countless others, all with pitfalls. Klopsteg claimed, perhaps facetiously but certainly in frustration, that NSF had forty-five mimeographed pages of definitions to choose from. Engineering, in particular, would suffer over this ambiguity.[31]

In the meantime, NSF leaders acted to establish workable, equitable policies and specific mechanisms for moving the Foundation's mission forward, soon settling mainly on grants to academic institutions to support the work of individual investigators. By July 1952 the board had drafted guidelines for evaluating proposals that emphasized scientific merit but also considered personnel quality, institutional resources, and budgetary reasonableness. Secondary factors, such as comparative merit and geographic or disciplinary distribution, could be considered to encourage the growth or improvement of a small college's research program.

While only a few proposals were rejected outright as unworthy, most of the 30 percent considered meritorious received less money than they requested. Indeed, in the lean year of 1952, NSF granted only one support dollar for every thirteen requested, which meant an accumulating "formidable mass of unsupported research of outstanding caliber" in addition to a flood of new submissions. Within these harsh fiscal limitations, though, Waterman pledged grantees full "freedom of inquiry and integrity in research."[32]

ENGINEERING IN THE FORMATIVE YEARS

During NSF's early years the National Science Board evaluated individual proposals as well as overall program areas. In June 1952 it approved its first research grants in engineering sciences—three, out of ninety-six applications. NSF awarded Daniel Drucker of Brown University $10,000 to conduct research in three-dimensional photoelastic techniques for two years. John G. Trump of MIT won $16,400 for a two-year study of fundamental processes in high-voltage breakdown in vacuum, and J. A. Sauer's proposal was approved at $15,500 for a year's work at Pennsylvania State College on the mechanical behavior and structure of linear high polymers. All were investigations of fundamental questions in their respective disciplines within the engineering sciences—defined as those sciences concerned with making specific scientific principles usable in engineering practice.

At the same meeting, the board, in the polite language NSF continues to employ, "declined" to support seventeen engineering proposals, which ranged from Big Bend Oil Company's $262,000 submission to study natural gas liquefaction and Bennett College's desired investigation of multielectrode radio tubes—both clearly aimed at application—to more basic but perhaps poorly delineated studies, such as biaxial fatigue strengths of metals and ionization characteristics of gases. NSF also chose not to fund studies of prestressed reinforced concrete for pavements, a subject soon to garner much interest.[33] Before long, engineers began to complain that too large a proportion of engineering proposals were being rejected relative to those of other disciplines—a subject of increasing vexation. Funded grants that year totaled somewhat less than $1.1 million.

In justifying a proposed expenditure of $450,000 for engineering sciences in FY 1953, the Foundation emphasized the "underlying scientific facts and laws" not directly applicable to practical solutions that the engineer needed to investigate before considering specific human needs. Yet the proposal summary sheets that staff prepared to assist the board in project selection often included a section titled "Possible Applications." Research for its own sake still hoped for ultimate usefulness. Thus, studies of photosynthesis and ways to concentrate solar radiation to operate heat engines could put solar energy to work warming buildings. Knowing more about tribology (the chemistry and physics of surfaces and their interactions, or friction) could lead to improved lubrication of engines or the control of rust. Additional fundamental work in nuclear energy could provide the "solid ground" upon which to base military and civilian development in that new technology.[34]

Other examples of early NSF engineering activity suggest tentative new directions. One was cooperative effort, whose value time would prove. In 1954, NSF staff recommended a Princeton University project, "Thermal Behavior of Buildings through Model Structures." By replicating different thermal conditions, the investigators hoped to determine which environmental features of a building—orientation, materials, shapes of structures like overhangs—made the greatest difference in terms of its heat control. This study was the first sponsored as a cross-disciplinary effort between mechanical engineers and architects. A Swarthmore project on electronic instrumentation for neurophysiology combined the fields of biology and electrical engineering to understand electronic switching of extremely low potentials while adding to fundamental knowledge of brain activity. Each investigator on the team contributed "something out of his own specialized stock." Thus, interdisciplinary research in engineering, considered the wave of the future today, enjoys a history almost as old as the Foundation.[35]

During the mid-1950s, most engineering grants followed basic-science approaches. Research projects expanded understanding of basic phenomena, such as fluid flow, heat transfer, and reactions at interfaces. These studies were conducted in various materials in varying solid, liquid, and gaseous states at fluctuating and extreme temperatures. Individual investigations included research on stresses from turbulence, vibration, applied forces, and chemical instabilities, or on the properties and behavior of existing materials and the development of new ones—metal alloys, polymers, complex composites. Investigator-initiated, they followed no patterns of emphasis. Support for engineering could include travel for conferences or field research or the purchase or construction of complicated specialized equipment to be retained by the institution. Developing high-precision tools and devices, from computers to hypersonic wind tunnels, that were necessary to accomplish myriad basic research tasks was the special purview of engineers, for both their own research and that of pure scientists.[36]

All this showed how fully NSF and its engineering community had embraced the concept of engineering science as a statement of what academic engineers were, and should be, about. Engineers' reliance on scientific principles and methods to design and create useful products and processes, increasing for a half-century, had become the norm. Ever more complex technologies required this approach; and overcoming the embarrassment of the common belief that the best engineering of World War II had been accomplished by scientists encouraged it. Engineers' craving for the social status of scientists drove them to it. But they remained distinct. As Edwin Layton says, engineers considered themselves to be scientists, "doing science in their own distinctive way," which did not mean accepting scientists' "priority of theory over practice" or of the ideal over the real. Finally, the liberal postwar federal funding that favored basic science completed the transformation, first in the laboratory, as shown in NSF's engineering grants, then in the classroom. It was no accident that engineering was born organizationally within the new Foundation as "Engineering Sciences."[37]

In October 1954 the National Science Board first considered a subject that has occupied it with varying degrees of urgency ever since— that of personnel strength in science and engineering, which was, in turn, an education issue. In 1952 the United States had an estimated 625,000 technically trained workers—about 1 percent of the workforce. Of these, about 135,000 engaged in some kind of research, with fewer than 40,000 "capable" of basic research. Lack of advanced training was a primary factor. In November 1952 Morgen told the MPE Divisional Committee what educators already knew, that "only recently have engineers pursued the doctorate."[38] Military service, which cut the college student population, and abundant postwar employment opportunities, which made graduate study less attractive, topped a variety of recent reasons. Now, NSF's very existence could help increase those numbers, for in order to acquire a research grant, engineering departments had to have graduate students as research assistants. That required them to emphasize graduate education, which, in order to prepare students for "scientific" research, shifted curricula ever more toward the sciences and mathematics.[39]

The NSB's immediate concern was a draft report to the president by the Special Interdepartmental Committee on the Training of Scientists and Engineers (with input from NSF representatives). The committee felt it essential to "improve the quality and increase the supply" of these trained Americans, because they ensured "our way of life" and determined "our future security and growth." Moreover, while America lagged, the Soviets' output of scientists and engineers was increasing and "approaching comparability with us in both quantity and quality."

The United States, with NSF at the lead, must improve teaching and enhance opportunities in the technical fields, the report urged. Hearing this, the board authorized the director to seek a $1.5 million budget increase for education. While a variety of factors stimulated concern from time to time, educating engineers and scientists would be an enduring NSF topic. The engineering profession was conducting its own series of studies on engineering education at this time. Its 1955 "Grinter Report" urged greater emphasis on science and mathematics fundamentals as more useful to engineers than specific engineering courses for coping with rapid technological change.[40]

Engineering's Place

For all that academic engineers embraced science and likened themselves to scientists, the reverse was painfully not true. The written record, polite and bland, suggests an early-developing undercurrent of uneasiness among engineers that their citizenship in the NSF was not quite first class. For one thing, too few engineers enjoyed power within the Foundation. They were rare among the professional staff. When it came time to nominate new members for the National Science Board in 1954, the Engineers Joint Council and the ASEE's Engineering College Research Council each forwarded half a dozen names in a clear effort to increase engineering representation, but the proportions on the board did not change appreciably for many years. It helped, a little, to mollify the engineers when in 1954 Thomas Sherwood, dean of chemical engineering at MIT, joined the MPE's Divisional Committee, a statutory consultant group of staff and representatives, primarily academic, from the various disciplines. He became its chair in 1957.[41]

Perceived monetary discrimination against engineers particularly stung. What Saville, who doggedly monitored NSF developments for the EJC, specifically asked Waterman in November 1953 is lost, but the director found it necessary to respond: "I consider the support of basic research in the engineering sciences of equal importance to the support of the other sciences in the Mathematical, Physical, and Engineering

Sciences area." Saville wrote again in August 1955 of the EJC's "grave concern" that engineering had suffered both a percentage and dollar reduction in MPE's allocation of funds for FY 1956. According to his figures, the MPE budget had risen from four to five million dollars, but engineering's share had dropped from $700,000 to $650,000—a 5 percent loss. In a formal ("Dear Doctor Saville") reply, Waterman tried to explain what "appears at present to be a reduction": the allocation was "preliminary"; MPE's specialized research, like that in radio astronomy, supported engineers but was budgeted elsewhere; and great increases in support of education (including engineering education) had to be compensated for somewhere. Within MPE's disciplines, Waterman continued, funds would be distributed according to the relative merit of proposals. Whether the EJC was satisfied with such reasoning is not recorded.[42]

A similar round of correspondence in 1957 between Morgen, then of the Purdue Research Foundation and representing the EJC, and Raymond Seeger, acting assistant director for MPE, suggests that the tensions continued. E. E. Litkenhous, program director for engineering in the mid-1950s, likewise made clear his dissatisfaction with engineering allocations within MPE and suggested that 20 percent be an arbitrary minimum figure.[43]

Early statistics were, in fact, showing that engineering sciences proposals were receiving proportionally fewer "excellent" and "good" ratings from reviewers and, hence, less funding. Why? Were scientists on review committees more harshly critical than engineers, as Karl Spangenberg, associate director of Stanford's Electronics Research Laboratory, suggested? Was engineering, as by definition too applied, at an inherent disadvantage when compared with, say, astronomy or mathematics? Was it true, as rejected applicants charged, that at NSF, "If a proposal is directly useful, it is of no interest; if it has practical application, it is turned down"?[44]

Was it possible, as Waterman seemed to be saying, that perhaps the fault lay with engineers who did not submit worthy proposals? Randal Robertson, who became head of Mathematical, Physical, and

Engineering Sciences in 1961, later suggested this might be at least partly true. Young assistant engineering professors, he opined, just seemed to "cook up" projects on their own, without the advice and guidance of senior faculty that, say, young astronomers were provided. So, while "Eric Walker and the other engineering groups" would complain, even the MPE Divisional Committee recommended against funding some engineering proposals, asking, "Where did you get *those?*" Perhaps some of those older senior engineering faculty, having spent most of their careers in specific, practical engineering applications, were unskilled in generating ideas or drafting proposals focused on basic principles. Perhaps a part of the problem lay with the relative paucity of graduate engineering programs.[45]

The MPE Divisional Committee appointed a panel on engineering sciences to look into the profession's lack of support; its chair, Frederick Lindvall, chair of the engineering and applied science division at the California Institute of Technology and ASEE president, reported back to the committee in late October 1957. His was not a statistical analysis of engineering support, he said, but rather an attempt to "indicate the place of Engineering Science among the other Sciences." The panel knew that place—or where it *ought* to be: "Everywhere in modern technology"—in the "frontier subject" of solid-state physics, for example—engineers used "tools, techniques, and objectives" that were "indistinguishable from those of a scientist" to pursue new information and theory in the basic sciences. Moreover, engineering education, undergraduate as well as graduate, was increasingly "turning toward an intensive scientific core," with ever more emphasis on research.[46]

NSF was, Lindvall said, in a "unique position to strengthen" American engineering. With the Defense Department now supporting 85 percent of engineering research, it was nearly all developmental in nature. Furthermore, Defense security requirements restricted publication. Even in industry, proprietary interests interfered with the free sharing of information. Lindvall's panel recommended that NSF invest more "risk capital" in unknown but promising young engineering investigators, provide more equipment and facilities to colleges, and emphasize the interdisciplinary character of much of current engineering when assessing engineering's contribution within MPE.[47] Engineers felt all these things to be not only true but obvious; that they felt the need to say them revealed their insecurity in NSF's world of pure science.

The sense of being left with NSF's crumbs persisted within the U.S. academic engineering community to such an extent that during the summer of 1958 the American Society for Engineering Education formed an ad hoc committee to improve the relative position of engineering as well as the relationship between the ASEE and NSF. Two of its members, Morgen and Gene Nordby, were former Foundation program directors for engineering. The committee did not criticize engineering's representation within the agency except to note that Engineering Sciences was seeking its sixth program director in about as many years, while other MPE programs had enjoyed continuing leadership. Replacing the string of rotating academics with a permanent director would be "highly desirable," the committee said, without pursuing the question of whether engineering staff left because of inadequate moral or monetary support from the institution. The committee made the first of many recommendations that NSF organize a Division of Engineering Science, "ultimately to become co-equal with the present MPE and Biological Science Divisions."

The ASEE investigators also looked for evidence of discrimination against engineering as an applied, not a basic, science but found none. There was plenty of fundamental research to pursue in the engineering sciences, they said. Nor did they see engineering science limited by a "physics point of view" among NSF engineering staff, as was sometimes charged. As for actual funding, the committee had seen the data showing that proportionally fewer meritorious engineering proposals had been funded than those of any other MPE program, except one year, when its share was second to lowest. Appreciating the difficulties of allocating "*very*" scarce" resources, the committee members recommended a "strategy of improvement rather than complaint." While admitting—perhaps insisting—that the line between pure and applied research was difficult to draw, they urged investigators to frame their proposals in the broadest, most basic terms possible and upgrade their overall quality. Play by the rules and play better, the ASEE advised its own.[48]

In August 1957 the National Science Board also took a critical look at NSF policies and activities. Board chair Detlev Bronk, who was at the same time president of the National Academy of Sciences, asked the members whether they saw the primary research objective of the Foundation to be the "development of new knowledge, the strengthening of our national life and economy, or the development of education and personnel." Potter, the engineer, spoke up immediately for the pursuit of basic knowledge, which had been his consistent posture. The board agreed and further decided that excellence had to be the first priority for granting research support.[49]

The numbers did show overall growth in engineering research funding. In FY 1952, engineering sciences had won 3 grants for a total of $41,900; in FY 1957, 103 grantees received $1.4 million for engineering research, the first year that engineering surpassed the million-dollar mark. The next year the number of engineering awards actually decreased, to 88, although the dollar amount modestly increased. All together, 284 awards had been given in engineering at that point for a total of $3.4 million. With NSF's total research support just short of $40 million, engineering's share was 8.5 percent.[50]

Then, on 4 October 1957, the Soviet Union launched *Sputnik,* the first artificial satellite in space. *Sputnik* not only represented "an affront to long-held assumptions of American scientific and technological superiority," it hurled the world into a new era. NSF felt its effects profoundly.[51]

THE *Sputnik* EFFECT

When Eisenhower addressed the nation on radio and television about *Sputnik* on 7 November 1957, his subject was "science in national security." A month after the fact, Americans were still in shock, their smug complacency shattered by a small earth-orbiting metal moon that *we* had not put there. Only after carefully detailing U.S. military might in such technologically sophisticated areas as long-range ballistic missiles and "clean" (lower-fallout) nuclear bombs did the president congratulate the Soviet scientists on their "achievement of the first importance." While earth satellites in themselves had "no direct present effect upon the nation's security," Eisenhower admitted that their powerful propulsion systems alone implied competence in advanced military technologies. The United States would have to pledge its eternal vigilance, military power, and economic and spiritual strength to counter the Soviet threat, the president said, and he put scientific education and increased basic research at the top of his list of action priorities.[52]

Sputnik's broad effects were immediate and significant. The National Science Foundation's budget went from $50 million to $133 million in one year. Engineering grew from $1.5 million to $4.2 million within the MPE divisional increase of $9.5 million to $23 million. The Foundation put itself through a "critical self-examination," which it included in its 1958 annual report as a study called "The Year of the Earth Satellites—the Status of Science and Education in the United States." By then, one of those orbiting spheres was American, but the Soviet slap to U.S. pride still stung. Congress created the National Aeronautics and Space Administration (NASA) as a successor to the National Advisory Committee for Aeronautics to pursue the space race directly, but there would be programs for NSF to support "up to the point where engineering for inclusion in the [space] vehicle has to begin." The Defense Department set up a high-ranking position of director of research and engineering. Both houses of Congress established standing committees on science and space. Government scientists and engineers got salary increases, and the National Defense Education Act of 1958 (PL 85-804) authorized the first federal aid to engineering education since the 1862 Morrill Act. NSF requested and got clearer legislative authority in Public Law 86-232, signed 6 September 1959, which permitted the Foundation to support programs (primarily teacher training) to increase research potential.[53]

How engineering would fare under the heightened attention was not clear. In mid-October 1957 Waterman submitted to the president *Basic Research—a National Resource,* a report on the Foundation that looked toward a "desirable balance between applications of science to defense, health, and the economy on the one hand, and basic research activity—the 'defense in depth' for our whole technology—on the other." A month after *Sputnik,* on the day of the president's speech, NSF's director spoke in San Antonio, where he acknowledged American disappointment at not being first in space but insisted that the U.S. satellite program was "soundly conceived" and "well engineered." American science was strong and healthy. "BUT—and this is a large but—we will quickly lose this advantage unless we provide greater support to basic research, and unless we set about reaching and training an even greater number of students with aptitude for science."[54]

The President's Science Advisory Committee (PSAC) issued a report in December 1958, *Strengthening American Science,* which asserted that not only national security but also national long-term health and economic welfare, the excellence of science, and the quality of American higher education were "now fatefully bound up with the care and thoughtfulness with which the Government supports research." Inadequate or erratic support that emphasized hardware to the neglect of fundamental understanding and failed to encourage exceptional scientists and research programs could result in "impoverished science and a second-rate technology." As NSF had urged, the PSAC said that the United States must earmark for basic research more than the current 8 percent of its now $10 billion research and development (R&D) budget.[55] So the nation's top scientific leadership agreed that engineering applications like building space satellites depended on increased research in fundamental principles more than increased effort in applied research itself. They also agreed that the support of the federal government was crucial.

Meanwhile, at least in part because of the Soviet challenge, NSF took on increased responsibilities within the federal establishment. Among these, the Foundation took over research on weather modification (mostly physical evaluation of atmospheric processes) as mandated by an act of Congress in 1958, established an Antarctic research program to continue the work of the International Geophysical Year, and increased its emphasis on oceanographic and atmospheric research as well as high-energy physics.[56] Some of these programs clearly represented applied, not pure, science, and engineering figured prominently in them. It was significant that NSF did not decline these additions to its program, even though they pushed the boundaries of the Foundation's original mission.

The space age also gave new impetus to basic research on the frontiers of technology, seeking understanding of fundamental phenomena and complex systems to guide new directions in engineering design. Thus, engineers had intensified interest in studying heat transfer to explain and control such effects as ablation (surface melting or vaporization), which was caused by the drag and heating of space vehicles moving through low-density atmospheres at hypersonic speeds. An NSF grant for a University of California project on thermal radiation, which accompanies rocket engines operating at extreme temperatures, supported some of the first basic work in that specialized area of heat transfer since the 1930s.[57]

Space travel accentuated a need long recognized in airplane design—for strong but lightweight construction materials in order to increase payload and range. Several researchers, such as D. R. Carver of Louisiana State University, began studying the dynamic behavior of sandwich-plate construction, that is, layering a honeycomb of stainless steel, aluminum, or other material between two thin plates of metal to maximize the strength-weight ratio. In 1964 Robert Schmidt at the University of Detroit studied the bending and buckling of five-ply sandwich shells. (In 1990 college students used these same principles to construct strong, ultralight, polymer composite materials for the bodies of the feather-weight solar-powered cars they designed and built to compete in the General Motors–sponsored transcontinental "Sunrayce."[58])

As Bush and others later said, *Sputnik* was primarily an engineering achievement. American engineers, like scientists, benefited from the surge of new money and attention—for the moment even at a slightly faster rate than some others. But prosperity without recognition would not long satisfy these professionals, who had, as Layton says, so consciously adopted a self-image based on science and keenly sought its social prestige as much as its financial status. Engineers were ready to push their boundaries.[59]

CHAPTER 2

TOWARD A LARGER ROLE

> It is pretty clear that engineering, although it has been ac-
> cepted as a member of the [NSF] family, has been in some
> respects treated as kind of a stepsister to the basic sciences.
> —Eric A. Walker to House Subcommittee on Science, 1965

To counter the Soviets' head start on ascendancy in space, President
John F. Kennedy announced the United States' intent to put an American on the moon before the close of the 1960s. The nuclear tensions
between the superpowers continued. Rachel Carson's *Silent Spring* in
1962 ushered in an era of environmental awareness. The nation's attempts to meet these and other technological challenges gave the National Science Foundation a decade of unequaled affluence—its golden
age. Yet this very generosity revealed politicians' desire, and need, to
focus on real problems and solutions. While NSF's advocates for basic science rightly grew nervous over the portent of such pressures, even
they sometimes felt boxed in by their own restrictive definitions. Meanwhile, the Foundation's engineering community grew restive, feeling
undervalued and inconsequential in an MPE organizational setting

dominated by the "pure" mathematical and physical sciences. Engineers pressed for administrative and professional recognition for their innovative and important contributions to society. In 1964, on the eve of the Foundation's fifteenth anniversary, engineering wrested divisional status from a reluctant institution.

PROVING ENGINEERING'S WORTH: BASIC, APPLIED, AND INTERDISCIPLINARY RESEARCH

From the beginning NSF's engineering community had actively sought what it considered a fair share of the Foundation's fiscal and programmatic attention. To encourage more qualified applicants to apply for research support, engineering staff contributed frequent articles to the technical literature explaining NSF's operating philosophy and funding procedures. For example, Gene Nordby and Robert Faiman, MPE engineering program director and engineer, respectively, gave engineers useful information about grant policies and favored research directions in the April 1958 issue of *Civil Engineering*. They noted that NSF's ground rules for supporting engineering research were a little different from those in other disciplines. For one thing, engineering projects often did not follow the usual classifications—say, aeronautical, chemical, or mechanical engineering—but rather were best described by areas of interest, such as fluid mechanics or thermodynamics. Use of such "science" terms, of course, would have political value in NSF's basic-science milieu, which the authors hinted at when they admonished potential grantees to let "time and economics" be their guidelines: Basic research in engineering would "advance a field of knowledge and deal with new principles," not "meet a deadline with a usable and economic result." Also, while agreeing with the MPE Divisional Committee's recent recommendation for giving priority to established investigators, whose work at the frontiers of knowledge would most likely result in greater achievements, the authors wrote encouragingly that NSF "*may* support the young untried scientist" with an excellent proposal to keep the supply of researchers and ideas from eventually drying up.

Nordby and Faiman cited ongoing NSF-supported projects to exemplify the fundamental nature of engineering research intended someday to solve real problems: Rutgers University's measurements of the upward migration of frozen moisture in soil, "thoroughly buttressed by theoretical analysis," could guide future highway construction. Investigators at the California Institute of Technology were studying the mechanics of underwater dunes, which was important for flood control and the management of natural waterways. Researchers at Georgia Institute of Technology were doing pioneering work on the effects of radioactive waste on sludge digestion, finding that such wastes could remove toxicity, stabilize organic matter, and reduce bacterial populations (though no one had yet figured out what to do with the long-term harmful radiation left in the system). Thus, application could be near at hand, so long as seeking "an economically feasible process, device, or product" was not the direct aim of the investigator.[1]

As the years went on, engineers found the struggle to stay within the basic-science boundaries increasingly problematic. For example, in 1960 NSF researchers in chemical process-control engineering at the University of Wisconsin proposed to develop an electronic closed-loop system to control a continuous chemical reaction by causing it to self-adjust in changing environments, thus continuously optimizing cost or yield of product. After studying the mathematical and statistical aspects of controlling a complex catalytic reaction, they would build and operate a small self-optimizing reactor. While most referees judged this proposal "excellent" and of fundamental importance, one called it "*not* a basic study," while another thought it would "make little contribution to basic statistical understanding, but rather would be a device of great utility." Work of immediate practicability, but still considered basic engineering, was that of the NSF grantee at the University of Michigan who derived and experimentally verified the mathematical correlation between the physical properties of materials and their machinability.[2]

Engineering, meanwhile, was moving in new directions, generally managing to keep the focus fundamental. By all accounts, interdisciplinary research in engineering was a rising star (although sometimes, it was said, resisted by university faculty firmly planted in traditional patterns). In 1959 two grants in plasma dynamics joined investigators in aerodynamics, thermodynamics, electrodynamics, chemistry, atomic and molecular physics, and applied mathematics. Another growing interest was the new field of bioengineering—the application of engineering techniques and principles to solve problems in the life sciences. Here, NSF worked closely with medical and biological researchers at the National Institutes of Health. Advances in communication science were emerging from the joint efforts of electrical engineers, biophysicists, experimental psychologists, neurophysiologists, linguists, and others. They included basic theoretical groundwork in the prediction and filtering of random processes, mathematical models of communication, and the logic of various elements of communication and computing devices. All this new knowledge would advance future pursuits as varied as computer design, signal transmission in the human nervous system, speech and language analysis, artificial intelligence, and cryptography.[3]

Another emerging topic, also multidisciplinary by nature, was system design and analysis, the study of the interrelationships among the parts of a complex technology. Investigators were finding, thanks to the awesome and increasing power of the digital computer, that systems could be comparable in mathematical structure despite dissimilar physical form.[4] Indeed, numerical analysis was becoming a "necessary complement" to experimental analysis, and still another engineering field could be anticipated.

All the while, growing "proposal pressure" was becoming an increasing worry. There just was not enough money to go around, especially with the mission agencies' funding becoming even more restricted to "oriented fundamental research." For example, the Atomic Energy Commission conducted applied nuclear research and development but looked to NSF to fund basic chemical, electrical, and physical research on nuclear reactions and the construction of experimental reactors at universities. In 1962 MPE recommended asking Congress for significantly larger appropriations, arguing that "we must empha-size that science is a necessity rather than a luxury and that to allow meritorious research to go undone is a waste of a national resource." But the Divisional Committee, meeting that October, also recognized that selling Congress on NSF's worthiness would require that the Founda-tion "demonstrate the impact of physical sciences on our social struc-ture."[5] A concept of responsibility to national need was emerging.

MAKING THE CASE FOR ENGINEERING

Engineers looked on from their MPE cubbyhole as the social sciences, which Vannevar Bush had called "political propaganda masquerading as science," won NSF divisional status in 1960. True, some admittedly uneasy board members made the social scientists "promise to behave as if" they were "really scientists" by pursuing only basic research and "punctiliously" meeting scientific standards of "objectivity, verifiabil-ity and generality." (They also had to forswear controversial subjects such as race, religion, sex, and politics.) If social scientists, who had gotten their nose under the NSF tent only through the dubious loophole of permissible "other sciences," could win official recognition, engi-neers could only wonder, why not them? They were *authentic* scientists. And they were increasingly resentful of their perceived short shrift in research grants, fellowships, and representation on the National Science Board.[6]

But engineering also won a determined, energetic, and vocal advocate that year when Harvard-educated engineer Eric A. Walker, president of Pennsylvania State University and the American Society for Engineering Education, was appointed to the National Science Board. With his typical color and candor, he later wrote, "I simply want to report that I outdid myself in eloquence when I pleaded the case for support of engineering research at meetings of the NSF board," and that "Van Bush [whom he considered to be a firm engineering advocate and disillusioned with NSF] would have been proud of me." Walker brought to Waterman—not his first attempt to change the status quo—an ASEE resolution enumerating the emergence of new research areas, the inter-relationships of engineering with other areas of research, increases in basic engineering proposals, and unprecedented growth in graduate engineering programs as evidence that it was "not only appropriate but necessary" to grant engineering divisional status within the Foundation. The engineering division of the American Association of Land-Grant Colleges and State Universities put forth a similar argument. In what must have been a coordinated lobbying effort, numerous other engineers' professional societies also pressured NSF, judging by the number and variety of responses that Waterman used to justify the Foundation's lack of action.[7]

Although sensitive to engineering's understandable need for the "prestige" of independence, Waterman held out, citing the desirability of keeping engineering within MPE because of the disciplinary "over-lap" and the insufficient (though admittedly growing) volume of engineering proposals. The social sciences, unlike engineering, were a "quite different field" from the physical sciences, he reasoned. More pointedly, Waterman stated, "If the engineers wish the Foundation to support engineering to a greater extent all they had to do was see that more first-class proposals reached us and more deserving applications for fellow-ships." Waterman suggested that what was most needed was a "really competent senior engineer in charge of our rapidly growing engineering sciences program." Moreover, administrative reorganization of the Foundation was "strictly the business of the Foundation staff," not advisory committees or other outsiders.[8]

However, when physicist Randal Robertson, former director of the MIT Radiation Laboratory and the Research Group of the Office of Naval Research, became head of the Division of Mathematical, Physical, and Engineering Sciences in 1961, engineers found a leader sympathetic to their insistence on improved status. If the argument was increasingly that of choosing basic or applied research and deciding which category engineering fit into, the "soft-spoken, thoughtful, high-principled" Robertson found it easy to define basic research in engineering as "any scientific activity that strengthen[ed] engineering practice."

Besides, applied research had already made inroads—in the social sciences, which were accepted as "relevant" to societal needs, and in the congressionally mandated weather modification program of 1958. Even Waterman himself, in reminding Senator Quentin Burdick of North Dakota in 1961 that NSF's mission was limited by statute to basic research, suggested that Congress might consider allowing more latitude.[9]

When the MPE Divisional Committee met in April 1961, Robertson reported that the National Science Board had decided in January to take no organizational action on the ASEE and similar resolutions but that "both new concepts in engineering as well as interdisciplinary efforts would be especially encouraged." Robertson acknowledged the "apparent paradox" involved in basic research and engineering sciences and saw that the issue of divisional status for engineering had become a debate over how far NSF should go in funding "applied and engineering research."[10] He was right. If engineering, which was linked with problem-solving, especially beyond its most fundamental aspects, won additional support and organizational latitude, could NSF's basic-science mission be sustained? At what point would the institution lose its original identity? Did that matter? They were approaching a crossroads.

In the meantime, Robertson kept the director informed about engineers' mounting insistence that their overall relationship with the Foundation be improved. While Robertson did not favor creating a position for engineering relatively higher than that for other disciplines, he could foresee a need for a division fostering both engineering "as presently conceived" and "applied programs important to U.S. technological development." Many areas important to the national welfare, he said, were not being adequately supported by the mission agencies, so NSF could rightly intervene.[11]

By June 1961 Robertson was ready to recommend an alternative approach to Waterman—that all seven of MPE's existing "program offices" (which, he said, were informal titles with no organizational status) become "sections." Further, he suggested use of the name "Engineering Section" rather than "Engineering Sciences Section" to "emphasize the breadth of our interest in the field of Engineering." The change, a sort of consolation prize in one sense, took place in October 1961 after nearly two years of needling by engineering partisans. In another sense, the broadened meaning of the shortened title, just "Engineering," was a significant concession. The engineering sciences were an important part of engineering, but the profession's scope was rightfully broader than that, the change implied. Samuel Seely, on leave from the Case Institute of Technology, came to head the new section, which included formal "programs" in engineering chemistry, energetics, mechanics, materials, and systems. Early on, Seely expressed the hope that sectional policy could "be modified to permit the support of a limited number of high grade proposals involving creative activities with broad intellectual implications even though they possess some attributes of practical engineering."[12]

The Foundation's 1961 reorganization officially recognized that engineering touched not only multiple disciplines within the physical sciences but also the biological and social sciences, and it allowed practical programs beyond the "classical disciplines" in broad subjects like design and systems engineering as well as in areas of national need, such as fire research, biomedical engineering, urban planning, and transportation engineering. Waterman said the new section would "seek rational methods for bringing all the resources of science to bear on the problems of our technological civilization."[13]

In May 1962, apparently in response to numerous professional inquiries, or perhaps at Walker's nudging, Waterman articulated for the National Science Board his views of what NSF-supported engineering should encompass. He recommended, and the board adopted, a resolution that "intellectual pursuits at educational institutions intended to advance significantly the basic engineering capabilities of the country" were eligible for Foundation funding as basic research in the engineering sciences. "Such work must be of a true scientific nature and not routine engineering practice, and must meet the usual NSF standards of originality and excellence." Waterman seemed to be trying to concede as little as possible, still using the narrower designation of engineering sciences, still proclaiming the Foundation's willingness to increase its support of engineering "upon receipt of meritorious proposals."[14] The *Journal of Engineering Education,* publishing the NSB resolution in January 1963, applauded the expansion of the types of research that NSF would consider supporting and cited Waterman's interest in interdisciplinary and "creative" work in systems engineering, such as energy, information, and environmental systems.[15]

But the unsatisfied engineering community continued to apply pressure for divisional status. The appointment of nuclear physicist Leland Haworth as Foundation director in 1963 upon Waterman's retirement provided a new excuse and impetus. The American Society of Civil Engineers (ASCE), for example, wrote to Haworth in August 1963 urging that "basic engineering and technological research be given full recognition and uncompromised leadership in its support." The ASCE also pointed out that since other nations, such as Canada and England, funded engineering research directly through separate agencies, NSF could fittingly assume a more primary role in engineering support. The American Institute of Chemical Engineers (AIChE) in September appended to their congratulations to Haworth a plea for greater awareness of the link between engineering and human welfare: "The discovery of scientific facts is merely one step in the process and no scientific fact in itself means anything until someone can use it for the production of energy or material for use."[16]

AIChE president W. R. Marshall's editorial in the December 1963 *Chemical Engineering Progress* forthrightly attributed the "general disregard for engineering by the federal agencies" to the establishment of the National Science Foundation, pointing out that the NSF bill's original draft did not even mention engineering. Despite repeated urging by the Engineers Joint Council and others, he wrote, there was still only one engineer on the twenty-four-member National Science Board, and a dismissive attitude toward the profession prevailed. He hoped that recent appointments of engineers to positions in other government agencies, such as General Electric executive J. Herbert Hollomon as assistant secretary for science and technology in the Commerce Department, would signal a better trend.[17]

Haworth was in many ways like Waterman. He was a "stickler for rules, detail, fine print—everything by the book." He, too, enjoyed colleagues' and Congress's unqualified trust in his integrity. And he spoke the familiar philosophical language. In his first annual report, for FY 1964, the former professor, Brookhaven National Laboratory director, and head of the Atomic Energy Commission emphasized that continuing progress in science and technology was essential to the public welfare and that the interrelationships among science, education, and government, at all levels, must remain healthy. The new director delineated at length what he saw as the differences among basic research (the quest for understanding nature), applied research (pursuits toward practical, usually specific, objectives), and development (design, fabrication, and testing of prototypes or processes). Basic research was the essence of the Foundation: "The technological fruits of research are harvested from a mosaic of knowledge made up of a great many experiments and the understanding derived from them—just as a tree springs from no single root but from many rootlets reaching deep down to the source of nourishment." But when Haworth wrote that expenditures for development, important though they were, should not be made to compete with investment in basic or "broadly based applied" research, a slight breeze of change could be felt stirring.[18]

The intense, chain-smoking Haworth also took pains to reassure engineers of his commitment to their profession—provided they measure up. In an editorial in the *Proceedings of the Institute of Electrical and Electronics Engineers* in January 1964, Haworth called engineering "one of the strongest bulwarks of the nation it serves." But engineers needed a broad scientific background coupled with creativity in design and development applications, and far more must pursue advanced education, he said. As engineers knew, the empirical methods of the past could no longer serve advanced technologies. According to Haworth, only 1 percent of U.S. engineers held doctoral degrees, while as many as one-fourth of practicing engineers in 1963 had no degree at all. These statistics were, in fact, beginning to change significantly. As research funds multiplied, so did engineers' credentials: Masters degrees increased from around 1,300 in 1940 and 4,800 in 1950 to over 15,000 in 1968. Doctorates went from some 100 to 500 to 3,000 during the same period.[19]

During this time, however, Haworth continued to resist an organizational promotion for engineering. Walker, who was unrelenting in his insistence for recognition, remembered his frustration. He later noted that he argued, in "one of the best speeches of my career," that "if the nation's nine hundred astronomers have an NSF home, its million and a half engineers deserve one," and managed to get the board to approve, "without audible dissent," a resolution requesting a division for engineering. But nothing happened. When nearly a year had passed, he questioned Haworth, who replied, in effect, "I didn't think you really meant it." The NSB's official record does not reveal this formal action; the minutes make quiet note that the board's MPE committee "requested Dr. Walker to consider [further] some important policy questions he had raised with the Committee (especially rapid utilization of scientific discoveries), and to report to the Committee at the January meeting." Another MPE committee action "requested the staff to consider ways of rendering additional support to the engineering sciences and to give more publicity to the recent shift in emphasis toward the support of creative engineering." It is possible that Walker remembered a stronger statement because of the keenness of his wish for it; conceivably he was thinking of resolutions he had brought to the board from various professional associations. But minutes, which are not a verbatim record, can skew meaning, by accident or design. It was also known that Haworth, shy and deliberate by his own admission, agonized overlong on decisions. The passionate Walker, who did not hesitate to poke the "parochial" Foundation, later muttered that "physicists have always had too much say in NSF"—a thought not uncommon among Foundation engineers over the years.[20]

DIVISION ACCOMPLISHED

What the record shows is an out-of-the-blue, one-sentence entry in the director's report to the National Science Board in a January 1964 executive session announcing that he was "considering the possible elevation of the Engineering Section to the status of a Division." Had the mounting pressures become too much? Whether Walker's election as chairman of the NSB on 21 May of that year mattered as cause or effect, engineers applauded the improving climate.[21] In a wide-ranging board discussion, Walker urged increased Foundation support for engineering sciences—not only for their own sake but also for the good of the national economy and the "enormous impact on the Congress and the public." More basic research in designing with prestressed concrete, for example, could result in "tremendous benefits" for the building industry as well as help Congress understand the "significance of basic research." Moreover, as he had earlier argued, sometimes "engineers want to explore just because they 'need to know' and not because they have missions to complete." In fact, Walker's comments stayed reasonably within NSF's basic-science model; he did, however, correctly forecast the increasing presence of politics in NSF affairs.[22]

At the same meeting, board member Harvey Brooks, dean of engineering and applied physics at Harvard and a scholar of science policy as well as technical subjects, wondered aloud if "basic engineering research" was not a contradiction in terms. If NSF moved more into engineering research as opposed to basic, meaning "generalizable," research in engineering science, how would it interface with the mission-oriented agencies and with industry? He also opined that the engineering sciences were losing interest among physicists, chemists, and mathematicians because "their underlying principles [were] well established." The important problems now lay in applying these principles to "increasingly complex real systems." Among the sciences in a stage of transition between physics or chemistry and engineering he placed much of solid state physics, microscopic physical metallurgy and materials, plasma physics, and electron optics. "All these subjects derive their importance," Brooks said, "from the fact that they underlie some problem in engineering *design,* although they can and should be investigated as sciences in their own right."[23] However individual members judged engineering philosophically, at this session the board approved one engineering grant, "Mechanism of Dropwise Condensation" at MIT, and twenty-nine staff declinations. That May, the NSB approved 137 engineering grants at $2.6 million and declined 489 at a requested amount of $9.1 million.[24]

In April 1964 MPE chief Robertson prepared for Haworth a list of advantages and disadvantages of divisional status for engineering. Most of the pluses centered around deriving increased support from the engineering community—within NSF and without—especially in universities, where engineering already enjoyed equal footing with the physical sciences. On the negative side, an engineering division would complicate the organizational tree, possibly make support of interdisciplinary work more difficult, require more space and personnel, and perhaps raise questions by the public and Congress. Robertson did not amplify any of these points offered in a bureaucrat's balance, nor did he mention a likely intensification of funding competition or philosophical differences. Finally, at its next meeting in late April the NSB unanimously "DEEMED it necessary that there be a Division of Engineering Sciences within the Foundation and AUTHORIZED the Director to activate such a Division at such time as he deems appropriate." On 11 June Haworth announced his intention to establish a Division of Engineering (not Engineering Sciences) and its statutorily required divisional committee. On 1 July the deed was done; in September the National Science Board approved it.[25]

What had been MPE would henceforth be a Division of Mathematical and Physical Sciences and a Division of Engineering. The board approved Haworth's choice of the term "engineering" alone to "signify the Foundation's intention to provide more general support for research in engineering." As Haworth expressed it to the press, the pace of advancement was fast closing the gap between engineering research and its practical application: "Today's developments in science and technology, the demands of sociological and cultural change, and the needs of national defense require that basic research in engineering be expanded. The new division will help to serve this purpose." During FY 1964 the Foundation provided approximately $13 million for basic research in engineering. Pinpointing with remarkable prescience emphases that would dominate in the coming years, the new division solicited proposals in systems engineering design, such as energy and information systems; interdisciplinary work in such areas as biomedical engineering, fire prevention, and transportation; and "analysis and synthesis of processes and systems which contribute to mastery of the environment."[26]

According to Walker, one reason Haworth had given for not moving forward quickly with the Engineering Division was that he could find no one to run it. Walker promised to solve that problem and took John Ide to the director's office. With a Harvard doctorate in communications engineering (1931), Ide came from the directorship of the SACLANT Antisubmarine Warfare Research Center, a NATO facility for research in oceanography and underwater acoustics in La Spezia, Italy. He had done geophysical research for Shell Oil Company before moving to the Naval Research Laboratory during the war. Ide duly got the job. Although a physicist by practice, he consistently won high marks within the engineering community for his understanding and sympathetic advocacy for engineering's goals.[27]

The new Engineering Divisional Committee, composed of academic and corporate leaders who met with engineering and other NSF staff, quickly began to explore "new directions," especially those areas cited as economically important in a new Engineers Joint Council report, *The Nation's Engineering Research Needs 1965–1985*. While not many high-quality proposals had yet been received in such growing fields as oceanographic engineering, the committee members thought NSF should "not stand idly by" while other agencies moved in. At the same time, the Foundation should keep within its own mission to avoid unnamed "political pitfalls," which had plagued other agencies pushing beyond their traditional territory.[28] The Engineering Divisional Committee first met, coincidentally but significantly, the same month that the National Academy of Engineering was finally established after nearly a century of effort, led in significant measure in recent years by Walker, who became its president in 1966.

Engineering's proponents concluded early that its major thrust, beyond the solving of engineering problems, had to be in the area of graduate research and education—to improve engineering college faculties and train engineers to ensure the future of the profession. By 1964 graduate engineering enrollments were increasing rapidly and at a higher rate than for other disciplines in the physical, life, or social sciences. While this appeared to be good news, more students required more research support and more teachers. NSF Graduate Traineeships, begun that year, assisted first-year advanced-degree candidates selected by the institutions receiving the funds. The first year's $6 million all went to engineering.[29]

Research Initiation Grants were even more important. During FY 1964, when engineering research grants in general almost doubled even before engineering's administrative promotion, the Engineering Section introduced Research Initiation Grants. Their purpose was to encourage meritorious younger faculty at smaller engineering schools that were striving for increased excellence. The need for such grants became clear when 581 proposals came in—far too many for the $1 million set aside for them. By squeezing here and there, the section managed to fund 106 proposals for a total of $1.5 million, still only one in six. Even this support, which came at the expense of regular research and equipment grants, could not continue within existing budget limitations. But the concept proved a good one, to be adopted throughout the Foundation and endure in some form throughout NSF's history. Engineering mechanics received the largest share of the first money but only about 16 percent of what was requested. The other major programs that benefited were systems, energetics, chemistry, and materials.[30]

The Engineering Division's internal annual report for FY 1965 was a fat one. With a good impression to establish, the division had much to say about its maiden year, which had enjoyed growth in all programs, including grants for traditional research, research initiation, specialized engineering equipment, conferences, and travel. One hundred and twenty-nine Research Initiation Grants totaling $1.5 million were, in their second year, awarded to seventy-one institutions—at a gratifying grant-to-proposal ratio of one to three. Or was it that fewer institutions applied, discouraged by the early, poor success rate of proposals? While merit continued to be the primary funding criterion, the review panels, heeding President Lyndon Johnson's democratic philosophy, deliberately sought the broadest possible distribution of awards and managed to include institutions in thirty-six states. For the first time, specialized research equipment appeared as a line item in the engineering budget. The initial budgeted amount of $500,000 grew from various sources to nearly $1.3 million, including half a million dollars for hybrid computer grants to the universities of Minnesota and Wisconsin.[31]

Five engineering programs enjoyed multimillion-dollar research budgets in FY 1965. In all, 541 proposers won grants worth $15.2 million—about a third of the number and a fourth of the money requested. Engineering mechanics came out on top, by far.

Program	Proposals		Grants	
	Number	Amount	Number	Amount
Engineering Chemistry	214	$ 8,733,239	88	$ 2,759,480
Engineering Energetics	233	11,032,055	71	2,741,035
Engineering Materials	250	10,579,699	77	2,557,650
Engineering Mechanics	412	18,602,751	139	4,115,150
Engineering Systems	199	11,786,944	64	2,745,830
Special Eng.	25	849,237	7	72,500
Foreign Travel	222	200,928	79	50,100
Conferences	18	140,915	15	72,795
Nuclear Reactors	2	375,750	1	97,000
Total	1,575	$62,301,518	541	$15,211,540

Summary of Engineering Division, Proposals and Grants, FY 1965[32]

The engineering report highlighted that

- engineering chemists were excited about work at Princeton University on separating fluid mixtures by parametric pumping, describing the technique as loosely analogous to the energy converted by a child's judiciously timed "pumping" on a swing. The process held promise for desalting water and for separating industrial and pharmaceutical chemicals.

- research in plasma dynamics, important in areas such as energy from nuclear fusion and spacecraft propulsion, was the big-ticket item in energetics.

- in materials, all manner of investigations in metals processing—metal flow through conical dies for manufacturing, flotation separation of minerals for mining, cutting and grinding techniques for hard and brittle materials—bore technological and economic significance.

- water behavior and management, from hydrological studies of rainfall runoff to the biomechanics of blood flow, particularly engaged the engineering mechanics community.

- systems engineers were studying hybrid computers (a marriage of the best features of accurate digital computers with fast analog machines) for work in control theory.

- advances in lasers, with their powerful coherent light sources, were leading to progress in holography. At the University of Michigan, for example, work in the X-ray range of the spectrum was making it possible to obtain three-dimensional micrographs from varying perspectives of crystalline and amorphous materials under huge magnification.[33]

HOT PURSUIT AND NATIONAL NEEDS

While the Engineering Division was clearly proud of its recent advancements, it also was prudent enough to look ahead. Questions of finding balance were foremost. Were current NSF engineering grants well proportioned among the engineering disciplines? What new areas of engineering should NSF be supporting? Like the rest of the Foundation in the self-examining 1960s, the Engineering Division wondered about its role in such societal problems as transportation, the urban environment, air and water pollution, and construction. Should it actively promote selected topics or just wait for proposals, as NSF had traditionally done? "It seems undesirably cautious to assume the purely passive approach," noted Engineering's 1965 report. Yet aggressively seeking "new" engineering could conflict with NSF's "basic" research mission, to say nothing of the interests of industry or other federal agencies. And what about long-range planning for research needs?

Indeed, feeling others' eyes upon it, the new division carefully considered engineering research within the Foundation's mission for basic research. The engineers noted that by defining basic research in each field as "the pursuit of new knowledge of fundamental importance to that field," NSF could reward more excellent academic proposals. However, the breadth of this definition confused even its own clientele, not to mention the broader scientific community. In the end the division settled on NSB's usage of the investigator's final objective as the factor determining whether the "product" was to be knowledge or use. But if NSF were to "promote" particular research more actively, especially in areas of national socioeconomic importance that did not match the familiar academic pattern, such as fire research or the effects of earthquakes on the infrastructure, inconsistencies in definition could be problematical. If research on basic principles for, say, a new type of motor succeeded in theoretical analysis, must funding for a prototype to test the theory be refused as "applied" research? Must an engineer or scientist be denied the right to "hot pursuit" of a promising lead that looks useful?[34] By 1965 legislators, citizens, and scientific bureaucrats as well as researchers were asking such questions.

One questioner was Hollomon, NSF engineering advisor and now assistant secretary of commerce. He wrote to Walker in April 1965 to complain that NSF was ignoring industrial and societal needs with great economic impact in favor of traditional academically defined physical sciences: "In the main, it seems to me that the NSF has not regarded as science such matters as the understanding of the transport system, application of computer technology to the simulation of communication systems, the relationship between housing and construction and design and engineering." Hollomon argued that these subjects were of basic national importance and did not involve any direct application to industry, commerce, or practical affairs. He thought it incumbent upon NSF to initiate and catalyze such research. In short, he called for a "rebalance" of NSF resources for Johnson's Great Society to have a chance at becoming reality.[35]

In summer and fall 1965, Engineering's Divisional Committee, which was renamed and more broadly authorized by executive decision as the Advisory Committee for Engineering, again studied program direction and distribution. Based on grant listings, Ide categorized about three-quarters of NSF engineering research as engineering science, with the remainder "slightly to the left of this," or "a little more applied than what is generally considered to be engineering science." Some "important applied or semi-applied" research could not be funded. At the same time Hollomon and some congressional committee members were charging that NSF was supporting only about $1 million of research valuable to the civil economy.

While Engineering's committee members protested that nearly all Foundation-funded research sooner or later addressed civilian economic issues, they differed on how to proceed. Should they simply document the relevance of their work more fully? Should they support broadened research on problems directly relevant to American society? Finally they agreed, in a formal "consensus statement," to ask the new National Academy of Engineering (NAE) to undertake a study of the nation's long-range research needs in engineering and to provide guidelines for future planning by federal funding agencies. But NAE could provide no immediate advice either; it would take some years before the newly founded organization would have sufficient members or resources to respond to such requests.[36]

The central debate was shifting. Where once the question between basic and applied research related to solving specific technical problems, now it encompassed the responsibilities of scientists and engineers to the earth and to humanity. As for accumulating knowledge, that would remain key at NSF, but soon it would no longer be enough.

NSF UNDER REVIEW

By 1965, the yet-small, relatively obscure National Science Foundation had grown appreciably in size and scope. Although it was still devoted to support of investigator-initiated, basic scientific research, NSF's character was subtly changing as new programs and responsibilities emerged—especially the so-called "Big Science" emphases of the 1960s. In 1965, NSF's fifteenth year, the agency, the White House, and the Congress all conducted critical reviews of the Foundation's progress.

NSF's retrospective appeared in the agency's annual report for 1965, where Haworth proudly pointed to the history of the Foundation in relation to U.S. national purposes. Echoing Bush, Haworth noted that when the "explosive" technological impact of World War II "crash programs" and the war-interrupted educations of young scientists and engineers "bankrupted" the "national store of unexploited fundamental scientific knowledge," NSF, with its basic-science mission, helped replenish the stock. While that operational doctrine remained "substantially valid," growth and maturity had broadened it, he said. Foundation grants continued to support the highest-quality research, however, and to allow maximum intellectual freedom with minimal administrative oversight.[37]

While engineering did not loom large in Haworth's report, an alert reader could find it there. Although Haworth did not state it directly, engineering components vitally affected the progress of many of NSF's most visible Big Science programs. The research centers at the Kitt Peak National Observatory in Arizona and the National Center for Atmospheric Research in Colorado were two examples. Project Mohole, an ambitious effort to penetrate and study the earth's mantle through sophisticated

undersea drilling past the earth's crust and the "Mohorovicic discontinuity (or Moho)," was another. This work, overseen by a Foundation
inexperienced in managing huge, multifaceted projects and performed
by a contracting firm whose closeness to Johnson exceeded its relevant
experience, would not prove a success. In the end—for political and
technical reasons—Mohole was NSF's only program to be killed at the
specific direction of Congress. But in 1965 Haworth still spoke of it as
an investigation of great promise for understanding the planet's inner-
most mysteries. That year Mohole consumed $24.7 million. U.S. inter-
national efforts operating under NSF auspices included engineering
participation in the continuing Antarctic Research Program begun under the International Geophysical Year, 1957–58, and the U.S.-Japan
Cooperative Science Program.[38]

The director mentioned two trends with special implications for
engineering. The first was interdisciplinary research. Haworth applauded the movement toward linking the natural sciences, social sciences, and engineering for the common good and thought that there was
not enough of it. The second trend, inferred from the first, was a recognition of the pressing need to use the pure and applied sciences to deal
with social issues in an increasingly complex society. The 1960s—a
decade that included campus protest, the increasingly unpopular war in
Vietnam, agitation for civil and gender rights, a "counterculture" movement, political assassinations, environmental pollution, and more—had
no lack of such issues to address.

Haworth, who continued to see NSF as the repository and pro-
tector of basic research—"the fountainhead of new ideas"—also recognized that by 1965 the national climate for broad federal involvement
had distinctly changed. Government support for science and engineering had become institutionalized and widely accepted. The social sciences had been brought into the federal circle, and in 1965 so had the
arts and humanities. Even federal support for education, long a state
enclave, was being widely favored as progressive and necessary. The
argument was that a strong, healthy nation demanded an educated popu-
lace, and educational opportunity was an entitlement in a democracy.[39]

Donald F. Hornig, director of the White House Office of Science and Technology, gave his fifteen-year assessment of NSF at a dinner celebrating the one hundredth meeting of the National Science Board. In wide-ranging comments that were both perceptive and challenging, Hornig praised the Foundation as "a great force in American science," but he also charged that it had "not yet fulfilled the leadership role its early supporters envisaged." NSF controlled only one-seventh of the federal research effort. On the federal role Hornig agreed with the President's Science Advisory Committee, which had stated in 1960 that either the government would "find the policies—and the resources—which will permit our universities to flourish and their duties to be adequately discharged—or no one will."

Hornig also pondered philosophical funding issues, that time-honored NSF concern. Grant awards based on merit alone clashed with democratic ideas of fairness and broad institutional development, he said. Under an elitist system, how could the government provide equal opportunity to all students with "substantially equal native ability"? Could government plan science and direct it toward identified national goals without conflicting with the idea of allowing scientific advance to spring from the minds of talented investigators? What happened to the quality of teaching when university research gained too much emphasis? Hornig thought it critical to "both broaden the base and sharpen the apex of our educational pyramid" by reserving some fellowships for award purely on merit while earmarking other programs for people and institutions "which show the latent capability to develop." Finally, the president's science representative agreed with the increasingly popular opinion that basic research and application of that knowledge to societal goals were properly inseparable, each strengthening the other. To allow applied research to follow basic, however, would require a redefinition of the programs of the Foundation. Hornig did not pursue the familiar issues he raised. But Congress was ready to do so.[40]

The House Subcommittee on Science, Research and Development, chaired by Democrat Emilio Daddario of Connecticut, undertook the third, "critical but constructive," and generally friendly probe of the

National Science Foundation in 1965. In a proper but previously unexer-
cised oversight role, the subcommittee asked the Legislative Reference
Service of the Library of Congress to review NSF's purposes, history,
and statutory authority so that Congress could assess its performance,
consider its future role, and determine what tools it would need to do
its job in coming years. Daddario's committee learned that U.S. preeminence in science was "now universally accepted," in significant measure thanks to NSF's influence, which was "far out of proportion to its
budget." Other agencies pumped more money into scientific research,
but the NSF ensured its quality by setting high national standards. The
report criticized the Foundation, however, for not providing federal
policy leadership and mismanaging certain large projects, such as the
Green Bank radio telescope, which had suffered construction delays,
and the Mohole Project, which was floundering over runaway costs, lack
of full-time, committed scientific leadership, and differing scientific
objectives requiring varied and expensive types of ships and undersea
drilling equipment. Finally, the report criticized NSF's relative lack of
support for engineering.[41]

Three months of congressional hearings followed. Haworth, the
lead witness, called primarily for more active "leadership" by NSF in
"promoting," not simply supporting, basic research. Waterman, too,
appeared before Daddario's subcommittee on 30 June 1965, still prais-
ing the virtues of basic research.[42] When Walker testified, however, he
spoke specifically and frankly about the frustrations of engineers, who
felt like "stepsisters" to the basic scientists within the NSF community.
He voiced engineers' regret at having settled in the "forming days" for
an agency that fell short of including engineering, both in name and
practice. The true nature of engineering was misunderstood, Walker
asserted. Engineering synthesized, not analyzed. It used knowledge
(together with money, personnel, and materials) to satisfy human needs,
not to accumulate more knowledge. But even theoreticians tested their
theories; why should engineers not be permitted the same right? He
urged Congress to reassess the definitions of science and engineering

and the growing gray areas between them toward the formulation of a fair and productive public policy.[43]

Thus, engineers, having tried with disappointing success to meet science on science's terms, moved to assert that engineering deserved recognition on its own merits, its differing values being equally valid. During the early 1960s, however, engineers' demands strayed only modestly beyond the "basic" research of the engineering sciences—the prevalent emphasis in U.S. engineering schools by then. Although, according to Layton, engineering science—"less abstracted and idealized" than science—is "much closer to the 'real' world of engineering," practicing engineers had begun to complain that many engineering faculty had become too theoretical and scientific to know, or care, about practical engineering experience.[44]

The profession was again on the cusp of change. This evolution in self-identification fits with Layton's observation that, ironically, as engineers became more scientific, they "tended to reject science, at least as a means of defining themselves and their social role." By the 1960s, some engineers went from "disenchantment" with science to an "outright repudiation" of it, claiming that engineering was the "broader and more challenging" profession. The NSF rhetoric, with a few exceptions like Walker, had not yet reached this intensity, but the engineering profession was catching up with its leaders, who chafed for a more central, self-defined role.[45]

It was politics, however, as politicians sought to mollify social discontent, that applied the forces of change to NSF's policies and even its charter. Congress was, in fact, poised for change. Effecting it would take three years.

CHAPTER 3

APPLIED SCIENCE AND RELEVANCE
The RANN Concept

> I believe that the time has come for the Foundation to assume
> a new posture, one that seeks the amelioration of
> the human condition through the wise use of our scientific
> strength, and one that recognizes that technological change
> is not only inevitable but can be adapted and managed to
> contribute, on balance, to the quality of human life.
>
> —William D. McElroy, *Science* 166:1252

When President Lyndon Johnson signed into law an amended charter
for the National Science Foundation on 18 July 1968, he ushered in a
new era—one that widened the Foundation's jurisdiction beyond its
basic-science mission and moved it into concerns directly affecting
society. At the same time, turnover in Foundation leadership brought
different outlooks and goals. The statutory and personnel changes carried profound implications for NSF's and the nation's engineering
communities. Engineering figured prominently in the resulting new
Foundation programs, especially Research Applied to National Needs
(RANN). Not everyone welcomed the new directions.

THE DADDARIO AMENDMENTS

The 1968 NSF act is still popularly referred to as the Daddario Amendments, for the thoughtful Connecticut representative who guided the legislative effort. Emilio Q. Daddario was a Yankee lawyer and decorated veteran of two wars when he won a House seat in 1958. He became a charter member of the House Committee on Science and Astronautics, the new standing committee formed in the wake of *Sputnik*. Interested in science since his school days, Daddario grew increasingly appreciative of the sweeping impact of science and technology on modern warfare from a stint in the Office of Strategic Services, where he analyzed the German war machine. He became science committee chair George P. Miller's choice to head the Subcommittee on Science, Research and Development when the California Democrat, himself a civil engineer, set it up in 1963. One of the Daddario committee's explicit charges was oversight of the National Science Foundation.[1]

Daddario proposed his first amending legislation for NSF in March 1966, armed with the Library of Congress's study of the Foundation, *First Fifteen Years,* and the National Academy of Sciences' report to the House science committee, *Basic Research and National Goals,* along with the subcommittee's own in-depth report, *The National Science Foundation, Its Present and Future,* which was based on the extensive 1965 hearings. A new law was necessary, Daddario wrote in *Science* in April, to "make NSF more sensitive to the shifting winds of our national scientific climate and the government's role therein." He meant in particular a congressional role, believing that federal science agencies had become too "pitched" toward the executive branch. Within the executive branch, he admonished NSF to "step forward and speak with the loud voice of a senior partner" lest it be "reduced to the nodding mechanisms of a junior colleague or the notetaking silence of a staff operation."

Daddario had no wish to diminish the Foundation's original basic research and education mission, he said, but it was time for NSF to direct "some research—basic or otherwise, and including engineering—to help bring the scientific base for new and emerging technologies required in the national interest to the point where their development can proceed through other federal agencies and industry."[2] Some of his colleagues, however, were less gentlemanly, more pointed. Science writer Daniel Greenberg noted as early as 1963 that Congress had become "querulous" over the mounting costs of research, and members' "own inability to grasp the significance of scientific matters" led them to demand evidence of likely payoff to justify the rising expenditures before they wrote more checks.[3]

As finally approved, the Daddario Amendments (sometimes called the Daddario-Kennedy Amendments to acknowledge Massachusetts senator Edward Kennedy's brief but key leadership) significantly altered several aspects of the NSF mission and structure.[4] Among them were the explicit inclusion of the social sciences and computer development as fundable disciplines; an expanded role for the National Science Board in promoting and reporting national policies for scientific research and education; and increased functions in international scientific cooperation and data collection. Four assistant directors, besides the director, would be presidentially appointed—a provision unanimously opposed by the NSB and Haworth. The statutory requirement for weather modification research was rescinded.[5]

Not only (although especially) for engineering, the path-turning change in NSF's charter was the specific sanctioning of applied research, which had been the tenth recommendation of the Daddario report. Section 3(c) of the new act authorized the Foundation "to initiate and support scientific research, *including applied research,* at academic and other non-profit institutions. When so directed by the President the Foundation is further authorized to support, through other appropriate organizations, applied scientific research relevant to national problems involving the public interest" (emphasis added). Thus, with three small words, tucked innocuously between commas, the National Science Foundation took on a new character.[6] In a March 1967 report, the House science committee had tried to soften the impact of this change by emphasizing that support for applied research was "authorized" but "made permissive . . . at the discretion of the Director." Support for applied work should not come at the expense of basic research, still emphatically NSF's primary mission; however, research aimed at solving major societal problems or engineering studies carried into "early phases of application" should be fundable.[7]

A review of the legislative history shows a tenor of testimony distinctly different from that accompanying the bills preceding NSF's formation in 1950 or, for that matter, from the chorus of calls for expanded basic research following *Sputnik*. Republican representative Charles Mosher of Ohio, a member of Daddario's subcommittee, spoke on the House floor in favor of applied research at NSF in July 1966 in nonpartisan arguments that became familiar with frequent uttering. He saw the distinction between basic and applied research as "artificial." To him the Foundation's present authority to support basic research in engineering was "somewhat contradictory since engineering involves in essence, the application of the laws of science—in other words, applied research." NSF's new authority could appropriately be used to fund promising research in this "ill-defined gray area" and allow a scientist the right of "hot pursuit" from basic research into logically following, potentially practical avenues. Or, as Donald Hornig, director of the Office of Science and Technology, had put it a few months earlier, "It shouldn't be necessary to stop work if there is any sign that it will become useful."[8]

Haworth, who refuted charges that he, like Waterman, backed only "pure" science by pointing to his own record in practical research, offered the Senate graphic evidence of the need for change. Under the present NSF act, he testified, the Foundation could give a fellowship to a student who could use the money to carry on applied research but could not support that same research under an NSF research grant. Both in House testimony and in an April 1966 letter to Daddario, the director stated his opinion that the Foundation had "reached the limit of what can be defined as 'basic research,' particularly in engineering." NSF needed a "clearer mandate to support work directly relevant to some of the problems of our society," many of whose solutions were engineering related. The meticulous Haworth later claimed that the applied-research authorization was added to the Daddario bill as a result of a conversation over a drink in the representative's garden where the NSF director had demanded that "they've got to make us honest."

Haworth was especially concerned with graduate engineering education, which was "commencing to swing back from a predominant emphasis on engineering sciences, divorced from empirical experience, to a more balanced approach." Increased emphasis on applied research by engineering graduate students was necessary to "develop engineers with greater competence to adapt new scientific knowledge into engineering practice for the welfare of the country" as well as to allow the "hot pursuit" of promising leads.[9] Bruce Seely's research confirms both Haworth's facts and their implications As research opportunities and the university reward system encouraged academic engineers to place more emphasis on knowledge and theory than on doing and design, they became more like scientists and less like real-world engineers. The boundaries between science and engineering became blurred, while a gap opened between academic and nonacademic engineers. By the 1960s, they "almost stopped talking to each other."[10] Now external forces were calling for real-world problem solving, and engineering was adjusting its approach, just as NSF was being called upon to do.

National Science Board commentary on Daddario's procession of NSF bills betrayed the conservative, scientist-dominated board's nagging discomfort with the encroachment of applied research. Basic science, Philip Handler, the new NSB chair,[11] insisted, was "among the most significant of human activities." Further, "since it seems perennially necessary to reconvince nonscientists that 'nothing is more practical than basic research,' there should be one agency in Government devoted exclusively" to it. Paraphrasing *Science—The Endless Frontier,* the board feared that over time a popular disposition toward applied science would incline Congress toward a "relative diminution in the basic science portion of the budget." Internally, board members fretted that "it would be very easy for NSF to mortgage its reputation on applied research ventures, where the criteria of scientific quality are less clear cut and . . . [more] open to criticism."

Nevertheless, the board, "on balance," accepted the legislative mandate as "desirable on several grounds." First, the dividing line between pure and applied science was unclear and arbitrary—and already crossed, as with Project Mohole and weather modification. There was a "genuine need" for applied work to "help bridge the gap between fundamental findings and socially useful systems or concepts." Also, program restrictions were causing some investigators to steer their research, and hence their educational impact, away from "legitimate applied interests" to those with better chances of funding. Haworth called this a "skewing" of American research with potential detriment to the needs of society.[12]

Among whose who questioned the legitimacy of NSF engineering and applied science was Frederick Seitz, president of the National Academy of Sciences, who worried that increased NSF involvement in applied avenues would reduce its support of basic research to "a starvation level." Yet he, too, was impatient over the fine lines that engineering, in particular, had been forced to draw. It was more important, he said, that proposed research be "soundly conceived and carried out imaginatively" than that "the investigator be motivated more by a search for truth than by an interest in the solution to challenging, difficult, and important problems which may have practical implications."[13]

The Engineers Joint Council applauded an increased role for NSF in addressing urgent national problems. As it was, the mission agencies had little tradition in such areas as urban development, while there was no potential profit motivating industry to pursue solutions to social problems. Augustus Kinzel, speaking for the National Academy of Engineering, noted that in 1962 engineering at NSF got only 7.4 percent of total federal expenditures for basic research; in 1965 the number was 6.6 percent. He considered this "decidedly less than good judgment" for a government trying to increase the economy and standard of living.[14]

Concurrently—and hardly coincidentally—with the legislative debate, the Department of Defense (DOD), growing skeptical over the payoff of $10 billion invested in research, perhaps a fourth of it basic, questioned the Bush premise that basic research was the pacemaker of technology. *Project Hindsight,* an eight-year effort that studied the respective contributions (viewed as discrete "events") of science and technology to twenty major weapon systems, concluded in 1966 that basic research had thus far provided but minor practical help to the military since the war. These findings were shocking but not unique. Writing in 1960 about the history of engineering, James Kip Finch gave numerous examples, from arch bridges to the vulcanization of rubber, to argue that "scientific explanation has followed rather than been basic to progress." Assistant commerce secretary Herbert Hollomon baldly asserted that "new technology flows from old technology, not from science." A final, gentler *Hindsight* draft, completed in October 1968, recognized the "less measurable" benefits of basic research and admitted they might "show up" in technology only years later. The report stood by its earlier findings, although in late 1970, a navy spokesman made the first of his conclusions on *Hindsight* that "a comprehensive base of science seems necessary to advance technology." Still, in Greenberg's not wholly sympathetic view, not only the "egregious outrage" of the results but the very asking of the question was a blow to the "ideology of pure science."[15]

NSF countered *Hindsight* with *Technology in Retrospect and Critical Events in Science (TRACES),* which appeared in 1969, after the Daddario Amendments. NSF's study showed the "overwhelming importance" of basic research to five recent technological innovations, crediting understanding from basic research with about 70 percent of the "key events" in their development. The Foundation report argued not that DOD's conclusions were wrong but rather that it had not looked back far enough in time; about 90 percent of the basic research on which the applications rested had been accomplished a decade or more earlier. Insisting that *TRACES* was a "complement," not a "rebuttal," to *Hindsight,* NSF affirmed that knowledge for its own sake was still essential to practical breakthroughs.[16]

Thus, growing support for a broader Foundation emerged within the technical community and in the political arena. But so also, at many levels, did philosophical conflict and ambivalence. The purists were loath to yield to any dilution of NSF's basic-science mission, and generally they dragged their feet, opposing the implementation of change. Yet they admitted, as Haworth did, that to curtail the contributions of engineering or applied sciences as not neatly fitting the traditional pattern made no sense and robbed the nation of valuable technical advances.[17]

In the end the NSF bill of 1968 passed by voice votes. Discussion was amicable. Relations between the National Science Foundation and Congress remained mutually respectful, even admiring. Daddario's committee had done its homework and understood and respected the Foundation's work over the years. Members of the NSF universe, many of them experiencing in Daddario's "responsible and intelligent" leadership their first contacts with the legislative branch, returned the compliment. But this was still, in Greenberg's phrase, a "new politics of science," a far different situation from NSF's first years, when scientists, despite mild government-imposed accountability, more or less determined their own research rules and basic-science agenda.[18]

Problems of environmental pollution, transportation, water supply, housing, and population pressures—the typical examples given—seemed to easily justify applied research on national needs. All would require interdisciplinary, collaborative research among the natural sciences, social sciences, and engineering—all explicitly included under NSF's new umbrella. As Daddario himself had argued, the time was "past due" for NSF "to assume a more positive, dynamic stance" and not passively wait for "talented outsiders to suggest appropriate projects" to fund. "The problems of living in today's environment are reaching proportions which are truly monumental," he wrote. "They will not be solved without an equally monumental lift from science and technology."[19]

ENLARGED MISSION, CHANGED MEANS

Thus, NSF had a new mandate after nearly two decades. What would it do with it? Before the Foundation had much opportunity to decide, two abrupt changes from different directions interfered with the anticipated flow of progress. One was that the Foundation's overall budget suffered a precipitous drop in FY 1969—its first decline ever. Congress's appropriation of $400 million for NSF operations was a decrease of about 20 percent from the $495 million of 1968. Engineering's dollars, too, began a rapid decline after four notably prosperous years. While not entirely a surprise to astute observers, who saw mounting frustration on Capitol Hill over problems like the mushrooming costs and poor management of Project Mohole, the cutback multiplied the recently increased pressures on the "new" Foundation.[20]

Anticipating such a cut, NSF deputy general counsel Charles Maechling, Jr., had warned Haworth in 1967 to use "some caution in wholeheartedly endorsing" the applied research clause in the then-pending Daddario bill. It could well, Maechling thought, become "detrimental" to federal support of basic research, not "additive" to it, as earlier assumptions had had it. Increased defense spending for the war in Vietnam, which would not likely end before the presidential election, was reason enough for pessimism. A budget deficit was likely, and all federal science programs—indeed, Johnson's Great Society itself—would suffer. (The new president, Richard Nixon, would soon show his own fiscal priorities; NSF's Golden Age was over.)[21]

The Engineering Advisory Committee, reporting to the National Science Board for 1968, saw the financial outlook differently. It predicted that NSF's engineering budget cutback, from an amount already too small for major impact, would be a "profound shock" to engineering schools, its effect disruptive and long lasting. Committee chair George Housner, a civil engineer from Caltech, foresaw a "major opportunity" for the Foundation, however, in the area of applied research and recommended that NSF prepare such a program and seek support for it "over and above the present funding of its Divisions." The committee presented a formal resolution to that effect to Haworth on October 1968, its focus on interdisciplinary efforts in university research and education.

Indeed, the committee emboldened itself to suggest that "in view of the present development and status of pure scientific research, . . . the emergence of many problems of great importance to society, and . . . the temper of the times," NSF should enlarge its engineering and applied research activities over a five-year period to "approximate equality" with its pure research programs. If anyone really expected an expansion that radical, the evidence does not appear in the record, but it was clear that the advisory committee's action was instrumental in moving forward the concept of applied research, with which these engineers comfortably identified, in an NSF climate not particularly friendly to change.[22]

The other significant change involved turnover in Foundation leadership, twice in short order. Furthermore, anticipating his departure after the 1968 election (though hurt at not being reappointed), the ailing Haworth left the congressionally created assistant director positions vacant. Although that gesture was generous to his successor, it left decision making and action to languish in the interim.

NSF's third director, William D. McElroy, who arrived in July 1969, was cut from different cloth than his two predecessors. The former director of Johns Hopkins University's McCollum-Pratt Institute and chair of its biology department eagerly embraced the Foundation's expanded mission and the political activity that accompanied it. By all accounts, the vigorous, outgoing McElroy embraced life. His work on the mechanisms of fireflies had brought him renown in the field of bioluminescence. Science was "fun." He equally enjoyed sports, partying, politics (although Nixon's choice, he had worked for Johnson and Humphrey in 1964), and the challenges of effecting change. He was a savvy and determined activist (some said "operator"—even "bulldozer"), announcing early his intent to see NSF's appropriation rise to a billion dollars annually and its popular image one of relevance and usefulness to society. As activists do, he ruffled feathers in his own agency. He stayed just two and a half years but decisively led NSF into the new era prompted by the Daddario legislation.[23]

When McElroy left to become chancellor of the University of California, San Diego, in 1972, H. Guyford Stever, a National Science Board member, assumed the directorship. He shared his predecessor's advocacy of activism and relevance, if not his leadership style. An orphan, Stever had earned a scholarship-supported degree in physics from Colgate University and a Ph.D. from Caltech. He had served as department head for mechanical engineering, marine engineering, and naval architecture at MIT and was president of Carnegie-Mellon University when called to NSF in 1972. Thus, he brought with him not only unusually eclectic intellectual baggage but also a strong engineering background.

In outlook Stever was a generalist, like McElroy. He was impatient with fearmongers who predicted that basic research would suffer at the hands of expanded applied investigation. Both scientists who worked best in solo pursuits of knowledge and those more productive in the presence of specific goals should be encouraged, the consensus builder insisted. Basic science remained NSF's primary mission, and while it was true that growth of fundamental research had slowed, the exponential expansion of the 1960s could hardly have continued indefinitely. What was more, "Science has got to come to grips with the current problems of the Nation and society."[24]

Another change, outside the Foundation's doors but significantly affecting it, was the so-called Mansfield Amendment of 1969, passed as an attachment to the military procurement bill for FY 1970. It forbade the Department of Defense from funding any basic research that did not directly bear upon military purposes. Senate majority leader Mike Mansfield, finding it inappropriate that since World War II more than half of the "government's contribution to science" had been channeled through DOD, where it nearly disappeared in the huge cold-war defense budget, sought to redirect sponsorship of basic research to civilian agencies, especially NSF. Intended for that role in the first place, the Foundation would be expected to pick up support of the bulk of the projects that Defense dropped. When other mission agencies inferred themselves to be under the same constraint as DOD, NSF found its resources even more stretched just when it stood at the threshold of broadened expectations.[25]

Still, NSF's broader mission was clearly timely. An unsettling antiscience climate was developing in both popular and intellectual circles. Books like Charles Reich's *The Greening of America* and Lewis Mumford's *The Myth of the Machine,* both of which assailed technological achievement as harmful to human life and values, were proliferating. Citizens were protesting the effects of technology, from nuclear waste to smoggy skies. How would the technological community respond?[26]

Engineers, were, in fact, showing increasing social consciousness, just as society was, despite the fact that "as a politically and socially conservative group, they generally reacted strongly against the counter-culture style of protest in the 1960s, which seemed to flow from a set of values so antithetical to their own."[27] Articles in professional journals probed engineers' ethical and political responsibilities as well as their unique opportunities for contributions to human welfare. Whereas earlier engineers' reform spirit had dwindled by World War II, as Layton claimed, now it was reviving. Whether creating a new introspective climate or responding to growing criticism, they began addressing hard questions about the quality of life—about pollution, overpopulation, traffic congestion, and more. Chemical engineers, for example, wrote opinion pieces on what they could and should be doing about dirty air and water, on protein synthesis "to feed a teeming world," and on alternative energy sources. (In 1968 they were also proud of advances in pesticide production "for individual comfort and greater crop yields"—a view that would change.) Mechanical engineers pondered "this spreading wave of disquiet" over the "deleterious side effects of technological progress." Where smoking chimneys were once seen as signs of prosperity, in concentrated population centers they now symbolized perversions of advancement. The ASME went so far as to prepare a long-term goals statement that emphasized "the importance of conscience as well as competence" and named leadership in the public interest, to make "technology a true servant of man," its top priority.[28]

NSF's Engineering Division, in an unusual comments section in its 1966 annual report, declared that technology's link between science and the problems of society had become a ubiquitous topic at engineering college visits and society meetings. NSF should therefore encourage innovative pursuits all along the basic-applied research spectrum. Engineering also called upon the Foundation to "take unaccustomed initiatives" (like soliciting particular proposals), to incorporate "important nonscientific elements" (like political goals) into its mission, and to promote interdisciplinary efforts.[29]

The NSF Engineering Advisory Committee agreed. "We believe," the members wrote in 1969, "that NSF and its Engineering Division have both the legislative mandate and an obligation to the nation to . . . seize [these] opportunities to serve more effectively the people's needs." They meant encouraging the universities, through innovative research and graduate teaching, to address compelling problems of society—especially the longer-term aspects of issues that the mission agencies were addressing in the short range. "The hope for economically optimum solutions to many of the nation's problems lies, after all, in *really* new ideas, developed through basic science, and tested in an NSF-Engineering program through the feasibility stage." Calling in 1967 for added attention to earthquake research, biomedical engineering, and computer development and application, the committee had already anticipated key areas of growth, all with major social components.[30]

To drive their point home, the engineers offered instructive examples of opportunities already missed. For one, the United States had led the world in understanding the science of ultrahard materials; from their knowledge of the physics and chemistry of high pressures, Americans had produced the first synthetic industrial diamonds. But in failing to exploit logically following technological development in ultrahard composite grinding- and cutting-tool materials, the country was yielding its lead to Japan and the Soviet Union. The same was true in the fields of cryogenics and lasers. Other countries were taking American fundamental findings into real applications while the United States lost out both in international competition and in practical problem solving at home.[31]

Handler, not a friend of applied research, simply argued for more money. In a study for the board's Long Range Planning Committee, he recalled the "superb" results from U.S. support of scientific inquiry in the 1960s: "American science led the world in virtually every discipline" and had generated a "huge technological capability." But the growth had now stopped, with funding constant for four fiscal years and inflation eroding its purchasing power by about 25 percent. Handler deplored such shortsightedness in the name of concern for human betterment, global economic competition, and national defense. The Soviet Union, he warned, was expanding its science efforts by 10 percent per year.[32]

INTERDISCIPLINARY PROBLEM SOLVING: IRRPOS

In this restless climate the National Science Foundation sought its new footing. The normally deliberative Haworth, still in charge, accepted his new duty and stepped out toward it with such vigor that even Daddario was dismayed. Following a normal practice, Haworth prepared a release, titled "Important Notice," for the universities announcing NSF's solicitation of proposals "for scientific research whether basic or applied." Daddario read in its tone a "distortion" of Congress's "*permissive authority* to support *relevant* applied research" into a mandate for it. Two decades later Daddario remembered the incident as an "overreaction" by NSF that was perceived as harmful to basic research even as it gave no particular help to applied. Why did Haworth not see this? Was he too untuned to change to lead it effectively? Possibly. But he *had* requested the additional authority. Haworth was later remembered as being "all for" applied research and long "perturbed that we had to hide the Engineering Division under language that tried to make it sound like basic science." Perhaps the baldness of his notice simply eluded the literally correct Haworth; the widowed and lonely workaholic was also ill for much of his tenure. In any case, the resulting tension was unfortunate, the more so for being unnecessary.[33]

That false start aside, Haworth did follow up on Congress's permission and the Engineering Advisory Committee's October 1968 resolution and appointed Randal Robertson, then associate director for research, to the task of establishing an NSF program to address societal needs. Robertson was eager to lead the Foundation beyond the confining restrictions of its basic research mission. The MIT-educated physicist (Ph.D., 1936) had spent most of his career in a nonacademic setting, twelve years of it at the Office of Naval Research. There he had promoted research "relevance" and the concept that basic research was research fundamental to the field in question, thus avoiding an "abstract . . . criterion of scientific purity."[34]

For his part, the outgoing director may have sensed an opportunity to extract a little extra money from the pinched FY 1970 budget by proposing a new program that would leave basic research funds intact. In November 1968 he requested a budget add-on of $15 million (later reduced to $10 million) for interdisciplinary research on problems of society. Those words soon became a program title, with the acronym IRPOS; Haworth subsequently insisted on changing it to Interdisciplinary Research *Relevant* to Problems of Our Society (IRRPOS) in a attempt to suggest less-sweeping promises.[35]

IRRPOS lasted less than two years, with barely a year of actual operations. Reasons for its creation as well as its demise provide context for the RANN program—Research Applied to National Needs—which would soon engage and then engulf the Foundation. Engineering swam in the thick of it. The IRRPOS program first became public during the Foundation's authorization hearings before Congress in March 1969, a new requirement in the budget cycle under the 1968 NSF act. Unfortunately, that was also when Daddario first heard about it. Haworth's lapse of judgment in not having Daddario briefed in advance was beyond discourtesy and proved costly to the new NSF effort. After a relatively hostile grilling that implied that IRRPOS was too nebulously and unimaginatively conceived to accomplish its objectives, Congress appropriated $6 million for the program.[36]

As the IRRPOS planners waited for funding to materialize, the mantle of Foundation leadership shifted in July from Haworth to McElroy. Firmly supportive of NSF's bedrock commitment to basic research but also unburdened by philosophical problems with its expanded mission,

McElroy soon told Robertson to draft plans and proceed with IRRPOS even with a minimal appropriation. In late October 1969 the director announced IRRPOS internally and to the universities as a "new and significant step in the Foundation's activities" that would "encourage the academic community to do research directly relevant to vital problems of our times." McElroy made it clear that proposals must feature "an explicit indication of societal relevance and potential social impact," including economic and policy-related factors. Purely scientific research, even if interdisciplinary, would be funded through existing programs.[37]

McElroy placed the management of IRRPOS in the newly established Office of Interdisciplinary Research, to be headed by "young, bright-light" Joel Snow, within the research directorate. He created an interdisciplinary, interdivisional committee to help develop the IRRPOS program and integrate it into existing activities. An external advisory committee would help identify problems and suggest research strategies. Covering all bases, the politician McElroy touted the new program in a mid-December 1969 news release sent to numerous political leaders, legislative and executive, and to the press. In March 1970 he informed the National Science Board in detail; not for the last time under McElroy, NSB members grumbled about being the last to know.[38]

When McElroy's eight-person internal committee reported back in December 1969, its chair, William Consolazio, admitted, in his separate cover memorandum, that NSF staff as a whole, including some committee members, had "serious misgivings about the new look and is, putting it mildly, resistant to change." NSF's statutory authority and mission could encompass IRRPOS research, the committee decided, within the limitations of available resources. But the Foundation must consider the importance of the problem, the potential for solution, and its socioeconomic payoff as well as the project's "public and political visibility"—all new criteria. Moreover, "a philosophy of active programming" would have to be "grafted onto the traditional essentially passive NSF *modus operandi*."[39]

Deputy assistant director for research Edward Todd was also cautious. With his characteristic bluntness, he warned that a new emphasis upon problem-orientation, "even if confined to one-fourth or less of the total program, represents a profound change" in NSF's operational philosophy, tradition, and practices. The Foundation's clientele—the universities—would be understandably confused, especially in light of the Foundation's basic-science image, which both NSF and Congress had so long cultivated. *Chemical Engineering Progress* agreed that IRRPOS was a "remarkable departure" from NSF's original mission but, fueling traditionalists' fears, applauded it, concluding that "it thus appears that NSF has finally learned that 'Science Ain't Everything.'"[40]

The academic community appeared to respond to IRRPOS with alacrity—for the sake of its new money if not its concept. Testifying in February 1970 before the House Subcommittee on Science, Research and Development for the FY 1971 appropriation, Snow reported that before the IRRPOS program was even announced, almost three hundred informal inquiries and more than fifty preliminary proposals had come in. Formal proposal traffic, however, later proved to be much slower. NSF made twenty-one IRRPOS grants by June 1970, mostly in the fields of the environment, urban problems, and energy. Despite the lateness of the funding start, IRRPOS managed to commit $5.98 million of its $6 million appropriation. The Foundation requested $10 million again for the second year of IRRPOS, and although federal funding for research was then dropping off markedly, even losing ground as a percentage of national economic activity, the Office of Management and Budget (OMB) increased the FY 1971 amount to $13 million. That number survived Daddario's stern scrutiny and the rest of the budget process.[41]

Meanwhile, the NSF Engineering Advisory Committee continued to monitor the progress of IRRPOS, to which it rightfully claimed parentage. The engineers were pleased when Snow reassured them that he "fully expected engineers to be the largest segment of investigators on proposed programs." Among their February 1970 resolutions was one recalling the engineering community's "long-established tradition of solving problems facing our society or segments thereof." The advisory committee then offered itself, augmented by "suitable additions" from other advisory committees or the professions, and a social scientist, as an interim ad hoc advisory committee for IRRPOS until a permanent body could be formed. McElroy, who replied that he had his own IRRPOS review panel in mind, promised to appoint two members of the Engineering Advisory Committee to it. He said he would keep the social scientist idea "under careful study."[42]

At the same time, however, the Engineering Advisory Committee "strongly and unanimously" recommended a "giant increase" in NSF support for engineering. The scope of national problems depending upon engineering solutions was so great that NSF should go beyond funding research to coordinate and focus national attention on these problems. Economic competitiveness, too, required more money for engineering, in areas ranging from enzyme engineering to telecommunications research. Interestingly, while speaking specifically about societal needs, the committee requested this budgetary support through the Engineering Division, not IRRPOS.[43]

For all the enthusiasm it levied in some quarters, IRRPOS stumbled along with appreciable difficulties. Staff was inadequate in numbers, and those pulled from NSF's traditional divisions sometimes resented attention they considered, perhaps unjustly, stolen from their programs. Despite promises of innovation, the program ran conventionally (responding passively to proposals), except that the review process was necessarily faster and less formal in order to get projects moving. That made quality difficult to sustain. The Foundation's administrative mechanism of assigning projects within NSF's bureaucracy by the interdisciplinary criterion rather than relevance was seen as a "major impediment" to applying science and technology to social problems. That is, highly relevant but intradivisional proposals would not fit the IRRPOS mold, to their disadvantage.[44]

In any case, if with palpable uncertainty, "even the National Science Foundation, once the federal establishment's lone bastion of nothing but basic research," was paying attention to "fundamental science's attractive neighbor, utilitarian research."[45]

RESEARCH APPLIED TO NATIONAL NEEDS

Whether IRRPOS would have prospered in the long run cannot be known, since in December 1970 the Office of Management and Budget suddenly offered McElroy a $100 million budget increase (NSF had modestly requested a repeat of $13 million for FY 1972) if the Foundation could design a new program to harness science and technology in the service of national needs.[46]

The budget bonanza came attached to costly OMB conditions. NSF must terminate its institutional development programs; fellowships and most other educational programs must also be phased out. Of the increase, about half would be earmarked for other agency research that had been cast adrift by the Mansfield Amendment. NSF would be expected to solicit Congress for additional money for basic research (a carrot more willingly pursued). Finally, simply reformatting IRRPOS would not do. OMB insisted on a new program, managed separately and aggressively to identify and sponsor research on social problems.[47]

McElroy, who had anticipated the 1970s as "the Transitional Decade," did not hesitate. Here, after all, was a chance to redeem NSF's budget for basic research. Ignoring the wishes of the powerful Office of Management and Budget, which spoke for the administration, would have been difficult in any case; turning down a bigger empire was probably beyond fair expectation. Indeed, the National Science Foundation had already been listening to the changing voices. In defending its budget request for engineering, NSF as early as 1967 began talking about the growing trend toward engineering "relevance"; the next year the words "national needs" leapt from the page.[48]

From our present vantage point McElroy's willingness and prescience seem not so remarkable. Nixon, like Johnson before him, was more interested in the usefulness of scientific endeavor than fancy theories or "pure" research. While Eisenhower had summoned science to exploit and then to tame the atom, and Kennedy's challenge to walk on the moon within a decade had been met in 1969, for now there would be no new ringing calls to esoteric frontiers of knowledge. Confronting the multitude of mundane but pressing problems of sustaining quality life at home had to take precedence. Knowledge without intent for utility was becoming difficult to justify. Also, both Johnson and Nixon resented the escalating clamor of public protest. Supporting "ivory-tower" academics on whose campuses "foul-mouthed" students denounced their curricula as irrelevant, the war in Vietnam as immoral, and industrial technology as deleterious to society and the environment had diminishing appeal. Further, they could point to statistics to argue that the nation had no need to encourage growth in the numbers of engineers and scientists; rising unemployment among engineers was reported at 2.7 percent in 1970, 3.2 percent in 1971. Presidential science advisor Edward David argued before the House science subcommittee in 1971 that NSF resources should be shifted from educational and institutional programs that produced an increasing quantity of unneeded science graduates to those that utilized scientists and engineers more effectively.[49]

The Foundation had already requested guidance from the National Academy of Engineering's Committee on Public Engineering Policy (COPEP) as it began feeling its way into the unfamiliar territory opened by the 1968 NSF act. The latter responded with two major studies in 1970. *Federal Support of Applied Research* stressed that while basic research responded well to the familiar "bottom up" (wait for proposals) style of program administration, a sensitive "top down" approach (where "the sponsor must join in defining needs and priorities") would be more appropriate for applied research. COPEP also strongly recommended that 20 to 30 percent of an augmented NSF budget be designated for applied research, a specificity that NSF was not eager to embrace. By supporting high-quality applied research, COPEP judged, NSF could fill important research gaps, especially research with long-range goals, which "market or mission forces are not sufficiently motivated" to pursue. The NAE study emphasized that federal support of applied research should include the creation of "bridges," such as transfer mechanisms or liaison personnel, from basic research to development.[50]

In *Priorities in Applied Research: An Initial Appraisal,* COPEP attempted to "marr[y] in one priority-ordering effort" the "technical feasibility and promise" of research with "social need and utility." In brief, the committee members recommended supporting work that was interdisciplinary and problem-oriented. Interdisciplinary efforts would reduce the parochialism of the "not-invented-here syndrome" prevalent among some academics, they said. As for national problem-solving priorities, understanding the structure and dynamics of the biosphere should be NSF's first. Other program recommendations, distilled from seven hundred suggestions by academy members, included developing techniques for applied social research (such as simulation modeling) and research on materials, construction and transportation systems, and electronics (especially computers).[51]

The COPEP reports, along with new deputy director Raymond Bisplinghoff's preliminary notes for McElroy, proved invaluable to Snow, who was named to head an eight-person task force with the unenviable charge of hammering out a budget justification, programming strategies, and a management structure for the new effort in scarcely a week's time. The timing of the National Science Board meeting and the FY 1972 budget cycle left no room for leisurely planning. On 17 December 1970, the board, with apparent discomfort but no written dissent, had approved in one unamplified sentence "the Director's general organizational and program plans for expanding Foundation support in applied areas, as authorized by the 1968 amendments to the NSF Act." What McElroy had proposed was a broad outline to bring together NSF's existing problem-oriented programs (including IRRPOS) and new ideas into a directorate for research applications. To help sell the program, he projected that a good deal, as much as 40 percent, of such national problem solving would fall into the basic research category; however, some "very suspicious" board members "dragged their feet and tried to sabotage it" at "every opportunity."[52]

Sometime during that intense, hectic Christmas week, the task force hit upon a name for the new program: Research Applied to National Needs, or RANN. McElroy considered the phrase "research applied" a stroke of genius because it could imply that either basic or applied research would then be directed to the solution of identified problems. On 2 January 1971, a Saturday, he and NSB chair (since

1970) Herbert Carter, a chemistry professor from the University of Illinois, gave their approval to the name and the program concept, although Carter would have preferred "Research *Applicable* to National Needs" as a more realistic commitment.[53]

NSF brought Research Applied to National Needs before the House Committee on Science and Astronautics during NSF's authorization hearings. Handler, previously board chair and now president of the National Academy of Sciences, objected—not specifically to RANN but to the gutting of educational and institutional programs, which he saw as a reversal of NSF's own policy. (Congress agreed and restored those funds, most of which the president subsequently impounded.) Elsewhere Handler complained that RANN's predecessor, IRRPOS— then in existence one year—had "not yet solved any major problem of society." Snow found this line of reasoning "peculiar," because "a scientist always figures that it takes years to do research, then years to expect the results to mean something." The National Science Board also drafted a letter to the Office of Management and Budget and the president's science advisor deploring these specific losses, especially during sharp reductions in universities' other flexible funds. Facing increasing skepticism on Capitol Hill, NSF staff issued a blitz of press releases outlining the major points of every Foundation leader's positive testimony.[54]

In the end NSF got a $109 million increase overall, including more than $40 million for basic research (even after subtracting funds to cover the Mansfield Amendment "dropouts"), which was the largest increase in NSF scientific research project support in any single year. RANN obligated $34 million in FY 1971. For FY 1972 Congress allowed RANN an amount "not to exceed $59 million" from its request of $81 million. In contrast to the hesitant Congress, the president, through OMB, welcomed RANN, asserting that NSF "should draw on all sectors of the scientific and technological community in working to meet significant domestic challenges."[55]

The National Science Foundation announced the creation of the Directorate for Research Applications on 2 March 1971. McElroy named

Alfred J. Eggers, Jr., as assistant director for research applications and head of RANN. Eggers, an experienced technical manager and expert on supersonic and hypersonic flow theories, was among many high-level "transplants" from the National Aeronautics and Space Administration (NASA) at that time. The engineering mechanics graduate came eager to build bridges from the physical sciences to engineering to society. Snow, who had had to dismantle his own IRRPOS program to design RANN, was named deputy assistant director for science and technology.[56]

The employees and functions assigned to the Office of Interdisciplinary Research (that is, IRRPOS) were transferred to Research Applications. So were specified programs, such as weather modification, and other projects (mostly engineering) from the Research Directorate, including earthquake, enzyme, and power systems engineering along with fire research and urban and regional systems. Also shifted were priority assignments from the Executive Office and other relevant proposals in hand, including some originally submitted as institutional grants, which were significantly revamped at NSF invitation—and OMB insistence—to meet the national needs criteria. So RANN began with an adopted family.

Planners of the goal-oriented program then set about to develop coordinated research efforts based on problems, not disciplines, within three technical divisions—Advanced Technology Applications, Social Systems and Human Resources, and Environmental Systems and Resources—and the Office of Exploratory Research and Problem Assessment. They established as criteria the importance, urgency, and timeliness of the problem; likelihood of payoff in economic and social benefits; potential for scientific and technological leverage on the problem; scientific readiness and capability (federal, academic, and industrial); the need for federal action; and the possibilities for a unique NSF contribution. No matter how they were configured at any one time, RANN research projects tended to group around three topic areas: the environment, urban problems, and energy. Later, productivity became a significant concern as well. Problems within them often overlapped.[57]

Engineers, who for the most part readily identified with applied, problem-oriented research, welcomed NSF's wider umbrella—though not without reservations—and anticipated being more centrally engaged in Foundation affairs. For example, members of the Institute of Electrical and Electronic Engineers, recognizing that solutions required "both political and multidisciplinary technological approaches," pledged themselves in 1972 to "take the lead in helping apply electrotechnology to social problems." They outlined a dozen specific problems, involving technologies from communication to medicine, with suggested technological solutions needing additional research.[58] They were singing RANN's song.

Eggers, in the job not quite a year, appeared before the House Subcommittee on Science, Research and Development to justify RANN in late February 1972. He opened by noting that RANN's purpose was "precisely in accord" with Nixon's State of the Union message on "harnessing 'the discoveries of science in the service of man.'" He spoke of selected national needs, RANN's "gap-bridging" between NSF's basic research programs and the R&D of the mission agencies, and the multifront approach his program was taking. For example, with support from RANN (and the U.S. Army Corps of Engineers), the Chesapeake Bay Consortium (the Johns Hopkins University, University of Maryland, Virginia Institute of Marine Sciences, and Smithsonian Institution, in cooperation with several federal agencies) was conducting research to gather baseline data on the great estuary and evaluate changes from human causes. Of particular interest were pollution from power plants, nutrient loading from waste discharge, and accelerated erosion and sedimentation, especially as these factors affected wetlands, shorelines, and shallows.[59]

Science & Government Report in July 1972 took note of the new RANN program and reported on remarks overheard "among oldtimers in Washington science-policy affairs" that the Nixon administration was pressuring the "traditionally 'pure'" NSF into applied efforts for "quick political payoff." The *Report,* published by the iconoclastic Greenberg, decided that RANN was "a long way from either subverting the

Foundation or bringing the blessings of science and technology to the nation." After outlining several RANN undertakings, the newsletter opined that RANN was "emerging as a cautious and fairly reasonable response to widespread interest in 'relevant' research." Since all major industrial countries were trying to stimulate nonmilitary research, the *Report* found it "difficult to see why RANN should be regarded as anything but a threat to obsolete ways of relating science and government. If anything, it might be faulted as too timid a threat."[60]

Bisplinghoff, whose brainchild RANN substantially was, agreed when he gave a RANN progress report to the American Association for the Advancement of Science in December. The former NASA official and MIT engineering dean anticipated that major science policy questions of the 1970s would deal with applying research and development to the solution of national domestic problems, whereas the current R&D infrastructure had been developed over thirty years in response to military, space, and atomic energy objectives. He saw NSF's RANN program as a "modest beginning" to the organizational changes that must be made to serve national needs effectively. Already, Bisplinghoff noted, the problem-directed research effort, spanning basic and applied work, had produced new insights, wider understanding, and a body of generalizable principles.[61]

From the beginning Research Applied to National Needs operated under peer scrutiny. The Research Applications Advisory Committee had been organized in October 1970 to advise the Office of Interdisciplinary Research; even before its first meeting it became the RANN evaluating group instead. First reporting to the National Science Board in April 1972, the group's members praised the RANN concept and were sympathetic to the difficulties of effectively and imaginatively coordinating problem-oriented research, especially with skimpy planning time. They offered abundant criticism and advice as well, which the RANN staff solicited and tried to implement.

Because change threatens, the RANN advisers admonished NSF leaders to devote more attention to interpreting RANN to the scientific community, within and beyond the Foundation. They saw a "strong strain of critical resentment" to RANN, partly because scientists misunderstood its purpose and methods. They warned RANN managers not to let the programs become too diffuse and thus endanger significant achievement in any one area. RANN's early projects were uneven in significance, probably because they were inherited from previous programs. The RANN committee cautioned against zealous oversell and overcommitment without long-term funding assurance. To achieve socioeconomic benefits, sufficient attention must be devoted to the transfer process by which research findings would come into use. All of these points appeared again and again throughout RANN's existence.[62]

The Office of Management and Budget, too, kept a watchful eye on the fledgling RANN. In February 1972 the budget agency's Carlyle Hystad interviewed NSF staff to test their responses to the new program and the effectiveness of its unusual approaches. He found a few people unhappy with the top RANN management, but the divergence of their views (too much management involvement, not enough guidance; too industry-oriented, not enough concern for industry) suggested to him that a prudent middle path was being taken. OMB was satisfied with management quality but agreed that the main problem was numerical; there simply weren't enough staff members for the workload. As a result, RANN was slow to obligate its funds, which could jeopardize its position on the next budget round. Overall, OMB gave RANN a good report.[63]

In spring 1972, Eggers asked the NAE Committee on Public Engineering Policy to "update and refine" its earlier recommendations on problem-oriented research for the current (FY 1974) budget planning process and long-range policy formation. Eggers asked COPEP to identify "prime technological targets of opportunity" and recommend specific, ranked programmatic initiatives particularly appropriate to the Foundation. The COPEP planners organized a workshop of themselves, RANN Advisory Committee members, and other specialists in September and reported back to NSF in November 1972. Their final report,

Priorities for Research Applicable to National Needs, was published in
1973.

These engineering policy reviewers identified two new programmatic priorities: "institutional functioning" (meaning primarily the
functioning of delivery systems) and "conservation and patterns of consumption"—both social science issues. For the first, they emphasized
that improving the human components of institutional operations and
achieving technical innovations in the natural sciences and engineering
were "equally necessary to resolve the problems of society," each being "treated in context with the other." They called for "more adventurous
and imaginative research" to "anticipate and outflank" the limitations
of dwindling resources. Finally, they proposed that 10 percent of
RANN's budget be reserved for technological opportunities and promising exploratory development that might serendipitously appear apart
from RANN's own planning.[64]

As for the National Science Foundation's own engineering
community, it proudly took credit for RANN. According to the Engineering Advisory Committee's 1971 annual report, the engineers were
pleased that many of the Foundation's programs and future plans reflected their own thinking—especially in matters of "interdisciplinary
interaction between engineers and social scientists" (IRRPOS) and
addressing the "pressing needs of society" (RANN). They applauded
NSF's interest in greater cooperative research between universities and
industry and the "infusion of R, D, and E into our mature industries."
Nodding toward the institution's traditional view, they recognized the
continuing need to support basic research and saw the time as "ripe" for
"intensive engineering application of the science base developed so
rapidly and effectively over the past 25 years."[65]

The engineers' report did get around, however, on page nine,
to admitting their concern for "a possible weakening" of the Engineering Division because of "spinoffs" like RANN. (Todd called RANN
takeovers "bureaucratic buccaneering." Transferring work on superhard
materials to RANN when NSF was struggling to build its new materi-

als science program, for example, was costly both to staff morale and productive, balanced research, he charged.[66]) Engineers, like others at NSF, feared fragmentation of their programs among several directorates and complained of not being sufficiently consulted in the planning of RANN. Muted disquiet continued even after RANN's Lewis Mayfield, deputy director of the Division of Advanced Technology Applications, predicted that at least one-fourth of the projected $80 million budget for RANN would support engineering efforts, and Stever told them he thought the National Science Board's attitude toward engineering was becoming more friendly (though "some confusion still exists" as to its proper role).[67]

About the same time that NSF established the RANN program to harness American technical expertise to national needs, Congress was pushing its own approach. Edward Kennedy intended with his bill, S. 32, the National Science Policy and Priorities Act of 1972 (and subsequent years), to refocus national research and engineering priorities from a defense orientation to the solution of civilian problems, create jobs for unemployed scientists and engineers, and revitalize the economy. By requiring NSF to develop national policies on problem solving and to include more representatives of industry, the technical community, and the public on the National Science Board, the bill would have gone far beyond Daddario's mild permission to engage in applied research in the national interest. Neither the Foundation nor the administration liked the legislation, and it came to nothing; rather, Congress took the more appealing tack of reestablishing by law the White House Office of Science and Technology, which Nixon had abolished in January 1973.[68]

Thus, in a world very different from its founding in the postwar period, NSF was driven to apply the nation's science and technology expertise to the solution of the myriad problems besetting America's cities, land, and people. Neither the Foundation nor its scientists and engineers denied the severity of the problems or that many were rooted in science and technology. They wanted to help, they said. But to expand NSF's programs to include applied research, even when basic research also enjoyed a significant budget boost, and to direct research toward specific ends were unsettling new directions. Resistance to accepting, much less implementing, such changes was strong within the established NSF culture. Having achieved success and gratifying, if unspectacular, growth in a basic-science-only mode, many NSF insiders saw no reason to alter their assumptions or operations. Further, they had not planned for or even foreseen the possibility of change in any strategic sense for twenty years. Satisfied and insular, they insufficiently appreciated the potential impact of changes happening outside the agency. IRRPOS reflected some internal desire to respond to external needs, but it was a small, add-on effort that strayed little from the conservative Foundation's traditional aims and practices.[69]

The force driving NSF, of course, was politics, although the deliberate application of that force was relatively new. Until Daddario's mid-1960s assertion of Congress's oversight role, the Foundation, small and quiet, had been left more or less to its own devices, more or less politically invisible. Now, however, politicians, answerable to increasingly restless and vocal constituents who, like themselves, often had little comprehension of science and engineering, found it logical and appealing to demand technological answers for technology-based problems surrounding them.[70] The president could dangle big program dollars and appoint new, amenable agents; the Congress could legislate action toward change. And they did. The agency, dependent on both, responded— as it had to. How well, and how it all affected engineering, remained to be seen.

CHAPTER 4

THE RANN YEARS—
FRUITFUL, BESET, AND BRIEF

> Even the staunchest of NSF's governing body will have to
> admit that but for RANN, NSF would not be the billion-
> dollar agency it is today.
>
> —Wil Lepkowski, *C&EN* 58:19

RANN's promoters began their multifaceted program with enthusiasm
and commitment, buoyed by the possibilities for making a difference
in a society riddled with crises. They oversaw the achievement of significant successes along a broad front of problems and anticipated a
growing influence in Foundation and national affairs. During RANN's
brief existence, the program enjoyed high-level support, but from the
beginning this applied-research endeavor was also viewed with suspi-
cion, even hostility, by NSF's traditional constituents, who felt threatened
by it. Engineering both prospered and suffered from its close affiliation
with Research Applied to National Needs; engineers championed the
RANN cause but also looked out for their profession's independent
interests.

RANN Running

Looking back from the mid-1980s, NSF Engineering staff member Syl McNinch recalled specific RANN accomplishments that still offered economic and social benefits. One simple but significant example was the RANN-developed wind deflectors on long-distance highway rigs, which continue to provide truckers with 3 percent to 10 percent fuel savings, amounting to hundreds of thousands of dollars annually. Joel Snow touted a University of Maryland study on environmental contaminants emitted from smokestacks, which convinced the Environmental Protection Agency that implementing the Clean Air Act would have to take into account pollutants from outside sources in addition to those locally produced. Another RANN research breakthrough was the discovery of powerful carcinogenic nitrosamines in a series of widely used industrial chemicals.[1]

Although later reorganizations shuffled RANN's research priorities somewhat, the three major and persisting emphases were energy, the environment, and urban life. Examples of specific, problem-focused research in each of these areas illustrate the operations as well as the scope and variety of RANN's accomplishments. The selected problems, widely deemed important and well documented, all required, or could convincingly justify, multidisciplinary approaches, including a significant role for engineering. They further represented contrasting degrees of popular glamour, technical sophistication and innovation, and success—for RANN and for society. Individuals' basic-research efforts also continued in these areas.

Solar Energy. While the geopolitics behind the Arab oil embargo did not crest until 1973, those who were alert could anticipate an energy crisis for other reasons. Demand for electricity had been growing rapidly for years and was expected to continue as society became more consumer product-oriented. Energy consumption was doubling every twenty years. At the same time, environmentalists were warning that alternatives must be found for dirty fossil fuels and dangerous nuclear power.

Accordingly, RANN supported research to identify, evaluate, develop, and verify practical alternative energy technologies to supplement future national energy sources. To avoid duplication or conflict with research in the mission agencies, the Foundation established a policy in mid-1971 that it would act primarily as a catalyst, transferring its results and responsibility to the appropriate development agency at the proof-of-concept stage.[2]

NSF funded research in several ordinary and exotic energy resources (wind, geothermal, coal conversion), in methods of power transmission (superconducting and cryogenic underground cables), and in energy systems. But harnessing the unlimited, reliable, clean, and "free" energy of the sun captured the imagination most. While the first known solar-activated device, a water pumping machine, was invented in 1615 and a few mechanical applications were attempted in the nineteenth century, little sustained work on solar energy had been done until the 1950s. An 1876 selenium barrier-layer photovoltaic cell had proved to be a true solid-state conversion device, but its overall conversion efficiency was about 0.6 percent; Bell Laboratories managed to raise that efficiency rate to 11 percent by 1955.[3]

In April 1972 Nixon named NSF the lead agency for federally sponsored solar energy research. Stever reported to the National Science Board that RANN had initiated more than sixty energy projects in FY 1972, fifteen of them in solar energy. Plans for RANN's five-year, $196 million Terrestrial Solar Energy program were well along and included research on materials and components, coordinated systems studies on "the equally important economic, environmental, and social impact factors," and proof-of-concept experiments. Some of this work was being pursued concurrently in the Engineering and Materials Research divisions, Stever noted. Engineering, for example, was concerning itself with such technical problems as heat transfer, mechanical structures, energy conversion devices, and corrosion properties—all fairly basic pursuits. Significant NSF efforts aimed at developing solar temperature control systems for buildings and electric power generation from either thermal or photovoltaic conversion of the sun's energy.[4]

In September 1973 Stever outlined for the NSB a multiphase, $1.75 million research project to advance the systems technology, economic feasibility, and social acceptability of utilizing solar energy for the heating, cooling, and hot-water needs of buildings—an interest among investigators since midcentury. The Massachusetts Institute of Technology had built four solar houses between 1939 and 1959, and there were others. By the early 1970s, about 25 percent of U.S. energy went for indoor climate control, a figure expected to double by the year 2000. NSF selected three major industrial contractors—General Electric, Westinghouse, and TRW Systems Group—to do feasibility studies on systems for varying climatic conditions, performance and functional requirements, and markets. In 1974 these companies also bid to conduct the next phase of the project: experimental design and systems optimization. That year RANN allocated $14.1 million for solar energy, its largest single item.[5]

In late 1974 the National Science Board approved one of RANN's showiest solar energy projects, the experimental heating and cooling of an entire school building in Atlanta, Georgia. Richard T. Duncan, Jr., of Westinghouse Electric Corporation in Baltimore, headed the project, which, including previous funding, amounted to nearly $830,000 for twenty-two months. Faculty members and graduate students of the Georgia Institute of Technology helped design and install the monitoring instrumentation and gather and analyze data. The George A. Towns Elementary School was the fifth school to gain a solar energy system, the first to use solar cooling. In Atlanta, where heating and cooling requirements were approximately equal, the system was expected to provide 70 percent of the school's heat and more than 60 percent of its air conditioning. In operation, sunlight striking the 10,000 square feet of flat-plate solar collectors on the school's flat roof heated circulating water to 160–215 degrees Fahrenheit. The heated water, stored in four 6,000-gallon tanks, provided hot water directly and powered either the heating or air conditioning system. The aluminized Mylar collector array was designed for a twenty-year life span.[6]

In 1976 Duncan and others admitted that cooling the Towns school had proved to be more of a technical challenge than heating, as expected, and that certain collector materials had had to be strengthened to meet the extreme thermal stress. Overall, they said, their design data and techniques had tested "valid and slightly conservative." But after repeated problems with piping deterioration caused by the interface of dissimilar metals and costly interim repairs, the school's solar-powered system was shut down in 1985. Failure, in general, could be attributed to inadequate design for fault tolerance, a not uncommon problem, and inadequately trained on-site maintenance staff.[7]

On a more fundamental level, NSF began funding Bernhard Seraphin at the University of Arizona in 1973 to work on chemical vapor deposition for fabricating photothermal solar converters that would produce enough heat to generate electricity by conventional means. He proposed a two-layer cell construction in which the top layer was a semiconductor material such as silicon, having high absorption for solar radiation and high transparency for blackbody radiation from the heated unit. The bottom layer was a metal film with high reflectance. His goal was to identify reliable, affordable optical coatings with optimum absorptivity. By 1974 Seraphin had successfully deposited silicon onto stainless steel substrates and silicon nitride antireflection coatings onto silicon layers. He wrote in 1976 that there was "absolutely no gamble with respect to the technological feasibility of solar energy conversion." Most problems, in fact, had more than one possible technical solution. Making that energy conversion economically feasible was the challenge.[8]

In early 1975 NSF transferred its solar (and other energy) projects, by then the largest component of RANN, to the newly formed Energy Research and Development Administration (ERDA), just as RANN's administrators had predicted and planned for.[9] Still, for those at the National Science Foundation who had conceived and nurtured RANN's energy efforts, it was difficult to feel the satisfaction of success in the face of a vanishing empire.

Research on solar energy enjoyed great congressional favor in succeeding years. (For FY 1977 Congress appropriated $290 million after the administration's request for $160 million. Even ERDA officials voiced doubts that they could spend so much efficiently.) Generous

funding continued under the Carter administration, but severe and uneven cutbacks during the Reagan years hurt solar research efforts. With
a new oil crisis in the Persian Gulf in 1990, solar energy again looked
inviting if costs could be pared or efficiency appreciably increased. By
then, terrestrial-grade photovoltaic cells, which absorb and concentrate
incoming light to agitate electrons and produce electricity directly, were
about 10 to 15 percent efficient, not too far advanced from the 1950s.
Aerospace-grade silicon cells, such as those utilized in the fastest-quali-
fying "Pride of Maryland" solar car in GM's 1990 "Sunrayce," were 19
to 20 percent efficient but forty times more expensive. Gallium arsenide
solar cells, up to 35 percent efficient, remained prohibitively expensive.[10]

Trace Contaminants. Worsening pollution of water, air, and
soil from industrial effluents, pesticides, toxic waste, and ordinary human activity had become increasingly alarming during the 1960s. Long
before the first Earth Day was observed on 22 April 1970, the National
Science Foundation had been interested in applying science and technology to the solution of environmental problems, even as science and
technology were causing many of them. The effort greatly accelerated
under RANN; a livable earth was easily identified as a national need.

Within the wide range of environmental problems that RANN
addressed, one that occupied political and much technical attention was
that of understanding, measuring, and controlling trace contaminants in
the environment. As congressional hearings proceeded on such topics
as mercury pollution in the Great Lakes and the costs and effects of
chronic exposure to low-level environmental pollutants, NSF engineers,
social scientists, biologists, and others joined forces. In 1971 they laid
out a plan to identify heavy-metal contaminants, such as lead and mercury, whether occurring naturally or as a result of human decision, that
were beyond an ecosystem's capacity to absorb. Fallout from nuclear
testing, tailings from mining and milling, and smokestack effluents were
obvious points of origin, but increasingly insidious was automobile
exhaust. RANN worked closely on the sources, routes, and targets of
such contaminants in cooperation with numerous other federal agencies,
from the U.S. Food and Drug Administration to the U.S. Geological

Survey, and about ten different NSF programs, including Engineering Systems.[11] Solving the problem was largely a matter of economics and, therefore, politics. Nobody preferred a dirty, unhealthy world, but how much would people be willing to pay to clean it up?

Interdisciplinary studies of lead contamination, which began under IRRPOS and continued under RANN, included those of University of Illinois researchers under the direction of zoologist Robert Metcalf and environmental engineer Ben Ewing. They proposed in 1970 to examine the "overall sociological significance of the ubiquitous pollutant lead upon human welfare," arguing, in terms to assure funding, that the problem of environmental lead had not only scientific and social relevance but was "ideally suited" to systems analysis by experts in agriculture, engineering, biochemistry, physiology, economics, and law. The Illinois program changed personnel and emphases but continued for some years, with an ultimate estimated cost of $2.9 million. Having determined, unsurprisingly, that urban areas were the "major reservoir" of lead, the investigators also called rural roadsides "critical." Their preliminary studies showed that natural lead concentrations could adversely affect crop growth under certain soil and environmental conditions.

Illinois coordinated its efforts with those of Colorado State University and later with researchers at the University of Missouri at Rolla. Investigators in a large environmental project at the Oak Ridge National Laboratory and many other universities also regularly shared information. The Colorado State team, headed by mechanical engineer Harry Edwards, set out to understand the atmospheric, chemical, and physical factors associated with the dispersion of automotive lead particulates. To answer the technological and societal question of whether lead antiknock compounds in gasoline should be continued, they worked to describe short-range lead-particulate dispersion from highways under varying meteorological conditions, to conduct wind tunnel studies on turbulent and advective transport (that is, via the mass motion of the atmosphere) in simulated city canyons, and to develop mathematical models of long-range transport. The researchers also produced instrumentation and techniques to measure exhaust behavior and lead travel through the atmosphere and the food chain. The Environmental Protection Agency's first regulations requiring the availability of unleaded gasoline came out in 1974 in response to these and numerous other studies.[12]

Researchers at the University of Missouri at Rolla studied how heavy-metal pollution from mining, milling, and smelting lead in the "New Lead Belt" area of southeastern Missouri was affecting the ecology of the surrounding Clark National Forest. New Ozark ore deposits, discovered in the 1950s and 1960s, had been exploited so heavily that by 1970 the area was the world's largest lead-producing region. Even after recovering associated ores, such as zinc, copper, and silver, about 88 percent of the mined material was metal-bearing waste that was disposed of in tailings dams and settling lagoons. Sooner or later these materials would find their way into the forest ecosystem. Smelter stack emissions deposited airborne lead-contaminated particulates and sulfuric acid mist over wide areas.

The RANN-supported researchers, under environmental engineer Bobby Wixson, found that both aquatic and terrestrial vegetation picked up and carried high-level concentrations. By 1975 Wixson and his codirector, Charles Jennett, could report that their disseminated research findings had helped industries develop improved technologies for monitoring and controlling pollution and given regulatory agencies the means to devise more effective standards and long-range strategies. In a mix of gratitude and chauvinism, they also celebrated RANN, pleased that the "practical, common sense working partnership developed in this NSF-RANN Project—with the University of Missouri serving as an unbiased intermediary between agencies and industry—has allowed environmentalists to work with industrial and agency representatives to define and resolve problems for social and economic benefits."[13]

Researchers from these universities were key participants in periodic conferences and other efforts of NSF's Trace Contaminants Program to seek broad input. Investigators also worked closely with numerous other NSF programs and federal and state agencies. Industrial trade associations, from the International Lead Zinc Research Organization to the Tuna Research Foundation, and corporations, from the Climax Molybdenum (Mining) Company to the lead antiknock compound manufacturer Ethyl Corporation, also established cooperative working relationships based on their joint interests. Evaluators gave this $1.1 million research effort high marks for its "timeliness and relevance" and for the "receptive attitude" and direct involvement of numerous principals, which combined to make this program one of RANN's most successfully utilized.[14]

Excavation Technology. As the land and air of American cities grew more congested and polluted, scientists and policy makers looked to the subterranean environment as a site for major urban systems, such as transportation, power transmission, and communication. They also considered underground fuel storage, commercial and industrial operations such as warehousing, and water and waste systems with an enthusiasm that seems naive today. Interest was almost astonishingly broadly based.

Underground tunneling to support such activities faced only two fundamental technological problems: developing tools powerful enough to drill through and break up hard rock or developing materials strong enough to support the roof of an excavation through soft material. Both required sophisticated materials and techniques that were not yet within reasonable bounds of efficiency or economy. A 1968 study by the National Research Council's Committee on Rapid Excavation, citing the rarity of innovations in tunneling technology, had concluded that a ten-year, $200 million research program could increase the tunneling rate by 200 to 300 percent with a 30 percent decrease in cost.[15]

The National Academies of Engineering and Sciences established a National Committee on Tunneling Technology and cosponsored with NSF a 1972 conference of experts in excavation technology to identify areas of research need and opportunity. The Federal Council for Science and Technology (FCST) Interagency Committee on Excavation Technology also offered vital input. By the next year, NSF was supporting tunneling research on a broad front. Rock cutting by training a hydraulic jet on a weak point, chemical weakening by application of surfactants, and fragmentation by focusing electron beams were all under investigation. A novel approach was to melt the rock. Here, as Los Alamos Scientific Laboratory researchers described the process, a penetrating bit, heated electrically to about 1,800 degrees Celsius, melted the rock before it and left behind a self-supporting hole lined with fused glass. Meanwhile, the melted material could be formed in various ways (glass rods, pellets, rock wool) for efficient removal. Computer modeling of the excavation process helped investigators understand in advance the "geologic environment" through which tunneling would proceed. Researchers also sought to standardize tunnel designs and procedures to reduce costs. Meanwhile, NSF's engineering division continued to fund basic engineering research in soil and rock mechanics.[16]

The American Society of Civil Engineers, through its Underground Construction Research Council, had submitted a successful research proposal on underground construction in 1970 to NSF's IRRPOS program. It continued under RANN. *The Use of Underground Space to Achieve National Goals,* the engineers' voluminous 1973 report—cosponsored by NSF, the U.S. Army Corps of Engineers, and the American Institute of Mining, Metallurgical, and Petroleum Engineers— evaluated the potential societal benefits of transferring many urban functions to subsurface space. Their "conservative analysis" showed that relocating public-works facilities underground could release "approximately $60 billion per year to other social and environmental needs," with "markedly improved" costs and efficiency. To make such transfers would require an R&D investment of $0.1 billion to $1.0 billion annually for five to ten years, a firm public policy to stimulate or require the moves, and the institutional apparatus to implement them. The investigators identified urban expressways and electrical energy distribution as the two most promising and feasible technologies to be sent underground. They anticipated 5,000 miles of subterranean freeways and 667,000 miles of power lines in ten years from a $1.55 billion investment, from which a $6 billion savings would accrue annually for release to other uses.[17]

Meanwhile, Charles Fairhurst of the University of Minnesota proposed research to improve the design of rock-tunnel linings by using the finite element method to calculate the influence of broken-rock stresses on tunnel support loads for various lining configurations. He also anticipated "back-calculating" the in-situ strength of actual loads in tunnel linings. Field tests and work on tunnel-boring machines at an underground copper-shale mine in northern Michigan would complement Fairhurst's theoretical and experimental work at the university. Fairhurst justified his proposal in terms of national and social needs, noting that cheaper, more rational tunneling techniques not only would benefit traditional mining and civil engineering operations but could also provide an economically competitive alternative to surface construction for many uses—a particular incentive in "adverse winter climates" like Minnesota's.[18] The University of Minnesota, in fact, housed a major research effort on tunneling technology throughout the RANN period.

National Science Foundation funding also supported the work of electrical engineer Stuart Hoenig at the University of Arizona for several years beginning in the mid-1970s. He worked on a device to predict the collapse of roofs in mines. This safety measure was important because, as ore is removed, the stress on the remaining rock increases and sometimes leads to a sudden, violent collapse, or "rock burst." Discovering that rocks under stress produce tiny electrical currents, Hoenig attempted to detect and measure these currents so that they could trigger warning signs when dangerous levels were reached. Further, by observing the electrical patterns (isocurrent contours) in stressed rock samples, he could predict the path of the rock fracture. While Hoenig's rock-burst detector was still in the prototype stage in 1980, his current-differential concept showed promise for nondestructive evaluation of ceramics as well as for mining applications.[19]

For all the interest and substantive new knowledge gained in excavation technology, American cities did not move underground to any great extent. Underground utilities may be common today, but extended subsurface freeways are not. The latter may have seemed an innovative idea at the time—and "innovation" was the buzzword for support in the 1970s—but the ability to bore and line a sturdy tunnel was not the whole solution. Removing polluting automobiles from the landscape was environmentally attractive, but where would the exhaust of thousands of underground gasoline-powered vehicles go? Possible psychological stress from being underground got more attention (which was little) than ventilation of subsurface space. By contrast, electrically powered mass-transit subway systems with multiple-car trains carrying thousands of people have proven their worth and are enjoying a renaissance in many cities.

RANN UNDER FIRE

RANN's problem-solving efforts continued through the mid-1970s, with the varying successes that would characterize any broad research program. Enthusiasm for research applicable to national needs remained strong in many quarters. Among RANN's continuing advocates was NSF director H. Guyford Stever, who in 1973 praised RANN as a pioneer in that big-picture "systems" interrelationship between the activity of a human society and the "technological fix." It was important, he said, to "break down some of the separation that has existed between basic research, applied research, development, and the potential users of the knowledge and know-how research and development generates." All these approaches, working together through universities, industry, and government, were essential for solving society's massive and growing problems. At a National Academy of Sciences symposium celebrating NSF's twenty-fifth anniversary in 1975, Stever noted that over these years, "Certain developments in science have made it clear that the science community cannot conduct its affairs as a pure search for truth apart from serious considerations of its human consequences."[20] Quentin Lindsey and Judith Lessler, who conducted a survey of early RANN projects to determine how well their results were being utilized in meeting national needs, concluded in 1976 that "RANN is beginning to live up to expectations."[21]

The RANN Advisory Committee, renamed and reconstituted at various times and not meeting regularly, continued to offer positive observations and recommendations that, perhaps inadvertently, implied an undercurrent of disapproval. Committee chair Lewis Grant, a professor of atmospheric science at Colorado State University, wrote to Eggers in February 1976 in praise of the depth and breadth of RANN, but he was unconvinced that either the general or scientific public fully understood it. He suggested that RANN's benefits needed to be quantified, just as the magnitude of the problems were.[22] Among the committee's recommendations was that "potential end users" be included on research teams, which should continue to work on midrange (two- to five-year) research in "gap-filling" and "bridging" roles with the mission agencies, universities, and industry. RANN should be "protected, developed, and expanded" by long-term and increased funding. Committee member William Vogely pronounced, "Having looked at RANN, we believe it is a viable concept; it must be preserved in any plan; and the 'warts' can be removed by a good plastic surgeon—major surgery is not required."[23]

Warts had indeed been spotted; criticisms were mounting. In November 1974 Wisconsin Democrat William Proxmire wrote to Stever on behalf of his subcommittee of the Senate appropriations committee about general "managerial difficiencies" [*sic*] at NSF, specifically challenging RANN's selection process ("Who is advising you on how to spend the taxpayers money?") and review of completed work ("If these projects are for critical national needs, why have the results not been more quickly evaluated and disseminated?"). Threatening a detailed budget review, Proxmire gave Stever two months to explain NSF procedures to "overcome these glaring difficiencies" [*sic*]. Stever's "NSF-has-made-some-progress" response sounded like a tacit acknowledgment of the validity of the charges, as did his expressing an "intent to strengthen" NSF's evaluation and management system.[24]

RANN was on its way to the block. The reasons were many. Beyond the stiff resistance to change common among maturing, prospering institutions, which had been a problem from the beginning, RANN never enjoyed the full favor of either NSF's broad leadership or its traditional constituency. Former NSB chair Eric Walker, who had promoted RANN, noted in particular the National Science Board's lack of support, especially under his successor, Philip Handler, his "exact opposite" philosophically. There was always at least a minority on the board unhappy with RANN. Speaking for himself in 1974, Handler admitted to a "nagging feeling with numbers of us" that in time "the tail would come to wag the dog." He objected to the "invasion of style" that accompanied RANN and its high visibility, which implied the making of too many promises to the public. McElroy's failure to bring the NSB into the earliest policy- and program-forming process had to bear part of the blame. The exigencies of time had made that opportunity for input and acceptance impossible, but winning over an excluded board after the fact to a program that they had tradition-bound reasons to distrust was probably a lost cause from the start. McElroy did his best to mollify the board with proposals to smooth its RANN-related procedures, but the change had been imposed, not negotiated, or even discussed, and the record contains numerous pointed exchanges. Even if McElroy had done everything right, however, it seems unlikely that the scientist-dominated National Science Board would have been comfortable with so radical a departure for NSF.[25]

NSF directors McElroy and Stever, who had been appointed by a president seeking prompt and practical results, strongly supported RANN (Handler said Waterman "would have opposed it with all his vigor") and brought in like-minded staff, but Richard Atkinson, who succeeded Stever in August 1976, did not share their enthusiasm. The Stanford University experimental psychologist and applied mathematician, who had developed mathematical theories about the nature of memory, early declared his belief that applied research "must remain a vital component of our Federal science programs." But he had in mind the responsibilities of the Defense Department and NASA more than those of the Foundation. His first Director's Statement, for NSF's annual report of 1976, used Vannevar Bush's *Science—The Endless Frontier* as a text, and RANN proponents could take little comfort from his declaration that "we must develop more realistic expectations about what can be achieved by large-scale research campaigns aimed at specific goals." Atkinson questioned both RANN program content and management. The enthusiastic and committed Eggers was unable to win

Atkinson's confidence, and he left NSF in July 1977; Robertson, also an early advocate, had long since retired.[26]

RANN represented a clash of cultures. NSF's basic-science constituency, both in universities and at the Foundation, felt threatened by it, both philosophically and financially. Pursuing a new goal of applied, problem-oriented research inevitably ran up against NSF's historic ideology that favored hands-off support of independent, curiosity-driven basic scientists. It felt like forced competition, unfair and unwelcome to the "natives." Worse, this change came on orders from the outside. Comfortable with a modus operandi familiar from two decades of success, NSF had difficulty adjusting to political imposition, especially when belatedly exercised. In her policy study of RANN, Mary Ellen Mogee notes that the reluctant NSF created RANN as a new organizational unit, which served to minimize dislocation of the existing organization and isolate the change. However, no matter how often or fervently NSF leaders reiterated their abiding commitment to fundamental research or asserted that OMB viewed RANN funding as a separate item, the doubters remained convinced that RANN existed at their expense.[27]

Detlev Bronk, a former NSB chair, was confident enough in 1975 to assert that NSF's "25-year tradition of primary devotion to uncommitted research is an adequate guarantee that Research Applied to National Needs will not drive out 'pure' research from the Foundation." Stever pointed to the positive growth of the basic research budget during the RANN years. He "agreed 100 percent with McElroy" that basic research was primary, even though it was essential to become more involved with "science in service to the nation"—that is, applications.[28] But these were becoming thin voices in a stiff wind. With RANN organized and staffed inadequately for the utilization and integrability of its results, the research-to-user link was weak, and it became difficult to answer the question, "What problem have you solved?"[29] Of course, that question would not have been put to basic scientists.

That Research Applied to National Needs underwent some form of reorganization every year of its existence indicated an uncertain, bumpy ride. As Mogee argues, NSF's slowness in comprehending the changing policy environment meant that both IRRPOS and RANN were created as "crisis decisions." Time for planning and organizing was brief at best, and there was no overall strategic plan in which to fit the new pieces.[30] That RANN was just "beginning" to live up to expectations four years into funding was more evidence to impatient political watchdogs of a poor start—even though its entire career was too short for a fair assessment. The frequent reorganizations also suggested ongoing tension as program leaders tried (too hard?) to be responsive to too many constituents—chief among them scientists (NSF's reason for being) and politicians (its means for being). Congress and the Office of Management and Budget, the National Academies of Engineering and Sciences, the overseeing General Accounting Office (GAO), and others all tried to impose their respective, often opposing views on what research applied to national needs should be and do.[31]

RANN endured numerous formal reviews during its short history, which also betrayed its tenuous footing. The OMB requested in March 1975 that NSF conduct a self-evaluation of RANN. The Foundation admitted to shortcomings in RANN's project selection and utilization criteria but claimed overall that by "empirical evidence" a "reasonable incremental contribution to 'national needs' has been made," even though RANN had "little control" over the implementation procedures of users. The tone was half-hearted.[32] A 1975 GAO report generally praised the RANN program and staff but recommended wider input on problem identification, tighter procedures for peer review, and more user involvement. (Eggers set up a RANN task force that produced a proposed program for research utilization in February 1977.) In March 1977 the House Science and Technology Committee released an investigative report supportive of RANN, but its statement that applied research under RANN should eventually surpass basic research expenditures was another red flag to the basic-science community.[33] RANN had become a political football, with its home team, NSF, short on spirit.

Then, when the so-called Simon Report, *Social and Behavioral Science Programs in the National Science Foundation,* called the quality of RANN's social science efforts "highly variable and on average relatively undistinguished, with only modest potential for useful application," all of RANN was tainted. Harvard economics professor Dale Jorgenson was among those protesting that this inadequate National Academy of Sciences study was unfair in evaluating RANN by objectives other than RANN's own, and a congressional staffer dismissed it as "another example of the Academy protecting its 'Old Boy's Club.'" But the damage was done.[34] In July 1977 an NSF science applications task force issued a report urging NSF to strengthen applied research and its transfer of results; but it offered "no vote of confidence to RANN."[35]

In the fall of 1977, after spending $468.3 million and enduring six and a half years of competing external pressures and palpable internal hostility, RANN disappeared by reorganization when Atkinson announced a new science and engineering applications directorate (which was never implemented; see chapter 5). That year applied research accounted for a modest 9 percent of NSF Research and Related Activities funding. By the mid-1980s, that figure would drop to 5.3 percent. In McNinch's words, "The RANN experience shows that basic research support programs at NSF have sufficient muscle to prevent any encroachment by applied research." It was clear that basic-science advocates considered work toward application an encroachment on *their* Foundation and included it without enthusiasm. It was not surprising that RANN had produced results that were neither good enough nor fast enough. Yet it was also true, as *C&EN*'s Wil Lepkowski later noted, that NSF's emergence as a billion-dollar agency in the next decade was traceable to the growth begun under RANN.[36]

ENGINEERING DURING THE RANN YEARS

While RANN was commonly, though too narrowly, viewed as primarily an engineering endeavor, engineering at NSF retained its separate organizational position. During the 1970s members of the engineering community endorsed RANN, but RANN was not theirs alone and it was not enough. They also looked out for their status, and not without wariness, as an independent NSF unit. Although pleased to see engineers appointed to high administrative positions within the Foundation, they were "tempered" and "troubled" by the "continuing disparity" between the words and the funds provided for engineering research in universities—still only about 11 percent of all allocations. University engineers, they complained, were "a constituency whose value is praised when program support is sought and whose need is minimized when funds are distributed." In particular, they objected to the puny $625,000 that NSF was providing for engineering equipment in 1971; $6 million would have been more reasonable, they said, since engineering research was by nature equipment-oriented and such equipment became quickly obsolete. In April 1971, just after the formation of RANN, NSF's Engineering Advisory Committee pointed out that slightly more engineers earned doctorates than did physical scientists in 1970 and passed a resolution urging that NSF change its name to National Science and Engineering Foundation in recognition of the academic emergence of their profession.[37]

Probably not expecting action on this proposal, the engineering advisors were nonetheless encouraged when the National Science Board, which was required by the 1968 law to produce an annual report, decided to focus on engineering in its fourth report. When this "ideal opportunity" to reach a broader audience was realized in 1972, the engineers cheered that their input had been "harmonious and remarkably free of the profound philosophic differences often uncovered when general concepts are reduced to specific words."[38] The report, titled *The Role of Engineers and Scientists in a National Policy for Technology* and written largely by John Ide, stressed that progress in solving industry's and society's problems would take "team efforts joining science and engineering with other professions and other fields of knowledge, in industry, the universities, and government." Making wise choices about the difficult priorities would require "credible public technological assessment," which included consideration of society's values.[39] This was RANN's tune, a popular one at the time.

But by November 1975, when Stever spoke to the National Academy of Engineering, he knew he must address mounting pressures from its members for greater recognition. He called engineering "ubiquitous in the scheme of things at the Foundation," despite the fact that NSF had not become a National Science and Engineering Foundation, "as some have long suggested we do." Organizationally, NSF kept long-range, basic engineering research in the Engineering Division, while RANN did the shorter-term engineering problem-solving. So the actual Division of Engineering funded only about a fourth of NSF's engineering-related work, Stever said, trying to convince his audience of engineers that his agency was not shortchanging their profession: "Engineering research has been involved in or had an impact on almost every field of research NSF supports, and engineering studies contribute significantly to our science policy activities." Stever mentioned energy and environmental studies, technology assessment, and productivity enhancement. Atkinson similarly tried to convince the American Society of Civil Engineers in 1976 that "truly, engineering expenditures already dominate the Federal budget for science and engineering."[40] They all knew this was not true at NSF.

And so NSF engineering remained administratively divided, supportive of RANN but increasingly focused on its own interests. Engineers felt the tensions of this period probably more than most. With RANN's reputation increasingly sullied and its effectiveness waning, they struggled over being ubiquitously associated with the concept of research applications—especially after striving so long for recognition as scientists doing engineering science. At the same time, engineers were beginning to wonder why they had to fit any arbitrary mold, and to assert the merit of their own professional culture. However one judged their contributions to the Foundation's or the nation's agenda, by the late 1970s, engineers, like so many others, saw that the troubled RANN era was ending, and they wanted their own identity and a say in what happened next.

CHAPTER 5

INTERREGNUM

Searching for a Place

Engineering is different in its style, its traditions and its university institutionalization from science. Therefore it is appropriate that the Foundation institutionalize it so that it is also visible and can follow the traditions and culture of engineering.
—Lewis M. Branscomb, "Principles Underlying NSF Policy Regarding Applied Science and Engineering"

Between the beginning of the end of RANN in the mid-1970s and the establishment of its independent directorate in 1981, engineering at the National Science Foundation lurched from one organizational home to another—four times in six years. Some of the time it lived in two jurisdictions at once. Perhaps the most striking feature of the history of engineering at NSF during the late 1970s was its institutional instability. As the RANN model became increasingly divisive and discredited—unfairly in some measure—its leaders searched for a new, more workable approach. Yet to abandon the RANN concept entirely was difficult politically and practically, even emotionally. Thus, these years witnessed a scramble to find a suitable fit for applied research, and for engineering, which was finally seen as entitled to its own identity. In its own way, the RANN interlude probably helped engineering's emergence. Politics could clearly be seen determining the Foundation's structure and priorities.

1975: THE DIRECTORATE FOR MATHEMATICAL AND PHYSICAL SCIENCES AND ENGINEERING

Much of engineering remained with RANN in the Directorate for Research Applications as long as RANN lasted, but the remaining engineering division programs in the Directorate for Research became part of a major NSF reorganization in the summer of 1975. Releasing the announcement in July, Stever simply said that the changes, "under active consideration for more than a year," would strengthen a busier NSF and provide an "effective management structure for dealing with current and future operations." Speaking to the National Academy of Engineering in November, he more candidly acknowledged the force of politics on internal administration—"closer scrutiny by Congress" and "tighter priority-setting by the Administration"—with budget cutbacks, questioning of research programs, and criticism of NSF management. Noting that he had not titled his talk "Science—In Search of the Golden Fleece," he humorously acknowledged Proxmire's earlier grilling but also suggested that as science and engineering cost more, became more visible, and affected society more dramatically, the research community should not be surprised by the "encroachment" of politics. Indeed, "mutual respect between science and the citizen" could well be healthier than science's previous "aloofness" and the public's "blind faith." Reorganizing the Foundation to improve management of research support was one NSF response to the "new climate."[1]

Thus, two existing directorates, Research and National and International Programs, were restructured "along more functional lines" into three—one of which was the new Directorate for Mathematical and

Physical Sciences and Engineering. (The other two were Astronomical, Atmospheric, Earth, and Ocean Sciences and Biological, Behavioral, and Social Sciences.) Engineering found itself in a configuration remarkably like that of the Foundation's first years—both in its immediate surroundings and the larger organizational framework. The umbrella had a broader reach for engineers. Assistant director Edward Creutz headed the new MPE's four divisions—Mathematical and Physical Sciences (excluding astronomy), Engineering, Materials Research, and Computer Research. The latter two had been peeled off from Engineering in 1971 and 1974, respectively; their work remained closely related, frequently overlapping. They would again return to the engineering fold, at least temporarily.[2]

The National Science Foundation seemed indeed to be circling back to more familiar, comfortable patterns of operation and image projection. For example, in late January 1976 NSF issued a series of press releases on the Foundation's congressional budget testimony for the coming fiscal year. In one, Eggers emphasized the critical need for applied research on the national problems of resource scarcity and environmental degradation and on such issues as the impact of environmental regulation on productivity, a major RANN focus by the mid-1970s. But in another, Stever urged, as did NSB chair Norman Hackerman in a third release, "strong investments" in basic research as "the keys to future economic and social progress." The director noted that NSF's proposed budget of $812 million, an increase of $80.4 million over FY 1976, was intended to "counteract the gradual decrease in Federal support of fundamental research in terms of constant dollars since 1968" (although NSF's inflation-adjusted support, unlike other agencies' and despite RANN, had grown over that period, according to an NSF report for OMB). Stever cited $233.2 million for MPE, an increase of $39.9 million. RANN would get $64.9 million, an overall decline.[3]

Even discounting the budget losses due to the transfer of RANN programs to other agencies upon proof of concept—as when RANN delivered its solar energy work to the Energy Research and Development Administration—it was clear that RANN's support was eroding. RANN was surely feeling a chill wind from the Foundation's summit. In office as acting director less than a month, Richard Atkinson informed the National Science Board in August 1976 that he planned to appoint a task force "to study the amalgamation of the Division of Engineering with RANN."[4] Atkinson's phraseology both reflected the common perception that Research Applied to National Needs was largely an engineering program and foretold the outcome of the study.

Reviewing the Foundation's budget history, Atkinson's Task Force on Engineering and Research Applications concluded that NSF's first fifteen years had clearly belonged to basic research, while recognition of the value of engineering for useful problem solving had grown through the 1960s. During the RANN years, emphasis on engineering relevance and application increased, but at the same time NSF's engineering community stressed the importance of understanding broad, basic engineering principles "common to or required for the solution of future technological and societal problems." Determining when basic engineering became applied, however, was proving increasingly difficult as the time lag between research discovery and application continually shrank.[5]

As the 1970s progressed, the interrelationship between Engineering and RANN had become more apparent—even in the language they used to justify their respective budgets and the similar research results they sought (but, some charged, did not necessarily coordinate). Both programs called themselves a bridge between basic engineering phenomena and economically useful systems. In 1975 Research Applications' deputy assistant director for analysis and planning, Harvey Averch, grew impatient: "The Foundation, I suppose, can argue either that it has multiple bridges for applied research or that Engineering is a 'bridge' to RANN. But in either case, the rationale for many of the activities proposed in Engineering as opposed to being in RANN is unclear." Many of their respective activities were virtually the same, such as Engineering's Food Engineering and RANN's Renewable Resources and Enzymes or Engineering's Advanced Automation (in manufacturing) and RANN's Production Research and Technology.[6]

The task force duly recommended the establishment of an Engineering and Research Applications Directorate. They said linking Engineering with RANN would respond directly to the intent of Congress and the White House, as stated in the National Science and Technology Policy, Organization, and Priorities Act of 1976 (the outcome of Kennedy's earlier S. 32 efforts), and the views of others, such as the National Academy of Sciences. All these agreed that problem-solving research merited "vigorous and perceptive" federal support in attaining diverse domestic and international goals. Besides responding to these "external influences," such a new directorate would expedite engineering and applied research within the Foundation.[7] But it was not to be.

1976–77: THE WHINNERY COMMITTEE EXAMINES SCIENCE APPLICATIONS

Atkinson may have intended major changes "to provide a more sharply focused scientific and engineering attack on emerging or existing national problems and to help build a better bridge between basic research and the solution of such problems," but he decided not simply to join RANN and Engineering as the task force recommended, reiterating his own apparent view. NSF's engineering community was by then less than eager to be linked with the besieged and failing program. Rather, Atkinson appointed a special sixteen-member, one-year science applications task force in December 1976 to look at the issue in the broadest sense. He charged it to "evaluate the nature, content, direction and coupling" of all NSF science applications programs, but "particularly" RANN and Engineering. Atkinson tapped John Whinnery, a professor of electrical engineering and computer science at the University of California, Berkeley, to chair the task force, whose members were specifically selected for diversity in technical competence and professional viewpoint, geographic and institutional distribution, and race and gender.[8]

Whinnery's hard-working committee met approximately monthly until summer. In April 1977 he reported to the National Science Board that while the group had not yet reached conclusions, he felt they would agree that NSF could not do everything, but "few members would argue for a purist stance in which no applications work is allowed in the

Foundation." In Whinnery's personal view, NSF's science applications programs spoke to important national priorities and had achieved "some very fine successes," such as in fire research, now transferred to the National Bureau of Standards, and earthquake engineering, which was influencing both design procedures and building codes. These two cases alone, Whinnery opined, justified the entire applications budget for several years.

He also commented perceptively on implications of the fact that applied research, by nature, touched the political arena more closely than did fundamental science: It involved setting priorities—a sensitive, value-ridden process. Further, political leaders would be more likely to favor "safe" programs with short lead times so that successes could be easily observed and measured. Whinnery asserted, however, that NSF should seek the highly advanced, often risky long-term projects. And its basic-research mission must not be weakened. Whinnery also noted that engineering had a "greater spectrum" of research motivations than other fields, which should be encouraged.[9]

In July 1977 Whinnery's task force submitted drafts of "Principles and Recommendations" and "Comments on Organization." Three principles "on which we generally agree" introduced the report: that basic research and NSF's role in it were "unique and absolutely essential"; that basic research, applied research (to meet longer-range, generalizable applications objectives), and research applications (user-focused, short-term activities) were interdependent and mutually supportive; and that a pluralistic approach to problem solving was commendable if it was first rate and properly coordinated. They especially liked interdisciplinary research and mechanisms to encourage the "hot pursuit" idea.[10]

But research applications (meaning RANN) needed "structured coordination" with the rest of the Foundation, more selective programming, closer cooperation with the mission agencies and with state and local governments, and improved transfer mechanisms to ensure utilization. Significantly, while recommending organizational and budgetary strengthening, the task force did not specifically endorse RANN. Indeed, it sounded like lukewarm enthusiasm when the Whinnery report judged that RANN's "positive factors outweigh the negative, so long as one does not expect too much too soon." Many of the criticisms, it noted, "appear to have arisen from unrealistic expectations."[11]

The task force and NSF staff went on to evaluate four different organizational models, two with variations, to improve the Foundation's efforts in science applications. They quickly discarded Atkinson's first proposal, to combine engineering and RANN, as not in itself contributing to better coordination with the rest of NSF. Likewise dropped were models to disperse RANN activities back into the research directorates (which would "expose" them all to the "same criticism now confined to RANN"), and to reorganize the entire Foundation into units grouped around basic, applied, and problem-oriented research, although a majority of the task force "seemed to favor" the latter approach. Some of these ideas would reemerge.[12]

1978: Applied Science and Research Applications Replaces RANN

In August 1977, Atkinson submitted a memorandum to the National Science Board that outlined a proposed directorate for science applications based on the Whinnery committee's model IIA, by which NSF applications programs could be coordinated and integrated with the research directorates' basic and applied efforts and more clearly presented to industry and academia. Later that month Atkinson reported to the task force that the board recommended calling the new unit Science and Engineering Applications (SEA). The title, he said, was "only to remind people that the directorate is concerned with the applications of both science and engineering." The Division of Engineering would not be moved to SEA. With formal board approval, Atkinson publicly announced the new directorate on 15 September 1977, which signaled the end of RANN. At some point before December, however, SEA was renamed, without explanation, Applied Science and Research Applications (ASRA), and engineering lost yet another chance at increased visibility. The new directorate was to promote, financially and philosophically, the interrelationship and full range of research—basic through applied; it would be concerned with high-quality applications of both science and engineering, in accordance with the 1968 NSF act. Focusing applied research on fewer, more carefully selected national issues would increase its impact, the ASRA planners hoped. The Division of Engineering, according to this plan, too, would remain in the Directorate for Mathematical and Physical Sciences and Engineering.[13]

Research Applied to National Needs finally officially ceased being when ASRA became effective, on 6 February 1978. Most of RANN's programs were kept in ASRA; some, like nonconventional food research and excavation technology, were to be phased out; and some were transferred to other NSF units. Weather Modification, for example, went logically to the Directorate for Astronomical, Atmospheric, Earth, and Ocean Sciences; Technology Assessment ended up in Scientific, Technological, and International Affairs.[14] Former NSF deputy director Raymond Bisplinghoff, in handwritten notes to Atkinson approving the organizational model chosen for ASRA, could not resist further comment. The RANN architect praised the visibility as well as the technical value of hot pursuit under RANN, but he now believed that the basic research that was needed to generate new knowledge relevant to emerging national problems should be paramount: "This is what NSF knows how to do. It is what industry and the mission agencies want." Bisplinghoff emphasized the adjective "emerging." The value of ASRA, in his view, would be in foreseeing future problems: "We should look *ahead* and not put too many bricks on already well constructed foundations."[15]

Within Applied Science and Research Applications, engineering could be found in several niches. One was in the Division of Integrated Basic Research (IBR), which supported long-term basic research with "high relevance to major problems." New in concept and organized by traditional academic disciplines to facilitate relations with the research community and the rest of the Foundation, IBR provided most of its support through selective supplemental funding of work in NSF's basic research directorates. Engineering efforts were as varied as the chemistry and engineering of solution mining, superconducting devices and materials, electromagnetic sensing systems and other innovative instrumentation, nondestructive evaluation, biomedical engineering, estuarine mechanics, and interfacial science.[16]

The importance of engineering in ASRA's Division of Applied Research (DAR) was implicit in its designation of "coherent areas" that encouraged interdisciplinary research on selected long-term, generic problems in such emerging areas as telecommunications and production research and technology. Increasing the rate of technological innovation was also a major objective. ASRA's other divisions, especially that of Problem-Focused Research Applications, which coordinated specified programs—such as earthquake engineering and chemical pollution research—were essentially a more tightly focused RANN.[17]

Jack T. Sanderson, who had replaced Eggers as assistant director for research applications in 1977, became assistant director for ASRA with warm recommendations from NSF management. The Harvard-educated applied experimental physicist and former assistant director of the university's physics laboratory had come in 1971 to NSF's Office of Budget Planning and Program Analysis, which he eventually headed. Although thought by some to be Atkinson's hatchet man, Sanderson's own interest, he said later, was "to make it work," through whatever change was necessary. He would prove to be a survivor.[18]

Change always affects someone negatively, or at least is perceived that way, and the latest shake-up at NSF was no exception. University of Utah investigators in minerals research got their members of Congress to complain to the Foundation that their support, currently under RANN, was being withdrawn. Sanderson, who was careful in all directions, replied that indeed problem-focused research had been redrawn around concentrated topical concerns, but ASRA's Division of Applied Research would maintain an "open window" for unsolicited high-quality proposals in other fields. (He also took pains to remind the complainants that the new directorate would be under severe budgetary constraints for FY 1978.) Basic research in minerals could also be eligible for support in the research directorates through IBR if it showed a potential for high relevance. Hot pursuit of promising leads would always be encouraged.[19]

ENGINEERING MEANWHILE . . .

What remained of engineering in the MPE directorate during the ASRA period underwent its own "intense self-examination" in 1978. Division director Henry Bourne again reported the "apparent paradox" of locating engineering in a part of NSF that dealt with basic science when the term "engineering" "strongly suggest[ed] an association with problem or mission oriented research"—a paradox he said universities also keenly felt. "In general," he explained, "the qualities that make research attractive for the NSF [basic] engineering program are fundamental content together with recognizable bearing on technical applications." Problems occurred, at one end, with science research administrators who were comfortable with work on only the most limited, classical engineering principles and, at the other, with mission agencies who took the existence of NSF programs as an excuse to ignore their own basic research in engineering. Referring to earlier criteria, Bourne noted that it was often difficult to decide whether engineering researchers thought first of a problem or the basic research needed to solve it. In any case, was this important? Their "complex motivations" did "not furnish useful criteria for distinguishing between basic and applied research." MPE's budget justification for FY 1978 put it simply that "new knowledge in the fields covered by MPE has both intrinsic interest and the potential for wide applicability to problems in the 'real world.'"[20]

By Bourne's retrospective reckoning, the Division of Engineering had enjoyed modest funding increases, about 10 percent per year, since 1972—not enough to initiate large new programs but sufficient for "reasonable health." As a percentage of the MPE and total NSF budgets, engineering had remained "amazingly constant," even though total dollar amounts had been "markedly" rising in recent years. NSF was providing nearly 54 percent of all federal support of basic engineering research in universities and a "significant fraction" of all such research by anyone, while industry's support of basic work was seriously declining.[21]

Other statistics confirmed the significance of NSF's engineering efforts. By 1978 the engineering division potentially served 247 engineering colleges in the United States, of which 192 offered graduate degrees. Approximately 16,500 engineering faculty members and 48,400 graduate students submitted about 2,200 proposals to NSF, 800 of which (not quite 40 percent) resulted in research grants. This was a significant improvement over the figures from twenty years earlier, probably reflecting the increasingly theoretical bent, or at least political astuteness and growing numbers, of academic engineers, who by then were writing proposals more congenial to NSF reviewers. The Foundation also continued to offer Research Initiation Grants, 61 in 1977, to young researchers (usually assistant professors in their first two years) and specialized equipment grants, 90 in 1977, totaling $2.25 million. Engineering faculty in numbers now exceeded their counterparts in physics, chemistry, and astronomy combined, while engineering graduate students more than doubled theirs. Undergraduate engineering enrollments were steadily rising, although it was still true that more of these students would eventually migrate to industry than remain in academe.[22]

Overall, the NSF Division of Engineering had seen many of its activities, such as enzyme engineering and materials, spun off to other programs, while others by now had become integrated into current technology, such as holography, optical computing, and the finite element method of dynamic structural analysis. At the same time, the National Research and Resource Facility for Submicron Structures at Cornell University, the division's first such facility and an important commitment signifying an evolving NSF view of the profession, concentrated personnel and state-of-the-art equipment in a single center to encourage superior, perhaps unique engineering research.[23] A good part of engineering, of course, especially its more immediately applied aspects, now resided in ASRA.

1979: Engineering and Applied Science Joined (EAS)

Testifying before the House Subcommittee on Science, Research and Technology for the Foundation's overall 1979 authorization, Sanderson requested $73.9 million for Applied Science and Research Applications. The entire program, he said, would be geared to couple basic and applied research more meaningfully and transfer the results to users for the general good—to do what the best of RANN did, only better. Interestingly, Atkinson closed his congressional budget testimony with "a strange remark for me to make," a plea for increased support for applied research. While ASRA currently controlled about 8 percent of NSF's research funds, Atkinson thought the number should be about 15 percent to 20 percent. In his view, "applied work prospers from being close to our basic research effort," while basic research was "in no way . . . hindered" by NSF support of applied, contrary to early fears. If Atkinson's comment was not "strange," for him it was unusual.[24]

In the broader format of the NSF annual report, Atkinson boasted the development during 1978 of new precision instruments that were enabling increasingly sophisticated research—such as advanced X-ray detectors to speed up the mapping of the crystal structure of enzymes—lasers, computers, and very large-array radio antennas whose resolving power equaled that of the largest optical telescopes in existence. Although he did not say so, it was engineering that was enabling these new strides in all manner of other research, including basic. In the next year's report, however, while Atkinson celebrated engineering breakthroughs on the first page, he noted that "by far the bulk of the research described here is important because of its relevance to basic scientific goals of understanding nature. Practical benefits will come later, if at all."[25] Engineers would continue to read mixed messages from the Foundation leadership in this period of uneasy transition.

Meanwhile, ASRA did not find a comfortable fit at NSF either. For one thing, the mechanism of supplementally funding problem-related basic research through the Division of Integrated Basic Research proved politically untenable as well as bureaucratically cumbersome,

Sanderson later admitted. Why not just give more money directly to the science directorates? Besides, ASRA inherited RANN's elemental problem of lacking a natural constituency, either in academe or the Foundation.[26] The name found no counterparts in universities, and engineering in NSF, still administratively bifurcated, had no compelling loyalty to ASRA, in which it had no specific identity. The organization chart would have to be redrawn—again.

Chemical & Engineering News called ASRA a "gingerbread contrivance of every possible way to name approaches to applying science to everything else." While ASRA was "an honest attempt to integrate research with the search for opportunities," in *C&EN*'s view it failed because of the unhappiness of engineers at NSF, "orphaned" in a directorate of mathematics and physical sciences. "Everyone knew that engineering, however basic, was nevertheless related to problem solving and belonged back with applied science."[27]

Beginning in early 1979, NSF considered, dropped, and then reconsidered a reorganization plan to give more emphasis to engineering and applied research—and to their relationship. A May 1979 staff study suggested that to combine Applied Science and Research Applications with Engineering could strengthen both. There was "increasing recognition" that many of the activities which they pursued separately were complementary and reflected a real-world continuum; they largely depended on the same research community. Combining their budgets would give Engineering about $70 million for FY 1979 and break down the "artificial distinction" between Engineering's basic research and ASRA's applied. Such a structure would also reinforce the university-industry couplings that both organizational units were seeking, such as the establishment of major centers of technology-oriented fundamental research.[28]

So once again the National Science Foundation shuffled engineering's deck. Atkinson announced in late May NSF's intent to establish, effective 1 July 1979, a Directorate for Engineering and Applied Science (EAS) to replace ASRA and absorb MPE's Division of Engineering. The reorganization would, he said, strengthen NSF's engineering programs by "giving engineering a single, more visible place in the NSF organization"; provide a broader base of science and engineering on which to build stronger programs in applied and problem-focused research; and recognize engineering's key role in the transfer of science into technology.[29]

Sanderson was named to head the new EAS directorate, his third (but not last) comparable assignment since 1977 in the continuing metamorphoses. He supervised the work of six divisions, which were a mix of traditional academic engineering disciplines transferred from MPE and varied aspects of applied and problem-oriented research that had organizationally zigzagged through the decade. The new grouping of efforts oriented to "national needs"—the remnants of RANN—was clearly a diminished share of the whole.[30]

Once again, as they invariably did, representatives of the various professional interests responded to the change as they perceived themselves affected. Champions of mechanics, for example, bombarded Atkinson with complaints that their basic engineering emphasis had no separate identity in EAS. Sanderson wrote numerous patient replies—to university engineering departments, the American Society of Mechanical Engineers, the Engineering Mechanics Division of the American Society of Civil Engineers—extolling the respected and significant place of mechanics in EAS's Mechanical Sciences and Engineering Group. But it was not a division.[31]

Bruno Weinschel, vice president of the Institute of Electrical and Electronics Engineers, raised a different issue—that of intensifying global economic competitiveness—and a different point of view. Unless the current negative trade balance trends could be reversed by increased U.S. productivity, he foresaw "serious social unrest" as well as economic losses. "I trust that this is not just a game of musical chairs but that with this reorganization there is more emphasis on *application*," Weinschel warned. "Basic science is extremely important to our long-term well-being; however, we have not been clever in recent years to reap the fruits of basic science." Nor were American "philosophies and educational systems," he worried, "geared to face this challenge."[32]

These two exchanges of correspondence show just how unsettled engineers themselves—not to mention the Foundation as a whole or the broader scientific community—were in the wake of RANN over what engineering was, where it belonged, and what it should do. Had it become too applied? Too basic? Sanderson strove to articulate an inclusive yet distinctive identity for engineering that was both acceptable to engineers and understandable to the public, especially the providers on Capitol Hill. Appearing before the House science subcommittee not many months after the formation of Engineering and Applied Science to argue for its FY 1981 authorization, he declared that EAS represented an attempt "to significantly strengthen both engineering and applied research." As he put it, "the engineering disciplines have always had a foot in both camps," being concerned with both new knowledge and its end use. To allow engineering to continue its vital bridging function between knowledge and application, between science and society, Sanderson requested $137 million for EAS for FY 1981, a 22.5 percent increase. He emphasized that EAS was sharply targeting its efforts to balance responses to high-priority national needs with "real growth" in programs "recognized as providing a unique contribution to the U.S. knowledge base in engineering and to U.S. technology and innovation," at a time when productivity was disturbingly lagging.[33]

Here was another external crisis, international competitiveness, to drive U.S. policy makers and the technical efforts they controlled through the power of the purse. Sanderson astutely exploited the prevailing concern, knowing that President Jimmy Carter had, in late 1979, announced his Industrial Innovation Initiative, which was intended to reverse the poor productivity trend. Another initiative was a cooperative university/industry automotive research program whereby auto companies would provide the "path to production" while government support would focus on social concerns, such as environmental acceptability, economic attractiveness, and efficiency. Small-business support and programs to promote access by state and local governments to "the Nation's wealth in scientific and technological resources" were other efforts gaining currency. Sanderson called EAS's budget "holistic," emphasizing the "interconnection and complementarity" of its many programmatic elements.[34]

But for all the positive publicity, and contrary to Sanderson's expectations, the Directorate for Engineering and Applied Science did not endure either. Engineering schools, which were being revitalized with teaching and research in new technologies, and the broader engineering community, which was worried about the growing competitive disadvantage, renewed their pressure for increased status and visibility for engineering at NSF. Skyrocketing demand would probably cause personnel shortages to continue and perhaps worsen throughout the coming decade. Even as EAS was on the Foundation's drawing board, Congress was seeking its own fix for America's technological shortcomings.

CONGRESSIONAL CHALLENGE: A NATIONAL TECHNOLOGY FOUNDATION?

Aroused by the mounting challenges of foreign economic competition, the sluggishness of American productivity and innovation, and scanty government-industry cooperation, George E. Brown, California Democrat and chair of the House Subcommittee on Science, Research and Technology, convened fact-finding hearings in June 1979. The engineering professions, meeting concurrently as the 1979 Engineers Public Affairs Forum, were noted participants in Brown's gathering, "Government and Innovation: An Engineering Perspective."[35] Following up, the concerned congressman lost no time in drafting legislation to create the National Technology Foundation (NTF), a new independent agency to be organized and managed similarly to the National Science Foundation, including a twenty-four-member, presidentially appointed board. It would absorb NSF's Engineering and Applied Science Directorate and small-business interests along with the National Bureau of Standards, the Patent and Trademark Office, the National Technical Information Service, and parts of other federal programs.

In response to the NTF possibility, Sanderson set up a meeting in December 1979 of engineering deans, Foundation EAS and education representatives, Brown, and Atkinson to "review the present status of engineering research and education and some specific areas where engineering can make a unique contribution to the future well-being of the country." A lawmaker's attendance at an executive agency meeting was unusual, but the director felt it "so important that the best case be made for engineering" that he bent precedent. Although Brown was to appear at eleven, Atkinson convened the others an hour earlier to brief them on the Foundation's views of Brown's proposal.

Those views were predictably skeptical. Atkinson's handwritten notes to welcome the group reminded him to emphasize the Foundation's, and his, commitment to strong support for engineering—and that although he was making the "best possible case" for engineering research and education in a possibly "tough budget year," other NSF constituencies would expect equal attention. Atkinson admitted "real concern" about the concept of the National Technology Foundation. It would, he thought, separate engineering from other sciences and could establish a competition for funds between NSF and an NTF "in the political arena where the decisions are less likely to reflect the best judgment of the research community." The fact was that the larger an agency's budget, the more visible and vulnerable to political scrutiny it became; NSF had long ago lost through growth whatever benign neglect it had enjoyed. Atkinson did not state the obvious, that NSF would lose a significant chunk of its programmatic and budgetary bailiwick if a technology foundation emerged.

The director's preferred alternative, "the best for science, engineering, and the country over the long term," was, unsurprisingly, to strengthen the technology programs of NSF and the mission agencies. (That he anticipated a technology foundation bill could be the clue to Atkinson's enthusiastic, self-admittedly "strange," earlier testimony supporting applied research.) He clearly perceived the NTF possibility as ominous. With the president, through the Office of Management and Budget, setting national priorities, which NSF was obliged to support; and with Congress having the final financial word, he and the Foundation had little power over the outcome, Atkinson lamented. So although he could not direct his colleagues to lobby, he did remind them that all citizens should make their views known. As it turned out, the White

House Office of Science and Technology Policy (OSTP) also had reservations about a technology foundation, which could "break important linkages that now couple many scientific and engineering projects." OSTP suggested the government could better serve technology by working to improve tax and antitrust policies and the economic climate rather than by creating a new agency.[36]

Brown introduced his National Technology Foundation initiative anyway—"for discussion purposes"—in mid-1980 and held hearings in September, where he clearly stated his intent to study the NTF concept but not push the bill toward immediate passage. Lewis Branscomb, the energetic chief scientist at IBM and chair of the National Science Board, led a parade of distinguished witnesses. Applauding the committee for focusing attention on the critical need to improve U.S. technological capacity, Branscomb opposed separating science and technology organizationally as not "wise or necessary." The "interweaving of research with near term and long term potentials for application, a mixture of basic and applied," was already a feature of virtually every NSF research directorate, he submitted, and a "major source of strength to this Nation." The acting NSF director, Donald Langenberg, followed, emphasizing the links among all the various philosophical and practical aspects of research in the sciences, the distinctions among which were "very tenuous and very fuzzy." He also stressed the increasingly vital relationships that universities were developing with U.S. industry and state and local governments.[37]

Brown warmly commended both NSF leaders. Then: "The question becomes why, in view of all this agreement and the excellent record which I thoroughly concur with—the Foundation is a great institution—why are we in such a mess? Why is it that things keep getting worse in the field of productivity and innovation?" Brown pressed Branscomb to see to it that those areas slated for transfer to the technology agency would receive "a level of attention more commensurate with the problems facing the country today." Branscomb could not, of course, speak for the National Science Board, or the Foundation, on specific items, but he did assure the congressman that the board had "one voice" on giving a higher priority to programs improving technological competitiveness. But, he concluded, there was "no crisp way the Government can orchestrate industry and science" to work together. "We have to lead those horses to the water, but they share the common pond. That is the pond of our scientific and engineering communities."[38]

On the other side, the lineup of those favoring the proposed National Technology Foundation was considerable and diverse. The U.S. Conference of Mayors, for example, thought that NSF had plainly failed to pursue social and urban problems, although it had "made a pass at it," presumably under RANN. If NSF was too "pure and pristine" to make that contribution, then someone else should. The National Academy of Engineering thought the NTF would encourage inventiveness, creativity, and the university-industry connection. Just as the National Academies of Engineering and Sciences were coming back together in friendly cooperation only after NAE had taken the essential step of breaking away, perhaps so would NTF's break from NSF serve their future collaboration. Myron Tribus, director of MIT's Center for Advanced Study in Engineering, while unsatisfied with the pending legislation, was "absolutely unequivocal" that the National Science Foundation, by its "practices, procedures, protocols, power structure, philosophy, peer review, political constituencies and personnel simply [was] the wrong organization to be concerned with enhancing our ability to deploy technology."[39]

The engineering professions also stepped up to endorse the intent of Brown's H.R. 6910. The board of the American Society for Engineering Education again unanimously recommended the establishment of an independent National Engineering Foundation or a renamed, refocused National Engineering and Science Foundation. Engineers had felt themselves "in the position of a neglected child" ever since the first non-inclusive NSF act, and engineering remained only a minor part of NSF, with but 10 or 12 percent of its budget. Dissatisfaction with NSF, they said, "[ran] very deep in the engineering community." The American Association of Engineering Societies, claiming to represent thirty-nine national organizations, considered the bill "a step in the right direction" and pledged engineers' help on technical and other relevant considerations. Representing industry, Lamont Eltinge, director of research of Eaton Corporation, agreed with Brown that a technology foundation should "not so much intervene in the short run market needs" as take a "longer term technology perspective" to fill gaps to regain "that competitive market excellence that we have lost in several areas."[40]

1981: A DIRECTORATE FOR ENGINEERING AT LAST

The attention generated by Brown's National Technology Foundation proposal awoke NSF to the political, if not technical, necessity for giving engineering a more central, visible, effectual role within the Foundation. There followed a flurry of board and staff activity to deflect the legislative challenge as well as improve NSF operations for maximum effectiveness and clout. In the midst of it all Atkinson left the Foundation, following his predecessor, McElroy, to the University of California, San Diego. The acting director, Langenberg, told *Science* that calling the NTF bill the "driving force" behind NSF's reorganization discussion was an "overstatement," if anyone believed him. The basic-research community still feared that change would threaten NSF's primary mission, while external critics and the engineering community suspected that change would not go far enough. Meanwhile, another year of severe budget limitations was in prospect. By the time that John B. Slaughter, provost of Washington State University, was designated NSF director in the fall of 1980, he found himself having to assure the press that he was "not managing a train wreck."[41]

The disquieted National Science Board had already created a discussion group to consider issues associated with the proposed National Technology Foundation. Its chair was Joseph Pettit, a former professor at Stanford, where his mentor, Frederick Terman, had as dean built a premiere engineering research program and encouraged the commercialization of its results, especially in electronics, as witnessed in Silicon Valley. Pettit was now president of the Georgia Institute of Technology, which had transformed itself from an insignificant engineering experiment station to one of the country's largest federal (primarily

143

military) research contractors. He had also chaired the graduate phase of the American Society for Engineering Education's 1960s goals study.[42] His own perspective sharpened from the group's work, Pettit reported back to the board in mid-June 1980. Since the NTF legislation implied that those NSF activities it affected were not adequately supported at the Foundation, he said, the group had searched for common agreement on what NSF did and should support, acknowledging that it could not do everything and insisting that NSF's methods, although necessarily different from those of a technology foundation, were not necessarily inferior.

Pettit's discussion group unanimously stated that "engineering, or engineering science, was fully compatible with other basic sciences with similar paradigms, reviewing procedures, refereeing procedures, [and] publications." Furthermore, science-driven applied research was clearly suitable for support under NSF's usual criteria. However, NSF should only fund problem-driven research that had generic application, rather than implementing a particular solution, and even then only when the mission of no other agency encompassed it. The group upheld NSF's responsibility to "support the production of knowledge which undergirds technology." All this led them to favor an independent directorate for engineering.

The words were spoken. During discussion of the report, the departing Atkinson agreed that perhaps the best way to respond to the NTF bill might be to create an engineering directorate, "to give clear visibility to engineering"—as well as more resources—and to distribute applied research throughout the relevant NSF disciplines. To monitor and coordinate interdirectorate and interdisciplinary applied research, Atkinson proposed establishing an office of applied research. Pettit's group authorized a staff study of Atkinson's proposition, to be presented to the Executive Committee in July and the full board in August 1980.[43]

For the rest of the year, the NSF community considered the implications and possible specifics of further change. The Duke- and Harvard-educated Branscomb, who had internally challenged NSF's distinctions between basic and applied research the year before and was providing an intellectual context for the entire exercise, drafted "principles" to undergird Foundation policy on applied science and engineering. Of first importance, he wrote, was an "infrastructure investment" in people and knowledge, logically centered in universities, to ensure the nation's long-term economic vigor and technological capability. As engineering research activities "stay[ed] ahead of the state of the art," they "necessarily push[ed] up against the scientific frontier," whether the work was motivated by the investigator's intellectual satisfaction or the "prospect of practical benefit." Thus, university science and engineering had an "intimate relationship, each supporting the other." Therefore, NSF should be involved with both. At the same time, Branscomb asserted that engineering was different from science and should be allowed to follow its own valid culture and traditions in a separate NSF organizational unit.[44]

Pettit's August 1980 comments to the board reflected his examination of specific NSF programs slated for transfer to the proposed technology foundation. He agreed with Branscomb that the "relationships between basic and applied research and between the scientific and engineering fields are mutually rewarding and supportive." Moreover, acknowledging political reality, he argued that since "the continuing demonstration of the utility of science for the public good appears necessary to provide justification for public support," it was essential for NSF to include applied research in order to retain "a broad base of congressional support" for basic research. A separate technology foundation was a poor idea for NSF.[45]

Once more, NSF's constituents responded to the rumors of change according to their particular interests. George Housner of Caltech worried that if EAS's Problem-Focused Research were lost in the proposed reorganization, projects on earthquake hazard reduction would be split up among other NSF units. Losing their identity would impair both efficiency and productivity. The politically savvy Housner sent his letter of concern to Brown, not to the National Science Foundation. Purdue engineering professor Moshe Barash wrote to Carter that engineering research and education were in dire straits. Instead, he continued, NSF seemed to be dismantling the very programs—Advanced Production

Research, for example—that were "our main hope for increasing industrial productivity" in the face of mushrooming foreign competition.[46]

F. Karl Willenbrock, dean of engineering at Southern Methodist University, was among those pushing for enhanced visibility and clout for engineering as a whole. To *Science* he stated that engineering at NSF suffered from second-class citizenship, and if one more cosmetic change was all that was coming, engineering would be forced to look elsewhere—presumably to a technology foundation. He told *Chemical & Engineering News* that NSF's difficulty was that "they consider engineering as the applications of what they do in science. That simply isn't the case. Engineering has its own tradition that isn't recognized by science."[47]

Walter Vincenti, an academic researcher and aeronautical engineer who studied historically the idea of engineering as knowledge, came to the same conclusion. Citing historians of technology and selecting out engineers as an "organizing" (devising, planning) subset of technologists generally, Vincenti emphasized that technology was not "derivative from science." Rather, it was "an autonomous body of knowledge, identifiably different from the scientific knowledge with which it interacts." While technology might *apply* science, it was not the same as *applied* science. The engineering knowledge needed to implement the art of design was, in his view, "enormously richer and more interesting" than that of applied science. Engineering went beyond science, too. Note that engineering schools maintain their own libraries, he wrote. The books, journals, and other materials found in science libraries were important to engineers, but insufficient.[48]

The NSF engineering discussion, as *Science* put it in September 1980, was "assuming the scale of a major science policy issue," and despite efforts by some engineers and scholars to frame more accurate concepts, the basic-versus-applied argument played out once again in those terms. The record bulged with new attempts to draft definitions that would once and for all differentiate the one from the other and the shadings between—although to no avail, and perhaps to no purpose beyond trying to include or exclude. As exasperated Henry Bourne, now

EAS deputy assistant director, sputtered in August 1980, "The 'recognized and specific need' of applied research might very well be 'to gain fuller knowledge or understanding of the fundamental aspects of phenomena and of observable facts' of basic research." In engineering, such definitions were "particularly troublesome," he said, and only relevant at the program level for setting objectives and research criteria.[49]

In Branscomb's opinion, an "artificial distinction" between basic and applied research was "unnecessary—and unhelpful—when the applied research is done by institutions [universities] that don't intend to do the applying." *Chemical & Engineering News* added the perceptive thought that "enough literature in science policy now exists to make a convincing case that the social climate really drives science—whether basic or applied." What was more, NSF had "never really taken the measure of that climate in establishing a successful, stable applied science program."[50] Langenberg, in preparing for the board a draft organizational structure in September 1980, further noted that lately NSF's applied-research debate was being "conducted in the increasingly intense light of concern over faltering national productivity and innovation"—another driving force that was economically based and politically carried.

Echoing Branscomb's call for new understandings, Langenberg went on to recommend a directorate for engineering and suggested that the point implicit in joining engineering and applied science in the current directorate—that applied science was associated exclusively with engineering—was "not conceptually correct." While the profession of engineering might be primarily concerned with application, engineering research ran the gamut from basic to applied in the same way that other disciplines did. It followed, then, that engineering should enjoy the same kind of functional structure as any other group of disciplines in the Foundation. Linking basic and applied research in all directorates, in Langenberg's judgment, would be "mutually reinforcing," even though it would require defining the "imprecisely definable" (what was basic, what was applied) for the purposes of accountability. The applied-research community might bewail its loss of visibility, of course, and the peer review and advisory committee mechanisms would have to adapt to serve the full spectrum of research needs.[51]

Nothing was ever easy, of course. In late October 1980 Donald Senich, EAS division director for problem-focused research, reminded the acting director that the 1977 Science Applications Task Force, after thorough study, had reported its "strong and essentially unanimous conviction" that the distinctions between basic and applied research and research applications were "crucial" and must be structurally respected. These distinctions reflected differences in organizational strategies, working styles, goals and purposes, and even personal temperaments. Further, dispersing problem-oriented research among the directorates could encourage a tendency to fund too many unrelated small projects, which would in turn reduce the social impact of them all. Since Congress had confirmed NSF's role in this area—as with the Earthquake Hazards Reduction Act—Senich proposed at least including a Division of Problem-Focused Research within the new engineering directorate.[52]

Theodore Wirths, NSF's director of small-business programs, cautioned that reorganizations had been following one another so fast that it was not possible to develop or assess their effectiveness—especially those affecting interdisciplinary research. Diffusing such efforts throughout NSF programs could mean that they would go to people with no familiarity, interest, or ability with applied programs; what was really needed, beyond careful planning, was "firm and sensitive" leadership. Overall, however, Wirths considered an engineering directorate to be "timely and appropriate." Success would depend on "new money," of course. And while there would never be enough to satisfy completely every discipline's "appetite for funds," he said, "we must really make a commitment to engineering that contemplates the unthinkable. This might mean that engineering might increase at a rate of 15 to 20% in our budgets for two or three years, while [basic] research increases at 5%." The latter might even remain level or decline slightly. "NSF's commitment is not likely to satisfy the engineering community if they conclude that they are our first priority *after* basic research is taken care of."[53]

At the other extreme was W. W. Havens, executive secretary of the American Physical Society, who stated flatly that "engineering research" was a contradiction in terms and that NSF should stick to science. To him, research meant "the exploration for and creation of new knowledge," while engineering required "a specific and concrete objective." Understanding the current pressure for "making engineers first class citizens in NSF," Havens thought the mission agencies could and should support "engineering development." Branscomb, also a physicist, replied, in "strong disagreement," that engineering research also sought new knowledge, albeit with "its own paradigm," which, if "somewhat different possibly" from science, was "equally valid and legitimate." Since NSF supported research but not development, they were "differing over the uses of words," Branscomb wrote. "The operational question is, should NSF support the activities of our engineering schools that give rise to the *capability* in other institutions (industry and national laboratories) to do engineering development at the highest levels of technical achievement and productivity? I believe we agree that it should."[54]

During the months of ongoing discussion, Judith Coakley of the EAS Problem Analysis Group got the task of recommending where to put what in a reorganization of Engineering and Applied Science. Her draft of mid-July 1980 created an engineering directorate whose goal was to strengthen fundamental engineering while enhancing its links with industrial and other societal applications. The new directorate's divisions would include Computer Science and Engineering and Materials Engineering, now back under the broad engineering umbrella. Coakley organized the other engineering designations—civil, chemical, electrical, mechanical, industrial—along historically familiar disciplinary lines. Sanderson favored this approach because it would facilitate communication with existing university departments; he knew the perils of not having an identifiable constituency and making too many waves at once. (He later approved his successor's function-based internal reorganization.)

Each new division was to encourage research spanning the spectrum of fundamental through applied engineering to the point of technology transfer to industry or government. Each would participate in industry-university cooperative programs; each would keep its small-business connections. Most programs in problem-focused and applied research were slated to go where their subject matter best fit; for example, earthquake hazard mitigation would transfer intact to the new civil engineering division. After considering several possibilities, NSF assigned cross-disciplinary and some problem-focused research to the new Office of Interdisciplinary Research (OIR) in the Engineering Directorate, to be assisted and advised by the Interdirectorate Interdisciplinary Research Committee. OIR was to identify, stimulate, and coordinate work in areas such as robotics, catalysis, resources availability, and natural hazards.[55]

Director-designate Slaughter and Langenberg, after consulting with academic, scientific, and engineering advisory boards and individuals within and beyond the Foundation, and considering current and projected resources, brought the reorganization proposal to the National Science Board in November 1980. Slaughter emphasized that the redistribution of applied research back into the research directorates—which the board, at its October meeting, stated its wish to do—would stress quality as the chief criterion, whether in basic or applied research, and would maintain interdisciplinary and problem-focused efforts. Slaughter's press release immediately following board approval declared, once again, that without altering NSF's commitment to basic research as its central mission, the reorganization to establish an engineering directorate would "broaden our base of support for applied research and give new emphasis to our support of engineering research and education." (In a concurrently debated concern, the Foundation declined to form a directorate for social sciences.)[56]

As always, critics spanning the spectrum stepped forward. *Science* responded to the news doubtfully, predicting that NSF's latest reorganization would probably "put the foundation uncomfortably in the middle between two major constituencies"—the basic research community on the one side and critics in Congress, industry, and engineering fearful of inadequate follow-through on the other. NSF's history had shown the likelihood of this scenario. *Science* also reported Slaughter's statement to the NSB that forming an engineering directorate would commit NSF to seeking increased resources for engineering. He hoped they would come from new dollars and not a redistribution of present funds, which could only alienate NSF's other communities. While the incoming Reagan administration seemed sympathetic to increased emphasis on engineering and applied science, it was making no promises in a tight budgetary framework. The Senate Committee on Labor and Human Resources remained "skeptical" that applied science would "fare as well as in the past when it was logically coupled with engineering," as it had been from RANN through EAS. "Acquiesc[ing] to the new arrangement," the committee warned NSF of its expectation that the Foundation would "demonstrate its continued commitment" to "the best applied research."[57]

For his part, the new director wrote in NSF's annual report for 1980 that "our experience with basic and applied research teaches us that the differences are not nearly as great as their names suggest." A project's potential contribution was of greater significance than the categorical name it went by, Slaughter believed. What was more, creating a directorate for engineering acknowledged the "unique linkages between engineering research and industrial productivity and innovation" at a time when it was imperative to address the sluggishness of the American technological response to international economic competition. The change, Slaughter anticipated, would make NSF more "agile" in responding to the challenges and opportunities of the coming decade.[58]

Engineering became an independent NSF directorate on 8 March 1981, with Sanderson at its head as assistant director for engineering. As finally constituted, it included four divisions—Electrical, Computer and Systems Engineering; Chemical and Process Engineering; Civil and Environmental Engineering; and Mechanical Engineering and Applied Mechanics—and the Problem Analysis Group. Engineering and Applied Science was abolished; its Applied Research and Problem-Focused Research programs and personnel were redistributed among NSF's four research directorates: Engineering; Biological, Behavioral and

Social Sciences; Astronomical, Atmospheric, Earth and Ocean Sciences; and Mathematical and Physical Sciences.[59]

So 1981 marked a pivotal moment for engineering at the National Science Foundation. It had finally—or again, if the roughly equivalent divisional status of 1964 is remembered—achieved full citizenship. NSF had recognized engineering, officially at least, on its own terms. It had been a long, frustrating struggle, whose many arguments varied in relative emphasis over time, but the issue of whether NSF should be supporting applied as well as basic research—and whether engineering equaled or was more or less than applied science or, indeed, was something different—perennially recurred. While the artificiality and "unhelpfulness" of distinguishing among such terms was gradually being recognized, considerable bureaucratic turf protection and long-standing wariness remained and would undoubtedly continue, especially as the various interests saw themselves in competition for a finite financial pie. Fundamentally, engineering within the historic basic-science culture of NSF was always alien in some degree. Tensions were inevitable.

In the end, the strongest imperative seemed once again to be that of national need. RANN was created to respond to critical energy, environmental, and social challenges of the 1960s, and the new Engineering Directorate was expected to help effect a favorable shift in U.S. global competitiveness. That is, what drove science was not abstract theories or definitions (though all sides used them), but society, as forced in the first instance by the White House and, in the second, by Congress. In practical terms the driver was politics, the vehicle money. The days of quiet internal decision-making at NSF were long over.

For all its problems, RANN, by moving boldly into research applications, may well have paved the political way for the Engineering Directorate, which meant the admission and approval of the differences between engineering and, say, math and physics, as longtime Engineering financial officer Paul Herer asserts. Why this change took such a tortuous path was partly due to the continuation of historical tensions and rivalries. Also, Richard Atkinson, who directed NSF during most of the transition period, was not notably sympathetic to applied research and engineering; some saw him as not exceptionally effective as an administrator. More immediately, it would have been an enormous leap politically, even emotionally, to admit the failure of RANN at once. Sanderson "knew going in" that the first reorganizations after RANN would not be stable. It would take time to find a plan acceptable to both the academics and the government funding sources, not to mention the affected technical communities. Hence the several intermediate steps. But the interval also gave engineering time, and probably a push, to establish its own claim to independent recognition. Unlike the rushed "crisis decision-making" that resulted in RANN, the formation of the engineering directorate was the product of broad philosophical and practical consideration. By 1981, unlike the 1960s, NSF was no stranger to strategic planning.[60]

For all that, the tensions were not over. Some basic scientists still resisted engineering's incursion into *their* NSF territory. Some engineering proponents still muttered that the only real answer would be to establish a separate, analogous national engineering foundation, or at least an agency whose name—National Science and Engineering Foundation—reflected engineering as an equal partner. Yet, the fact that the overall organization chart survived more or less intact for a decade suggests that an acceptable accommodation had been found. Brown quietly dropped his campaign for a technology foundation, which indicates that he, too—for the moment—was satisfied. Perhaps, as Sanderson mused, Brown, a "bright politician," had intended his bill to act as just the internal prod it proved to be.

As for the effects of all these changes while they were taking place, Sanderson said simply that of course "productivity goes to pot during a reorganization." And morale could not help but suffer when personnel were jerked from one philosophical emphasis or managerial unit to another, especially when it happened time after time in short order or when the people jerked did not wish to go. Those involved in applied research were particularly hard hit; after the initial halcyon days of RANN, theirs had been one defensive posture after another. And then they disappeared altogether from the boxes on the organizational chart.[61]

But engineering as a whole had prospered and looked forward to increased visibility and influence in the new decade.

Chapter 6

By Any Name, Engineering

> Engineering is the bridge between science and society. New knowledge in the physical sciences and mathematics is translated through engineering into new products and services for mankind.
>
> —National Science Board

Despite jurisdictional uncertainties and the general unrest accompanying repeated organizational change, the engineering research that NSF funded became increasingly sophisticated and productive. A few examples from areas marking significant development during the late 1970s and beyond—earthquake engineering, materials research, and computer development—illustrate the interrelationships among varied facets of NSF research and the accelerating pace of accumulating and using knowledge, to say nothing of remarkable specific achievement.

These case studies, which were chosen as significant, representative efforts in consultation with project advisors, NSF leaders, and program staff and confirmed by abundant documentation, represent a mix of engineering with other disciplines as well as a spectrum of contemporary issues and problems. The topics remain timely and important.

The 1989 Loma Prieta Earthquake occurred while these stories were being chosen. Although it was never the only source of support, NSF played an influential, visible, and definable role in these areas, in one case by congressional mandate, in another by funding default, and finally by enthusiastically seizing opportunity.

The research outlined in these stories occurred in and out of engineering's specific jurisdiction and changed locations over time, which was indicative both of engineering's administrative disorder and its intellectual reach. Finally, though beginning earlier and continuing later, these topics all saw major activity in the period of transition following RANN.

Earthquake Engineering: Building to Withstand

Earthquakes are short-lived phenomena, usually over in seconds. They rank below floods and winds in terms of total losses over an extended period. Yet, a single major quake in a populated area is "as bad a natural disaster as there is," causing catastrophic injury and loss of life, destruction of property, and economic and social dysfunction.[1]

The National Science Foundation supported earthquake research—both that of seismologists, who study forces deep within the earth, and engineers, who concern themselves with effects near the surface and structures above it—as early as 1961, when it funded individual projects in engineering mechanics. By 1977, when Congress mandated that NSF devote specific attention to earthquake hazard mitigation, the agency had already made itself a locus of such efforts.

Spurred by the great Alaska earthquake of 1964, for which monitoring instrumentation had been inadequate, the Foundation began a formal program in earthquake engineering in 1966. It funded the University of Illinois to build the first U.S. electronically programmed earthquake simulating machine, or "shake table," on which engineers could test some earthquake effects on scaled-down buildings. NSF also began contracting with the National Academy of Engineering to arrange for immediate on-site investigations of earthquake damage before rebuilding and repair destroyed this "perishable information." Funding that first year was a modest $243,000; for the next five years the level of support fluctuated between $659,000 and $1.3 million.[2]

Justifying to Congress its engineering mechanics budget for FY 1968, NSF stressed the need for its new earthquake engineering program. A recent survey of earthquake engineering projects not funded by NSF had yielded a list of only five, for a probable total of not more than $50,000 per year. The next year NSF engineers noted that current building codes "could only be described as crude and this would even be a gracious view." Stronger, more appropriate construction requirements would not be possible, of course, without much more, and better, technical data on which to base them. The million dollars spent annually was not only "quite inadequate to continue a coordinated effort," it also existed at the expense of other basic research in engineering mechanics.

Meanwhile, a major earthquake occurred somewhere in the world nearly once a year with property damage alone ranging from $500 million to $1 billion. NSF planned a "broad-scale attack" of research to cover topics including engineering seismology, soil mechanics, structural dynamics, socioeconomic problems, and potential effects on utilities, transportation, and communication systems.[3] H. B. Seed, of the University of California (UC), Berkeley, for example, was studying soil and foundation response to understand the liquefaction phenomenon that left buildings floating in quicksand, while James Yao pursued the dynamic response and reliability of adaptive systems of earthquake-resistant structures at the University of New Mexico.[4]

Earthquake engineering became a line item in NSF's budget in FY 1971 at about $2 million. The billion-dollar San Fernando earthquake of February 1971, as powerful shakes do, spurred further research efforts. That year investigators were ready with instruments to measure effects on soil and structures and computers to analyze those effects. Of great help was the NSF-funded National Information Service for Earthquake Engineering (NISEE), operated by Caltech and UC Berkeley to collect, organize, and disseminate information. A key NISEE activity was providing a clearinghouse for relevant computer software developed under research awards.[5]

Earthquake engineering came to occupy a significant and visible place in the Disasters and Natural Hazards Division of the Foundation's Research Applied to National Needs program. By FY 1973 RANN was seeking more than $6 million for earthquake engineering—to develop improved building codes, study seismic risk, predict soil and foundation behavior, and evaluate the socioeconomic effects and costs of earthquakes. The program soon included about sixty projects, principally at Caltech, UC Berkeley, UCLA, Illinois, and MIT. At NSF's RANN 2 Symposium in November 1976, earthquake engineering was a featured topic. Caltech earthquake expert George Housner stressed engineers' ever-present concern (unlike basic scientists') that "economical optimization" was as important as technical considerations of stresses and deformations under dynamic loadings. Since complete protection was neither affordable nor technically possible in all cases, they must seek a balanced approach.[6]

By 1976 the National Science Foundation could boast that RANN's earthquake engineering program had developed computer programs on soil response that were used by "virtually all" engineering consulting firms that did construction-related earthquake analysis. The American Association of State Highway Transportation Officials employed RANN-developed statistical material in its design codes for bridges. The Nuclear Regulatory Commission found NSF studies on tsunami behavior helpful when considering coastal sites for nuclear power plants. In 1978 IBM's new, award-winning Santa Teresa Laboratory in San Jose, California, incorporated NSF seismic and energy research findings into its design.[7]

RANN earthquake engineers also applied their design methods to other dynamic hazards, such as extreme wind and explosion, while more fundamental, longer-range engineering research on earthquakes simultaneously moved forward in NSF's Division of Engineering. Some jurisdictional conflicts were inevitable when the division got little money for dramatic programmatic increases while RANN enjoyed funds and visibility, but in general, said Michael Gaus, who stayed in the Engineering Division, the two benefited from each other. The older unit retained its responsibility to develop human resources in the field through Research Initiation Grants and other educational programs.[8] RANN also supported social scientists' work on the sociological, psychological, and economic issues associated with earthquakes to balance technological advances with human needs, and NSF's Geophysics Program in the Research Directorate also funded earthquake research, $2.4 million in FY 1972. Numerous other federal agencies, principally the National Oceanic and Atmospheric Administration and the U.S. Geological Survey but also the National Bureau of Standards, the U.S. Army Corps of Engineers, and even the Department of Housing and Urban Development were active in the field, but NSF was "by far the dominant factor in earthquake engineering research."[9]

In 1977, even as engineering at NSF was being jostled through repeated reorganizations, earthquake engineering gained momentum and status when Congress passed the Earthquake Hazards Reduction Act. The law required NSF and nine other federal agencies to establish and maintain the National Earthquake Hazards Reduction Program (NEHRP, called "neeherp"), which was to be coordinated by the Federal Emergency Management Agency and would include research to understand earthquakes and to develop technologically and economically feasible design and construction methods, improved earthquake prediction systems, model building codes, and emergency planning and education guidelines.[10]

California senator Alan Cranston had initiated the legislation in 1968, based on a report by Housner of NSF earthquake engineering research. Cranston's bill failed when, as Housner later remembered it, California's representative from Palmdale—"sitting right on the fault!" but apparently fearful that admitting danger would jeopardize tourism— killed it, arguing that it was not needed in his state. The National Science Foundation itself had opposed new legislation in 1972 following the costly 1971 San Fernando earthquake, preferring to act through increased appropriations, not mandated programs; flexibility would better ensure excellence, leaders argued. In 1975 it likewise declined additional statutory authority as not necessary.[11]

But in 1976—the worst year for earthquakes in four centuries, according to a House report—science subcommittee chair James Symington, Democrat of Missouri, pointed out that the degree of earthquake damage was linked to whether governments had or had not concerned themselves with prediction and preparedness. He called for hearings to direct U.S. policy. Representative George Brown of California, harboring none of the reservations of his Palmdale colleague, cited the state's great risk and championed the legislative effort, as Cranston continued to do in the Senate. Thus Congress—as it had from weather modification to RANN—demanded action on a specific, visible problem that would, to be sure, serve political as well as perceived national needs. The NEHRP Act authorized a total of $102.5 million over three years for the Foundation to fund research on earthquake engineering in universities. The act specified the same amount for the Geological Survey.[12]

Since obtaining scientific information on unpredictably and infrequently occurring phenomena is costly and difficult, the desirability of international cooperation quickly became apparent. In 1979 NSF and the Ministry of Construction and Science and the Technology Agency of Japan formalized their countries' occasional collaboration with the U.S.-Japan Cooperative Earthquake Research Project Utilizing Large-Scale Testing Facilities. The purpose of the joint program was to improve scientific knowledge and engineering practices directed toward earthquake-resistant building design. Their cooperative research involved structural tests on full-scale reinforced concrete buildings built to new safety specifications in Japan's Large-Size Structural Laboratory and tests of components and subassemblies and corresponding correlation analyses in both countries. In 1981 the two nations implemented a new three-year agreement for research on steel-frame structures. By then NSF also had joint programs with the People's Republic of China, Yugoslavia, Italy, India, and other countries. Most of the money was spent in the United States, but the research and its results were coordinated with those of the cooperating nations in the interests of improving knowledge and everyone's ability to cope with earthquake devastation.[13]

In the United States, earthquake research and engineering proved their worth following the 6.4-magnitude El Centro–Calexico earthquake in California in October 1979, which destroyed a six-story office building and two bridges and seriously damaged gasoline storage tanks, irrigation and water systems, and lines of communication. Within three days NSF's Donald Senich could report to his director that the Earthquake Engineering Research Institute (EERI) and numerous NSF-supported individual experts were on site studying structural damage and effects on the infrastructure. NSF support had made this earthquake-prone area "the most densely instrumented in the world," a synchronized strong-motion instrument array being part of a joint project between UC San Diego and the National University of Mexico. The resulting strong-motion records, "unique in quantity and quality," would significantly advance knowledge about earthquakes, especially when buttressed with long-term studies, Senich wrote.[14]

Earthquake engineers by 1981 were reporting advanced concepts that are still a focus of attention. With their guidance, architects had begun to design buildings that were "uncoupled" from the earth. This new base-isolation approach floated the building on a system of foundation bearings that intercepted damaging ground motion and prevented its transmittal into the structure. (In the orthodox method, a building strengthened to resist seismic forces often survived, but with its contents severely shaken and its occupants correspondingly endangered.) The most promising foundation bearings were elastomeric pads. Laminating alternating layers of rubber and steel made these bearings very stiff vertically; could support large loads while remaining flexible laterally to move with the motion. Thus, the entire stiff building would retain its structural integrity over this horizontal shock absorber.[15]

A further improvement, implemented in a New Zealand government office building, was to fit the center of each rubber-steel laminate bearing with a lead plug, which first resisted and then damped the force before being itself deformed. Such buildings, said James Kelly, an earthquake engineer at Berkeley, could slide laterally as much as thirty centimeters. Retrofitting existing buildings (or bridges) with base isolators, especially the historic, stiff masonry buildings common in earthquake-prone California, was possible technically and feasible economically, Kelly said, but the conservative, cautious United States did not construct its first building using shock-isolator devices until 1983.[16]

At that same time, however, basic NSF earthquake research was being conducted on only one limited-motion 20-by-20-foot shake table, a ten-year-old facility at UC Berkeley. And the smaller, older shake table at the University of Illinois was, by comparison, almost a toy, according to Gaus. Both needed major repairs. Other military and commercial tables were in operation at various locations, but Japan, for one, was way ahead in applying this technology.[17]

In the mid-1980s NSF endured an earthquake of its own, a political one. When NSF's Earthquake Hazard Mitigation activities finally settled in the Civil and Environmental Engineering Division of the new Engineering Directorate in 1981, they took up nearly two-thirds of the division's budget ($18 million of not quite $28 million). But for FY 1982, an austere year, earthquake research was slated for a much smaller increase than that of the directorate as a whole (1.8 percent versus 8.4 percent). Most of the earthquake funding went to siting and design work; less than $1.7 million supported societal response studies, an important area begun under RANN. Despite significant contributions to understanding and preparedness, especially since the 1977 congressional mandate, a growing awareness of the limitations of narrowly focused individual research led the technical community to consider a more coherent, interdisciplinary, systematic approach. It should include all mitigation efforts, whether directed toward life or property, and encourage concentration, not on the 120-story buildings that "everyone" wanted to study, but on unglamorous areas like the low, nonengineered existing buildings where the most serious, widespread damage occurred.[18]

So in November 1985, after some months of delays, NSF invited proposals from universities for an Earthquake Engineering Research Center, as it was then seeking other types of interdisciplinary, industry- and state government-cooperative engineering research centers.[19] When the originally planned "one or more" centers became just one, and at the State University of New York (SUNY) at Buffalo, California partisans threw down the gauntlet. "Everyone" knew that California, locus of American earthquake activity, was already the locus of "world-class" earthquake research. California senator Pete Wilson issued a blistering press release challenging the objectivity of the selection and demanding a full investigation by the General Accounting Office. California's existing personnel and equipment "clearly [had] the best chance to solve many of the riddles of earthquakes. We shouldn't bet taxpayers' money on anything less than our best chance," stormed Wilson.

NSF stood by its decision, "made on the merits" of the proposals, with "no bias and no political involvement." As various staff later explained it, NSF promised up to $5 million per year provided that amount was matched by state, industry, or other nonfederal funds. California's fund-raising was bogged down in state politics, while New York stood ready with existing monies. New York offered a better technical proposal, involving a broadly based consortium of some seventy investigators in nearly every state and exhibiting an understanding and enthusiasm for the systems approach. Relying on its reputation as the premiere research center, California "wanted to do the same old thing" and concentrate on initial damage, not such aftereffects as fire; not how to help society cope, but how to research a problem to get the results published. This was not the only time that the National Science Foundation selected an up-and-coming second-tier institution over an overconfident established one, with predictable political fallout.[20]

As the noise died down, the SUNY Buffalo facility, with its new and versatile shake table, quietly pursued research on engineering solutions, from state-of-the-art base-isolation devices to reinforced everyday masonry walls. Research in California and elsewhere proceeded as well. Comparing the effects of similarly powerful earthquakes in Armenia in December 1988 and California (the Loma Prieta) in October 1989 (6.9 and 7.1, respectively, on the Richter scale) showed the contributions of U.S. earthquake engineering research. In California, where design and building codes had utilized the accumulated knowledge, the losses, in both lives and property, were a fraction of those in the Middle Eastern republic.[21]

MATERIALS: NEW, MORE, AND BETTER

Americans consumed about ten tons of materials per capita per year, estimated the National Academy of Sciences in *Science and Technology: A Five-Year Outlook,* which was published in 1979. Tracing the rapid growth of research on alternative materials—begun by necessity during World War II, when many sources of essential natural materials were cut off—the academy noted how spectacular results had led to major new industries, such as plastics, synthetic fibers and rubbers, and nuclear power. NAS forecasted continuing significant change as new technologies affected what materials people used and how. NAS also observed, unhappily, that the U.S. constant-dollar investment in materials research had now declined after peaking in the 1960s. "Innovative risk-taking" had also declined, which "inhibit[ed] the movement of new methods and products from the laboratory to everyday use." Still, advances continued, "stimulated by the interaction of industrial technology with basic research in the universities, government, and industry."[22]

Materials research, which provided "the essential scientific underpinning for technological advances," had for years been a significant pursuit at the National Science Foundation, although it had lived there in many organizational homes. It began as a division in the Research Directorate in 1971, and after being a unit separate from, but parallel to, engineering for a number of years, materials work finally ended up in 1985 in the Engineering Directorate's Division of Engineering Science in Mechanics, Structures, and Materials Engineering. The Mathematical and Physical Sciences Directorate also maintained a division devoted to materials research. Their respective efforts both complemented and overlapped. In the late 1970s NSF provided about 55 percent of all federal support for materials research at academic institutions.[23]

Pursuing a multifaceted discipline with innumerable multi-disciplinary applications, materials researchers sought to develop new materials, such as polymers and composites; understand and alter the properties of existing materials; and find ways to bond unlike materials. To understand the relationships among the composition, structure, and properties of materials was the intellectual challenge; to develop materials technologies for social, economic, and national well-being was its practical follow-through for engineers.[24]

One late-1970s materials emphasis, among many that could illustrate engineering/materials connections, explored surface and inter-surface phenomena that were related to catalysis, corrosion, optical properties, and mechanical behavior. For example, at the Stanford Synchrotron Radiation Laboratory, Joachim Stöhr was using a powerful technique called Extended X-ray Absorption Fine Structure (EXAFS) to detect the rate at which electrons were given off from oxygen atoms adsorbed on an aluminum surface. By tuning mono-energetic X-rays through the characteristic X-ray energy for oxygen, he could tell how the oxygen atoms interacted with contiguous atoms and measure the distances between them with great accuracy. His results had important implications for catalysis and corrosion studies involving oxygen and other light elements and practical significance, as corrosion was costing the U.S. economy an estimated $70 billion per year.[25]

Another materials research area involved amorphous solids—that is, those lacking the periodic lattice structure of crystalline materials. Breakthroughs were occurring in glass ceramics (the oldest and best-known such material), amorphous polymers, metallic glasses, and amorphous semiconductors. The use of optical scattering to study glass, begun by the English physicist John William Strutt, Lord Rayleigh, in 1919, had evolved into the modern field of fiber optics, which was revolutionizing communications systems. In the early 1960s, Pol Duwez of Caltech had propelled molten globules of metal at high velocity against cold surfaces to produce glassy metal alloys, or metallic glass. In the late 1970s Allied Chemical Corporation, using a variation of Duwez's "splat-quenching" technique, made Metglas, a metallic glass ribbon useful in coils for transformer cores. Easily magnetized and demagnetized, Metglas transformers enjoyed appreciably lower energy losses. In the mid-1980s a Caltech materials research group was continuing to study exactly how rapid cooling (quenching) produced "new kinds of materials—glasses whose normally disorderly atoms now line up in crystals, metals whose normally aligned atoms become random—with interesting magnetic properties."[26]

The National Science Foundation also maintained inherited materials research. In 1960 the Defense Department's Advanced Research Projects Agency (ARPA, later called DARPA, then ARPA again) established a system of Interdisciplinary Laboratories (IDLs) at selected universities to upgrade materials research specifically and increase the number of graduates in the field. Stimulated by aerospace, electronics, and defense requirements, the IDLs had made significant advances in electronic and optical properties of materials but did little on other aspects on materials engineering. When ARPA discontinued its IDL support in July 1972 in the wake of the Mansfield Amendment, NSF assumed sponsorship of these facilities, calling them Materials Research Laboratories, or MRLs.[27]

The MRLs became the heart of NSF's materials research effort. One particularly successful early program was at the University of Pennsylvania. It supported nearly one hundred faculty members, postdoctoral researchers, and graduate students from the departments of chemistry, chemical and biochemical engineering, electrical and engineering science, metallurgy and materials science, and physics. Not only did Penn's program enjoy strong local management and university support, it had developed a first-rate, large-scale "thrust concept." That is, a core of researchers worked toward solving a commonly defined problem using a variety of theoretical and experimental techniques, around which related but more peripheral work was grouped. Penn made major contributions with its thrusts in molecular crystals preparation and synthesis, microscopic understanding of surfaces and interfaces, materials analysis (especially embrittlement of alloy steels), and chemical and extractive metallurgical processes to control corrosion.[28]

In the late 1970s funding for the modernization and replacement of instrumentation at the materials laboratories was especially critical. For FY 1979 NSF requested, for example, an additional $1.6 million, which included $200,000 for increased dedicated running time for more high-energy, high-beam-current experiments at the Stanford Synchrotron Radiation Laboratory and $100,000 to construct a second-generation hybrid magnet at the National Magnet Laboratory at MIT. Unfortunately, upgrading equipment could mean lowering support for research, as the National Academy of Sciences had noted in its 1979 report.[29]

Over the years the MRL program both contributed and required diverse scientific and technical competence, sophisticated equipment and facilities, and sustained cooperative effort. From their origin in 1960 to the mid-1980s, MRLs had produced about 3,000 Ph.D.s in materials science. The Foundation had supported a total of seventeen MRLs, including five new ones, between FYs 1972 and 1984, with the number varying at any one time as funding levels and proposal excellence varied. The average three-year continuing MRL grant was about $2 million, the highest $4 million. But with the MRL budget increasing only 8.6 percent over a twelve-year period, and the current $27 million only 4.5 percent higher than the previous year's budget, NSF's director had to recommend in 1985 that mature or marginal MRLs be phased out in order to redirect funds to strengthen the most innovative and healthy projects. Supporting individuals and "middle-sized" Materials Research Groups of five to ten investigators would help ensure that the strongest components of phased-out MRLs could continue, he said.[30]

Two diverse MRLs winning further Foundation support in 1985 were those at Carnegie-Mellon (CMU) and Case Western Reserve universities. Both had strong engineering components. At Carnegie-Mellon, research focused on scientific problems underlying the technology of joining materials. Investigators involved with the fusion-joining thrust studied changes occurring in alloys during the high-temperature welding process and the mechanical properties and behavior of the resulting joints. Metallurgical, materials science, and mechanical engineers

worked with mathematicians and others on varying approaches, ranging from computer modeling of solidification to experimentation on
crystalline thermodynamics. CMU's thrust in interfacial phenomena,
evolving from and overtaking the first, was also interdisciplinary. This
research emphasized the relationships between the microstructure and
the electrical, chemical, and optical properties of one material where it
interfaced with another. The Case Western interdisciplinary researchers concerned themselves with polymers—the relationship of their mechanical properties to the form and structure of polymer chains, failure
behavior of polymer composites and blends, and high-temperature corrosion. Calling on an external, largely nonacademic advisory panel, this
MRL emphasized its link with industrial and other technological users
of the research results. These two MRLs received $850,000 and $840,000,
respectively, for one additional year, some of which supported an up-
grading of facilities and equipment.[31]

A showcase NSF program with a significant materials component centered at Cornell University's facility for studying "the physics
of smallness." It began in November 1975, when the National Science
Board took under consideration the need to establish a "national research and resource center for microfabrication," to exploit "the most
recent—and future—technological advances in submicron technology."
The board anticipated a centralized facility housing equipment to design, fabricate, and analyze submicron components reaching ultimately
the 10- to 100-angstrom range (dimensions of the order of the wave-
length of light down to interatomic distances). Such equipment—cus-
tom-designed electron beam exposure systems (similar to scanning
electron microscopes), high-intensity soft X-ray replication apparatus,
and plasma and chemical etching instruments—was so complex and so
expensive that no single educational institution could afford it alone. Industry facilities, meanwhile, generally focused on specific short-range
problems, such as integrated circuits, not the fundamental issues in, say,
the integrated optics or superconducting magnetic-field sensors that the
new center planned to address. The microstructures engineering facil-

ity, monitored by the NSF Engineering Division and a panel of university experts, would gather researchers from various institutions with interdisciplinary expertise, including materials researchers, computer scientists, chemists, and mathematicians.[32]

Strong in relevant multidisciplinary research and existing equipment (if not conveniently accessible in its "centrally isolated" location), Cornell got NSB's approval in May 1977 and an initial grant of $2 million for FY 1977 to house and sponsor the National Research and Resource Facility for Sub-Micron Structures. Now called the National Nanofabrication Facility, it received $750,000 for each of the succeeding four years, for a total of $5 million. Researchers foresaw, by the close of the 1980s, computers with hundreds to tens-of-thousands of times their present capacity, instruments to measure dynamic processes in living cells, new structures for chemical catalysis, and tiny devices that could perform surgery inside human arteries. NSF awarded the Cornell center a five-year, $9.75 million grant renewal in 1981.[33]

In 1985 the National Science Foundation could justify the microstructures engineering investment. That year a team of Cornell and Bell Laboratories investigators developed the world's fastest semiconductor device, one that could switch a circuit on or off in 5.8 trillionths of a second. The semiconductor circuit, called a ring oscillator, was a series of tiny switches—each under a thousandth of a millimeter in size and each built up of thin (a few atoms thick), "exquisitely smooth" and uniform, alternating layers of compound semiconducting materials. Its submicron size and thin layers ensured fast, easily controlled passage of an electrical signal across the circuit.[34]

In January 1990 Ronald Balazik stated in *Minerals Today* that it was "important to recognize that society is entering an age in which materials and technology are synonymous." Indeed, there would have been no revolution in telecommunications or computerization without optical fibers and silicon chips. New materials improved economic competitiveness and military readiness by sharpening the performance of products, tools, and processing equipment made from them, as well as changing the needs for, and sources of, raw materials. But Balazik reiterated the growing concern that America had lost its lead in materials development. Although the federal government spent more than $1 billion on materials science annually, other countries were overtaking the United States in developing and using new materials. While the United States enjoyed a mature electronics ceramics industry in 1990, for example, the new field of strong and corrosion-resistant structural ceramics was dominated by Japan.[35]

Thus, materials engineering became an increasingly significant emphasis at the National Science Foundation, especially after it inherited the Department of Defense's dedicated laboratories. Despite administrative confusion, materials researchers asked complex and innovative questions, used sophisticated instruments and methods, and achieved results validated in numerous high-tech applications in the real world. The Foundation sought and boasted of ultimate usefulness, however basic and theoretical an individual investigator's research might be. Finally, centers like the materials laboratories and the Cornell submicron facility, beyond their own considerable achievements, would soon serve as models for future NSF research efforts, such as the Engineering Research Centers of the mid-1980s.

COMPUTERS AND ENGINEERING

Computer science, said NSF's annual report in 1977, "sprang from a remarkable fusion of abstract logic and electronics accomplished by mathematicians, logicians, philosophers, and engineers." Nourished by commercial demands and the continuing electronics-technology revolution, computer science offered researchers their first opportunity to deal with complex systems—whether physical, chemical, biological, or social.[36] That engineers swiftly embraced the computer and each new advance in the technology hardly needs saying. But engineers also took the lead in developing computer technology, from the earliest days to the proliferation of programming languages and computer graphics in the 1960s to the miniaturization of integrated circuits in the 1970s to the automation and robotics of the 1980s. Perhaps in no other field has change come so fast or so dramatically.

The National Science Foundation straightaway engaged the computer revolution. In 1965 the NSF Advisory Committee for Engineering had strongly recommended that the Engineering Division "have an important part, or even the major role, in the broad area of computers." When NSF first formed the Office of Computing Activities in 1967, however, it reported to the director. This office emphasized computer use in education and training and sought to make computing equipment available to universities. In 1974 the Foundation encouraged a broader focus by creating the Division of Computer Research, a function parallel with engineering, in the Research Directorate. Linked for several years with mathematics when the Directorate for Mathematical and Physical Sciences and Engineering was formed in 1975, computer research also appeared in the 1979 Directorate for Engineering and Applied Science and in the 1981 Engineering Directorate. These configurations obtained with variations until May 1986, when Computer and Information Science and Engineering consolidated all of NSF's computer efforts in an independent administrative unit.[37]

By the mid-1970s NSF investigators were concentrating on the computing process itself, not just getting answers from computers. During the RANN years computer models were used to predict urban traffic flow, dispatch police, and design structures. Although the systems were crude in comparison with the Computer-Aided Design (CAD) of today, the Foundation highlighted computer methods for building design as early as 1968. NSF-supported researchers A. G. H. Dietz and Alan Hershdorfer at MIT were then developing a computer language, which they called Build, that could help design four building systems— activities, spaces, surfaces, and structure—and analyze the relationships among them in use. Computer graphics of the day allowed the designer to view or modify information, "which is displayed on a cathode tube scope and which can be altered with a special light pen."[38]

Increasing miniaturization of electronic circuits had to be the most awe-inspiring and significant ongoing computer achievement, the one with the most widespread implications. Those in the know talked of large-scale integration giving way by the late 1970s to very-large-scale integration to interconnect ever more components on one microelectronic circuit, or chip—and with escalating performance and reliability and plummeting costs. Assistant director for MPE James Krumhansl noted in 1978 that the complexity of integrated circuits had approximately doubled every year since 1959, when planar structure was invented. During the 1970s researchers improved their fabrication techniques sufficiently to reduce the size of a transistor from 10 microns to 2.5 microns, thus increasing the possible number of transistors on a chip from 5,000 to 100,000. By 1978 single-chip microprocessors were technically and economically feasible.

In the 1980s work continued on ever-tinier electronic circuits, which were made possible by advances in semiconductor materials and submicron technologies; Cornell's nanofabrication center contributed significantly to the knowledge base. In 1982 NSF reported that employing new techniques, such as electron beam, X-ray, and ion beam lithography, could "multiply circuit-packing density so many times that today's small computers may fit on the head of a pin." That year's "super chips" contained as many as 450,000 transistors, with a million transistors per chip predicted by 1990 (a prophecy met and exceeded). Such new knowledge, superseding just-yesterday's new knowledge, found immediate application in integrated optic systems, communications, and robotics.[39]

NSF interest in computer-controlled flexible automatic production systems blossomed under RANN; by 1978 NSF was spending about $1 million on funding robotics research annually. Robots by 1982 could sense their environments through machine vision systems, ultrasonic range sensors, and tactile and force sensors. They could analyze data and make decisions to control and modify their own motions in increasingly imprecise surroundings. Such "intelligent" robotic systems proved their worth in manufacturing environments hostile to humans, as in the presence of great heat or chemical toxicity, or nuclear radiation, or where quality control was important, as in competitive automobile assembly-line manufacturing. The value to disabled persons of "smart" prostheses controlled by microelectronic devices that responded to tiny muscular signals was beyond calculation. These achievements were a long way from George DeFlorio's caution of 1963 in *Electrical Engineering* that "present results do not justify undue optimism for solution to most of the traditional and difficult problems involved in getting machines to behave like people."[40]

Optical communication was another computer-age achievement to which engineers contributed, as optical fibers, short-wavelength semiconductor lasers, cutting-edge electromagnetics, and ultrafast networks answered the demand for ever faster and more effective transmission of information. Again changes were revolutionary, beginning with the enthusiastic, if tentative, discussions of the potential applications of lasers' coherent light in the 1960s. In the mid-1980s Kristina Johnson, a University of Colorado engineer, made breakthroughs on optical computers, using light, not electric current, for fast and efficient computing on machines that could work on several problems at once. While existing gallium arsenide switches were incredibly fast (on to off in as little as .000000001 seconds), they required so much power for multiple-problem computing that the system could melt from the heat. So Johnson used a ferro-electric liquid crystal light shutter, a material synthesized by condensed-matter scientist Noel Clark and chemist David Walba, which was not quite as fast as gallium arsenide (on to off in .000001 seconds) but, since it required less power, ran 10,000 times cooler. It also boasted 100 times better contrast.[41]

Throughout this period, computer researchers made some of their most numerous, varied, and significant contributions to earthquake engineering. In a 1981 Foundation publication of recent research report abstracts, section after section described new engineering computer programs, with a sizable percentage of them devoted to static and/or dynamic analysis of structures under stress—dams, buildings, structural members, storage tanks. Others considered soil response to ground motion, such as the liquefaction phenomenon, or shearing and other rock behavior.[42]

The great majority of these computer programs relied on a method of calculation called finite element analysis, or FEA. NSF support, though absent during the pioneering period, was instrumental in the subsequent development of this method, which began with a concept of "regional discretization," first proposed in the 1940s but dropped for want of computers able to conduct the extensive numerical operations. In 1952 and 1953 M. J. Turner, head of the structural dynamics unit of Boeing Corporation, and Ray W. Clough, a Berkeley structural engineer participating in a Boeing summer faculty program, worked together to advance prevailing theories of structural analysis for complex airplane wing configurations. By 1957 Berkeley had a computer capable of carrying out the sequence of matrix operations necessary to solve practical plane-stress problems using Turner's triangular plate elements. Clough did these early calculations, named the method finite element analysis in a 1960 paper, and directed his graduate student E. L. Wilson to write the first computer program for finite element plane stress analysis in the early 1960s.

Basically, the FEA method defined a complex shape (say, an airplane wing or a large shell structure such as the "boatsails" roof of the opera house in Sydney, Australia) as a series of small, flat geometric elements hooked together, much like a microminiature version of a geodesic dome. Algebraic equations describing the connections among contiguous elements at their node points (vertices) could be solved for a "reasonable approximation" of the surface. Attempting this process, though conceptually much simpler than solving the complex differential equations of a more mathematical approach, was out of the question until the advent of the computer. Today, while mainframes are required for complex analyses, some commercial FEA is even done on personal computers.[43]

Thanks to the Foundation's encouragement, evident in early grants and numerous published results acknowledging NSF support, the finite element technique became widely used in textbooks as well as engineering practice by the 1970s for all manner of stress analysis applications. Although mathematicians disparaged the approach for its lack of rigor, NSF engineers pointed with pride to the contributions of numerical analysis techniques. By the late 1960s, for example, Clough was using computers and the finite element method to analyze a "wide variety of structures ranging from missile cases and airplane wings to apartment buildings." Other work at Berkeley, under Seed, Clough, and others, centered on soil and foundation responses during earthquakes. In 1971 Cornell structural engineer Arthur Nilson studied the bond stress-slip relationship between reinforced concrete and steel to improve the numerical technique for predicting and measuring such stresses and cracks. At Stanford's Materials Research Laboratory, researchers applied the finite element method to the study of the deformation and fracture of metals.[44]

Besides applications in structural mechanics, the finite element method was also useful in analyzing fluid flows, such as the oscillations of wave action in harbors. Hydrologists also used FEA to understand floods and water control, distribution of contaminants, and drought conditions. In the mid-1970s NSF supported finite element research at Stanford on movements expected and observed in clay temporarily displaced in urban excavations and at Colorado State University on the mechanics of avalanches.[45]

The results of projects funded both to refine the finite element method and to apply it to specific problems, as shown in an impressive array of scholarly publications and significant changes in accepted ways of doing things, illustrate how the National Science Foundation could push the direction and extent of American progress in targeted areas. This was also true for support of materials laboratories and earthquake research, as the Foundation honed its ability to focus and "leverage" its financially limited influence, a specific goal of NSF director Erich Bloch and other contemporary leaders.[46]

These three case studies show a remarkable and unanticipated degree of interrelatedness, underscoring the importance of interdisciplinary effort. They highlight the user link and the continued interest in practical applications and problem-solving, even as the mechanism designed to foster that approach, RANN, disintegrated and disappeared.

They also foreshadow the growing connections with industry and government at all levels and key engineering initiatives that would characterize the 1980s.

RESOURCES AND POLICIES FOR FULL PARTNERSHIP

Engineering is the liberal education of the technological age.
—John A. White, "Current and Emerging
Issues Facing American Engineering Education"

With the achievement of organizational equity within the National Science Foundation in 1981, the engineering community sought to make the most of the opportunity for increased efficacy and influence. NSF leaders for engineering first examined available and potential human resources in the field to ensure that the expanded mission could be met. This effort meant a close look at engineering education to learn how adequately engineers were being prepared to contribute to problem solving in a rapidly evolving world. Who would the engineers of tomorrow be? Who could they be? The task was to stimulate broader, more highly qualified participation in the engineering profession, which meant refining engineering's mission to serve current and anticipated challenges.

THE SETTING: VOLATILE TIMES

Engineering had come of age at the National Science Foundation at the best of times and the worst. With a directorate of its own and allies such as the able and sympathetic George Brown in Congress, the NSF engineering community anticipated a broader technical and political role during the 1980s. Engineering achievements were pouring forth at an accelerating rate, and the need for such accomplishments to meet the mounting challenges in the global and domestic arenas was never greater. At the same time, however, the Reagan administration's priority of lessening the federal presence in American life held ominous implications for NSF.

Director John B. Slaughter, reviewing NSF's historical environment as one of alternating periods of stability and change, declared in February 1982 that "rarely" had the Foundation "faced circumstances as dynamic and volatile as we face now." NSF's budget, severely cut in the new president's drive to reduce the size and scope of government, was left inadequate to meet "pressing needs," though with help from the Office of Management and Budget and the Office of Science and Technology Policy, it had "fared better than most" federal technical agencies. Reagan gave his philosophical approval to basic research for strengthening national defense and, to some extent, for meeting international economic competition, but the Department of Defense took most of the money directly. At the same time, inflation ate up apparent increases in budgets, the economy stagnated, and deficits mounted.

In view of the administration's emphasis on nonfederal support of research, Slaughter saw that NSF would have to broaden its contacts and influence with other sectors, especially private industry and local governments, in order to maintain research quality and to "catalyze" maximum innovation and productivity. Nor could beneficial international cooperative opportunities be overlooked. Education would be a key issue, as would development of the "tremendous pool" of underutilized female and minority talent—a waste that the nation could no longer afford. Slaughter stressed the need for concentrating on the long view to anticipate and cope with the looming fiscal and political realities. Thus he foresaw many of the hallmarks of NSF's engineering history in its most active, visible decade. Others—especially technical, political, and organizational responses to expanding threats to America's global economic posture—soon emerged.[1]

If the difficulties of a changed ideological climate were not enough, the National Science Foundation itself endured four changes of leadership in as many years. Atkinson left NSF in 1980, followed in close order by Slaughter, Edward Knapp, and Erich Bloch—only the last serving a full six-year term (1984–90). Lewis Branscomb, who chaired the National Science Board during those years, confirms that such rapid turnover could not help but affect the institution. With each new start, continuity and efficiency were lost as successive directors sought their footing with staff and board as well as with Congress and the executive branch. While Branscomb saw no significant differences among the four directors' views on engineering, they naturally varied in their styles and effectiveness in working with their constituencies. The Reagan administration, in turn, initially distrusted the National Science Board, no member of which it had appointed.[2]

EDUCATING ENGINEERS: A HISTORICAL AND GROWING CONCERN

While education, then in financial peril, topped Slaughter's 1982 list of strategic issues, the education of scientists and engineers had been a programmatic concern for NSF since 1950. To ensure the future of scientific and technical endeavor, the Foundation established a Division of Scientific Personnel and Education as part of its first order of business. In the wake of the Korean War and the deepening cold war, Eisenhower created the high-level Special Interdepartmental Committee on the Training of Scientists and Engineers "to improve the quality and increase the supply." Although Foundation leaders resented the committee's purported usurpation of their own education function, they agreed with Eisenhower's view that education was "essential" to the American way of life, future security, and economic growth—then inevitably seen in terms of the Soviet Union's increasingly menacing technological prowess. Education first became an NSF directorate in 1973 and was renamed the Directorate for Science *and Engineering* Education (SEE, emphasis added) in 1981, at a time, not likely coincidentally, when the Foundation was eager to convince legislators of its encompassing commitment, lest Congress create a rival technology foundation.[3]

Yet education seemed perennially in crisis—the first to suffer in periods of shortfall. When the Reagan administration cut NSF education funding drastically for FY 1982, arguing that education was better served at the state and local levels, Slaughter gave up and on 30 April 1982 abolished the directorate, gathering its remaining hungry programs into the lower-level Office of Scientific and Engineering Personnel and Education. Branscomb later speculated that the administration's ax-wielding might have been a blessing in disguise, since it forced the Foundation to abandon a collection of increasingly fragmented, politically driven, poorly functioning programs and replace them with a more carefully defined effort. The House Committee on Science and Technology thought so at the time. In any case, the devastating hit stimulated widespread review and reevaluation of NSF's education agenda and its results.[4]

The Debate over Curriculum. Two issues have historically engaged proponents of engineering education, long predating, but directly bearing on, the National Science Foundation. The earlier, and more persistent and fundamental, was the question of what constituted a proper engineering education. The second was a recurring worry over the adequacy of the supply of engineers. As early as 1874 the American Society of Civil Engineers heard T. C. Clarke advocate an educational program centered on general principles, leaving technical knowledge to be acquired in practice. But two years later Alexander Lyman Holley, founder and president of the American Institute of Mining Engineers, argued that engineering schools put too much emphasis on abstract principles. He thought actual practice should come first, with theoretical study later. At a meeting that Holley helped convene at the technology-rich Centennial Exhibition in Philadelphia in 1876, twenty-five prominent practicing and teaching engineers argued widely on the subject but agreed in the end that "real engineers" needed a broad education. It should include "essential underlying principles," a "considerable range of so-called cultural studies," and early practical training.[5]

This early attempt at synthesis was significant at a time of two distinctly different approaches to engineering training, at least in mechanical engineering. The traditional, individualized "shop" method emphasized practical training directed toward running or ultimately owning a machine shop. The key to this "culture," as analyzed by Monte Calvert, was applied science. The newer, more impersonal "school culture" was grounded in theoretical concepts in mathematics and science in a formal environment. This approach defined professionalism in terms of standardized academic achievement and trained engineers for employment in the large, bureaucratic corporations that were emerging to serve the rapidly industrializing nation. The latter culture prevailed by the early twentieth century, just as the engineering profession generally was becoming more science-based and college-oriented.[6]

Engineering, in fact, pioneered as a profession in concerning itself with education, beginning with the formation of the Society for the Promotion of Engineering Education (SPEE) in 1893 and continuing through the SPEE's numerous education studies during the first half of the twentieth century, described in the prologue and chapter 1. As noted, Mann's 1918 report emphasized fundamentals combined with practice, while in 1930 Wickenden called for a "compromise" position—a four-year undergraduate course (lengthening it was deemed "unrealistic") that combined "science, engineering methods, and social relations," with professional development continuing through either graduate study or practical experience.[7] In 1955 Linton E. Grinter, graduate school dean of the University of Florida, urged a strong curricular base in fundamental engineering sciences and mathematics, while eliminating "engineering art and practice" courses. Leading educators had promoted this approach for some fifty years, but now, with massive postwar federal funding for basic research and the new report legitimizing this path, even "traditional, mainstream engineering schools" began to change. Engineering funding during NSF's early history clearly reflected the growing ascendancy of basic research in engineering science and advanced it with the financial and social status of its patronage.[8]

Yet as the philosophical seesaw between theory and practice continued and intensified, the engineering curriculum was beginning to strain at the seams. How could it absorb ever more mathematics (both a functional tool for analysis and design and a conceptual tool for developing insights into the systematic orderliness of physical processes) and new fields of investigation, such as nuclear energy and computer science? Either some existing curricular material would have to be deleted or the training time extended.

By 1966, when Eric Walker looked back, he labeled the Wickenden report" the biggest disaster that ever happened to engineering education." Wickenden's call for a liberal science education for engineers along with specialization in a "deep, narrow, professional" discipline was a contradiction that had produced a "kind of schizophrenia" among engineers that still plagued them. Although Wickenden "knew" that "you just can't learn enough in the four-year engineering program" to merit "true professional status," he apparently "didn't have guts enough" to recommend outright the "post-scholastic" training he privately favored. In 1990, Walker still regretted that such an expanded program, akin to a physician's internship in combining advanced instruction with introductory experience in business and technical specialties, had never been implemented. Augmenting the inadequate baccalaureate, a professional degree unique to engineering would be more appropriate than an academic doctorate for many practicing engineers, he asserted.[9]

Walker was then chairing a major Goals of Engineering Education investigation for the American Society for Engineering Education, with Joseph Pettit of Stanford and Purdue's George Hawkins directing its graduate and undergraduate phases, respectively. This nationwide study, partly funded by NSF and in progress since late 1961, drew upon exhaustive questionnaires, site visits, interviews, informational documents, masses of data, five regional meetings involving more than nine hundred participants, and lively discussion in several formats (such as the 1966 NAE symposium where Walker criticized Wickenden) among engineering educators, practitioners, and employers. In January 1968 the ASEE published its final report.

The landmark goals study pointed out the difficulty of maintaining, within the confines of a four-year undergraduate program, a broadened liberal education *and* the specialized training needed to keep up with the accelerating pace and complexity of technological development. The widespread focus on fundamentals and engineering science, especially since World War II, had run up against increasing calls for a broader, interdisciplinary curriculum, including nontechnical subjects like economics, management, and the humanities. All this was needed, of course, if future engineers were to help solve society's problems, as the investigators anticipated—before RANN—they would be expected to do. Noting both the necessity for graduate study in engineering to ensure depth of knowledge as complex technologies multiplied and its "remarkable growth" (2.8 times as many master's degrees since 1950 and more than 4 times as many doctorates), the committee recommended that a specialized master's degree be considered the basic professional degree for engineers. It insisted on the "fullest possible integration" of research with the "educational purpose" of engineering colleges and substantially increased federal, and industrial, funding.[10]

The forthright Walker had encouraged provocative ideas, he later said, to attract attention and stimulate discussion, and ultimately to bring about change, but even he was taken aback by the fury that a preliminary version of the Goals Study, appearing in 1965, unleashed in the field. The degree proposal took the worst hit, even though Walker could point to several schools, such as Cornell, Rensselaer, and California, that were already adopting or developing a four-year bachelor of science and five-year master of engineering degree program, with the latter as the first "professional" degree. Moreover, he wrote, the increasing numbers of engineers with advanced degrees would eventually by their presence "determine the requirements of professionalism." Ralph Morgen, a former NSF engineering program head and now at the

Stevens Institute, approved the preliminary goals report ("we cannot give enough education any more in four years to justify a professional degree"), but most of his chemical engineering colleagues vehemently disagreed, citing arbitrariness, a potential diminution of the value of both degrees, especially to current holders, and a host of other likely adverse outcomes. The civil and mechanical engineering professional societies also protested to varying degrees.[11]

Terry Reynolds and Bruce Seely persuasively connect the goals study firestorm to the "changing nature of academic engineering, which had begun to lose touch with industry after World War II as it gravitated toward academic science." By the 1960s, academic engineers, with their lucrative federal research contracts, had "developed a culture and values more akin to the pursuit of knowledge for its own sake than to practical engineering."[12] Nonacademic professional engineers' explosive response to the goals study signaled another impending change in emphasis: The goals study controversies took place concurrently with the emergence of engineering (not engineering sciences) as a division in NSF and the developing Daddario amendments permitting applied research, industry's natural enclave.

As for industry's active concern, Thorndike Saville had condemned American industrialists in 1952 for their "indiscriminating if not ignorant" criticism of engineering education, even as they "conjured up every conceivable argument and legal device to excuse them from any considerable support of the educational institutions upon whose graduates their future is based." This charge may have been legitimate when Saville made it, at a time when the engineering curriculum was leaning heavily toward academic science, but industry was clearly worried about the training of engineers by the early 1960s. In 1963, Standard Oil chemical engineers Charles Rowe and William Spaulding, for example, foreshadowed coming shifts in attitude in arguing for early curricular specialization in order to prepare students for "practical application problems," the daily fare of engineers in industry. Such engineers required a "balanced compromise between that which scientific data tell us is true or feasible, and that which is practical in terms of operability, reliability, safety, and cost." The swing toward science in engineering education should be reversed, they wrote, to keep "*engineering* in the forefront."[13]

By the 1970s the debate was becoming an "acrimonious public quarrel" within the engineering community. R. C. Quittenton, academic vice president of the SUNY College of Technology in Utica, in 1978 bluntly accused engineering of "drifting from true engineering into pseudoscience, partly under 'guidance' from the NSF. . . . So-called engineers who do not like to work with their hands" were but pseudoscientists, he charged, as were faculty who had "never built a bridge, paved a road, [or] designed a sewage plant." Because pure science had more "status" than engineering in universities and granting agencies, struggling engineering colleges were becoming so "pure" they nearly "ceased to be engineering schools at all."[14]

In the 1980s, some engineers complained that students were being force-fed too much knowledge in the available time, without learning how to learn or how to organize and apply knowledge to problem solving. They got too much exposure to science and too little, too late, to "real industrial problems and techniques," although "cooperative" internship programs were becoming more popular. Hardly a new idea, cooperative courses went back at least to 1907 when "academic entrepreneur" Dugald Jackson established one for electrical engineers at MIT to link with General Electric in an effort to match academic supply with industrial demand. While a five-year degree requirement would help, no one seemed to know how to staff or finance such an expansion. As Bloch saw it, academic departments rarely taught systems engineering and were slow in moving into new fields, like biotechnology or innovative materials engineering. Nam Suh, engineering head in the late 1980s, agreed. To him, engineers needed training to identify a problem, conceptualize it, analyze it, and implement a solution. Of these tasks, American universities adequately taught only analysis.[15]

What all these engineering leaders seemed to be saying in these challenging times was that curricula concentrating on academic science no longer served the needs of most engineers, who for their part needed to look both inward and ahead to develop training useful to their unique profession and to the country. Without denying the necessity for fundamental principles, especially in new computer-driven fields, engineers should focus on the straightest road to their ultimate goal: to make and to do for human betterment. To the leaders of the 1980s, design and production had to be key educational emphases.

The Quantity Issue: "Eating Our Seed Corn." The qualitative concerns about engineering education continued, but by the early 1980s a lesser, but periodically intense, related theme had reemerged—the problem of numbers. On 8 February 1980 Carter wrote to the secretary of education and the director of NSF, "I am increasingly concerned whether our science and engineering education is adequate, both in quality and in number of graduates, for our long-term needs." The president, a former engineer, asked for a review of secondary and college-level education policies "to ensure that we are taking measures which will preserve our national strength."[16]

Responses to the president's memorandum were numerous, varying in scope of investigation and degree of disquiet but agreeing in principle on major elements of the new crisis in engineering education. The significance of the problem rested on two ubiquitous concerns. The first was national security. Should there be a national emergency or mobilization, the demand for qualified technologists would dangerously strain the supply—or worse, if the exigency were a surprise. Second, with some economists assigning as much as 35 percent of the growth in American productivity to engineering innovation, engineering education simply had to be a national priority—in the name of international economic competitiveness (the emerging "national need") if not the domestic standard of living.[17]

One of the most detailed assessments of the supply of engineers came in 1980 from the American Society for Engineering Education (ASEE) and the American Association of Engineering Societies (AAES), which jointly pronounced the situation unique and alarming. Numbers of undergraduates had reached a historic high and were still climbing, but with graduate enrollments and filled faculty positions declining (in contrast to Foundation observations in 1967 that graduate enrollments in engineering were growing more pronouncedly than in other fields), there were simply "far too many" students for the available teaching capacity. Huge class sizes precluded interactive instruction, a situation detrimental to both learning and teaching.

Meanwhile, higher-paying, more technically stimulating opportunities in industry meant that fewer engineering graduates were going on for advanced degrees.[18] (The National Academy of Engineering estimated in 1986 that it took a Ph.D. engineer about twenty-one years to surpass the total accumulated earnings of a B.S. engineer.[19]) According to 1981 NSB statistics, new engineering doctorates had peaked at 3,774 in 1972; in 1980 there were 2,751. Making up the shortfall of American teaching assistants were large numbers of students from abroad, eager for a place in this troubled educational system. If they were deficient in English language skills or taught from cultures dismissive of female intellect, the classroom environment further deteriorated. From 1972 to 1980 the number of foreign nationals earning engineering Ph.D.s increased from 773 to 982; simple arithmetic showed a decrease of 1,232 engineering doctorates awarded to U.S. citizens, a 41 percent drop. With many of the noncitizens eventually returning to their countries of origin, the writers of the ASEE/AAES report were driven to ask, "Where are the faculty coming from who will teach the classes of the 1990s?" We were, said one, "literally 'eating our seed corn.'" And if it was true that "the general health of engineering education [was] fundamental to the general health of the national economy," where was the United States headed?

Existing native-born Ph.D.s and university faculty, too, were responding to the lure of industry's lucre, while demographics suggested "an impending bulge in faculty retirements in the 1990s." Exxon Research and Engineering Company president Edward David noted that "even at MIT" some faculty positions had been vacant for more than four years. So engineers, especially the younger ones, who remained on campus struggled with uncompetitive salaries and low social status vis-à-vis their compatriots in industry, along with overwhelming class sizes and heavy course loads that gnawed away at their own research time. Meanwhile, new knowledge, technologies, and methodologies were continually emerging, but university facilities and equipment for exploiting them for either teaching or research were notoriously and perennially inadequate or obsolete—"too old, too small, and in some cases too dangerous."[20]

Marian Visich and John Truxal, professors at the State University of New York at Stony Brook, detected another aspect of the education crisis, one that NSF affected. They feared that the rapid growth in the 1980s of private industry support, which concentrated on elite schools, could engender two sharply divided groups of institutions— about 25 excellent schools and perhaps 225 others "seriously out of touch with modern engineering." Worse, "we don't see the National Science Foundation (or the federal government) taking the actions necessary to reverse these trends." If MIT's graduate students came primarily from undergraduate institutions that were threatened by the current cuts in funding, could the United States afford to abandon them? After all, these second-echelon colleges produced most of the nation's teachers, contributed most of the gap-filling research and state and local governments' technological capability, and led efforts to open engineering to minorities and to women.[21]

With these and other insights, the ASEE/AAES investigators recommended immediate and specific actions, starting with engaging "someone (the President?)" to convey to university presidents the critical importance of engineering schools to the nation's economic health. After that, faculty recruitment and development—including smaller classes and improved salaries, facilities, and equipment—had to be the "first priority." They urged a "couple of thousand" post-baccalaureate fellowships to encourage the best engineering students to pursue specialization in a technical field, management, or public policy. This strategy would "refill the faculty pipeline" and prepare engineers with needed industrial specialties.

The engineering educators went on to insist that every engineering school be guaranteed an annual 5 percent replacement of equipment and instrumentation to keep up with needs in such new fields as robotics, microelectronics, and acoustics, even at an annual cost of nearly $40 million indefinitely. The reality in 1980 was that most existing university equipment, whose useful life span was ten years and decreasing, dated from the 1950s. The ASEE/AAES urged increased cooperation among the universities, government, and industry and stressed that the National Science Foundation "must be mandated to support fundamental engineering research as a responsibility equal to its responsibility to support basic science research." While the recent creation of the Engineering and Applied Science Directorate was "a step in the right direction," the study group thought a reorganized National Science Board or a National Engineering Foundation should be considered. Engineering's advocates, appreciating the power of politics, were increasingly calling on the system to effect a political solution—a legislative mandate, if NSF could not, or would not, produce its own.[22]

But while these outside engineers were fairly moderate in tone, NSF's engineers were alarmist. In November 1980, the EAS advisory committee requested a meeting "as soon as practical" with the National Science Board to convey their sense of urgency about the "worsen[ing] crisis" in engineering education, which, if it continued, would "seriously undermine" the nation's ability to generate and use new engineering knowledge. The committee charged that reorganization without an "appreciable increase" in funding for engineering would be cosmetic and "not an appropriate response to the crisis." In fact, the advisers demanded an immediate 100 percent increase in the engineering budget, with more in the long run, for both research and education to redress NSF's lingering and serious imbalance in its support of engineering and science.[23]

In July 1981 the NSB executive committee, after studying NSF education efforts, admitted that it was "gravely concerned" that the nation was "far less well prepared to meet the needs and challenges of the future than is necessary for it to maintain world scientific and technical leadership." Everyone, from James Conant, the first NSB chair on, had recognized that the limiting factor in all science was the human one; thus, it was incumbent on the present board to continue to support the long-term development of a strong human resource base. But money—big money—would be needed, and NSF was then facing the gutting of its education program by an administration favoring other priorities.[24]

CONGRESS STEPS IN: THE MANPOWER BILL

Congress moved to force the administration to act. In April 1982 the House Committee on Science and Technology scheduled hearings on H.R. 5254, the National Engineering and Science Manpower Bill, the purpose of which was to ensure an adequate supply of engineers and scientists to meet the nation's anticipated needs. It would "provide federal leadership and coordination with state and local governments and private efforts on education." Committee chair Don Fuqua and subcommittee head Doug Walgren, Democrats of Florida and Pennsylvania, respectively, had introduced the bill the previous December, on the heels of—and in clear response to—the administration's "unwise and unwarranted" emasculation of NSF's education budget. Congress's final, augmented appropriation of $20.9 million for FY 1982 represented a reduction of almost $50 million from FY 1981. The displeased Walgren noted that the Foundation's education budget, which he said had taken up 60 percent of NSF's total budget in 1960, was 3 percent in 1980.[25]

S. L. Glashow, of Harvard's Lyman Laboratory of Physics, put the need for human resources as vividly as anyone. The issue was global competitiveness. From being only recently the "unquestioned technological hub of the world," the United States had passed "the torch of scientific endeavor" to other peoples, he wrote. "Steel, ships, sewing-machines, stereos and shoes" were "lost industries." Japanese cars were cheaper, perhaps better made. "Proud RCA" was distributing Japanese goods assembled in Korea. The French government controlled American Motors. "We even buy Polish robots." Advanced electronics and computers were still doing well but were facing challenges from Japan. Exhausting their "heritage of capital and raw materials," Americans would find their familiar technological society out of reach and would soon be left with "their Big Macs . . . and perhaps, [their] Federally subsidized weapons industries."[26]

Debate on the manpower bill illustrated the scope of the education problem. Ohio Democrat John Glenn, who had introduced a companion bill in the Senate, chastised the administration for appearing to "abandon altogether any serious effort" to upgrade American technical education just when action was crucial to "halt the erosion of our store of human and educational resources." Democratic senator Charles Robb testified that in 1979 his state of Virginia had granted only seventeen degrees in high-school mathematics education and nine in science education. While advocating state control of basic education, Robb urged substantial federal support for research and development in universities and closer, "more imaginative partnerships" between universities and industry, including the sharing of costly research equipment and staff.[27]

Reagan's science advisor, George Keyworth, justified the administration's policy of discontinuing federal support for many technical education programs, such as NSF's, to the April 1982 National Engineering Action Conference by charging that such programs were "rooted in the 1960s." For two decades of rapid economic growth, the government had broadened benefits and participation, and "we consumed our knowledge and resource base faster than we replenished it to keep the economic engine speeding along." The 1980s called for different strategies, he insisted: "Now we must focus on production—of new knowledge, of new use of that knowledge, and even of new institutions—to get our economy in balance." On how that would be accomplished without an educational component, Keyworth was silent, but the political reaction to the big-spending Johnson years spoke loudly.[28]

When he testified before Walgren's Subcommittee on Science, Research and Technology on 29 April, Keyworth "took exception" to the bill's "basic premise that we can centralize national manpower planning for the nation," preferring the Reagan approach of leaving unperturbed a "responsive demand/supply employment market." (Fuqua thought this interpretation incredible; the bill simply provided for an exchange of ideas and coordination of programs. Agreeing, Representative Margaret Heckler, Massachusetts Republican, pointed out that American industrial leaders were among the bill's strongest supporters.) Keyworth approved of the legislators' interest in "assuring and improving the quality—not necessarily the quantity—of our research university faculties," and in bolstering all students' precollege science and mathematics preparation.[29]

While troubled by specific provisions of the manpower bill, the National Science Foundation sympathized with its spirit. Branscomb was arguing by then, as were many engineers, to differentiate support for engineering education from that of, say, physics or chemistry, since these fields were qualitatively different. For the professional practice of engineering, he said, funds explicitly designated for education and technical capability, not research per se, were most needed, at both the undergraduate and graduate level. Branscomb was gratified that after years of neglecting the "health of the scientific and engineering community's infrastructure," the Office of Science and Technology Policy had recently suggested to Congress that federal action in technical education might be necessary as a "catalyst where market responses [were] slow or where the problems [were] particularly difficult for the private sector to solve."[30]

However, Fuqua's bill became lost in a maze of political maneuvering, its most Foundation-friendly provisions not disentangled until the 1985 NSF authorization bill. But the House science committee in 1982 did write into NSF's authorization bill statutory permanence for the Directorate for Science and Engineering Education (SEE), which Edward Knapp re-created on 1 October 1983, a year and a half after SEE's administrative demotion. NSF could not functionally or organizationally change SEE without explanation to Congress.[31] Once again, politics took the National Science Foundation by the hand, this time as the Democrat-controlled Congress tried to blunt the initiatives of the Republican administration to cut spending and federal involvement.

THE ENGINEERING MISSION: NEW EMPHASIS, NEW DIRECTION

Now in the public spotlight, the National Science Foundation saw the need to plan for the long term and to articulate how it would take the initiative in meeting the nation's needs. Foundation leaders had little difficulty in agreeing that the role of engineering would expand significantly. The demands of national security and fears of the economic prowess of competitors such as Japan and Germany dictated as much. So did the call for all manner of high-tech commercial and consumer products. Indeed, "not since *Sputnik*" had there been such worries about the quality of the U.S. science and technology base and the country's technological leadership.[32]

Contemporary data showed that until FY 1983, when it was barely overtaken by the Department of Defense, the National Science Foundation was the leading agency supporting basic academic engineering research, at $108.4 million (of an obligated federal total of $278.3 million). Viewed from another angle, more than nine-tenths of NSF's support of academic engineering went to basic research, which was the largest proportion of any federal agency. At the same time, however, NSF's underfunding of engineering research was a serious problem. Too few potential researchers were served, and these inadequately. Engineering proposals still had one of the lowest success ratios in the Foundation, in some areas as poor as one award for every fifteen applicants. And in real purchasing power, 1982 grantees were living with dollars adequate for 1974. Industrial support for academic engineering research was thin, although a 1980 funding figure of $87 million was four times greater than that of 1972. Meanwhile, international comparisons showed that the percentage of engineers in the American labor force had declined since 1965, while doubling in Japan and Germany. In 1980 Japan graduated 87,000 engineers, which was 9,000 more than the United States produced from more than twice the population.[33]

Driven by these concerns as well as by Brown's perennial reintroduction of his National Technology Foundation bill, NSF leaders determined that they must define and outline a mission for engineering to serve current and future needs. Accordingly, the National Science Board made NSF's engineering mission the main agenda for its annual offsite, issue-focused meeting in June 1983. The NSB wanted to examine what additional federal support was necessary to strengthen U.S. engineering capability and, of that, what portion was NSF's responsibility.

Further, was NSF's traditional process of funding unsolicited research proposals the best approach for engineering, or were other, more directive methods needed? To inform the board's thinking, its Task Group on Engineering Issues, chaired by William F. Miller, president of SRI International, asked prominent engineering leaders to identify and rank current or emerging issues critical for the coming decade. These outside experts, representing professional associations, industry, and government, expressed the gamut of now-familiar concerns, but over and over they pressed for investment in academic activities that would improve American productivity and competitiveness. The President's National Productivity Advisory Committee, as a typical example, wrote to urge a focus in universities, working with and through industry, on manufacturing systems technology and education as the federal government's most useful long-term contribution to America's economic health.[34]

To address the "challenges to the Nation's engineering strength and, derivatively, to its economy and security," board members agreed early that NSF should "expand and alter its traditional role in support of engineering sciences at academic institutions." Determining what that meant exactly was more problematic for a board that represented differing interpretations of common terminology and varying philosophies on how far change should go. Branscomb, the chair, was concerned that engineering at NSF had been expanding *without* altering or modernizing its traditional approach, with the consequence that engineering was looking "more like applied physics with every passing year." Rather, these disciplines were not the same in practice and should not be lumped together in academic research either.

Knapp allowed that NSF's role in engineering was "very important" but cautioned that it must be understood and judged in terms of both the engineering community's needs and the budgetary and organizational impact on the Foundation. Admitting that engineering had not historically enjoyed equitable Foundation support, Knapp wondered in a background paper whether a funding strategy of slow buildup or a big initiative (for, say, engineering "centers of excellence") would be more "effective"—meaning, in part, more palatable to NSF's other constituents.[35] Now in June Knapp asked for a board statement declaring its consensus that NSF "should look after the same health of engineering . . . that it has with respect to the scientific enterprise, if the Board believes that." Clearly, not all members were sure. Probably neither was Knapp, who had another concern. As he later voiced it, the "fundamental question underneath the surface" was whether engineering should be severed from the Foundation. If engineering could grow only at the expense of the sciences, it would be "very, very difficult" for those assembled to agree to a large engineering expansion. At the same time, no one was eager to relinquish NSF's engineering empire, however modest it was. Knapp preferred the board to state that the Foundation's role in engineering was analogous to, but different from, its role in science.

While some board members, by their own words, were "pussy-footing around," unable or unwilling to grant parity to engineering in the NSF context, Miller plainly predicted, as did others who were politically aware, that if NSF did not support engineering fully, some other agency would. So if the board agreed—as he thought the members did—that science and engineering (including the research and training aspects of each) belonged together, then it was "timely and ripe" for NSF to give its full backing to the "important *new* engineering mission." The nation needed more emphasis on engineering, and the Foundation was "the right place for this to occur." Therefore, urged Miller, NSF should "move forward aggressively without impinging upon our science."

Board members Donald Rice, president of the Rand Corporation, and Allied-Signal's president of Engineered Materials Research Mary Good argued that such a negative outcome as science being impinged upon was unlikely, since the Office of Management and Budget and the Office of Science and Technology Policy had "multiple faucets that get turned on and off at different rates." They suggested that NSF would fare better "if we're playing in the game than if we're trying to keep our specialized set of marbles off in our own little corner." Appreciating the reassurances, Knapp concurred: "You said it better than I did."

Various participants clung to worries about the budgetary implications of a new engineering emphasis, and Miller strongly recommended that the mission document not address this point. Such a posture from the NSB would sound defensive, he said. The board should address the policy issues; the director's negotiations would seek the proper budgetary balance. Branscomb questioned if budget was a main issue at all: "I could change the engineering program *fundamentally* in the direction we're describing without spending a single nickel more." Financial growth would help, of course, but what the program really needed was a "clear charter." Several members were still palpably reluctant, but the board accepted, "with no nays or abstentions (that were audible)," the task group's report "in principle"—subject to language revision, review in July by the Executive Committee, and final board approval in August.[36]

Without apparent awareness of scholars' continuing attempts to describe and explain the relationship, the board reaffirmed the "interlocking" and "mutually supportive" nature of science and engineering. Refining Layton's "mirror-image twins" model of technology (which Walker defined as the "fruits of engineering") and science, Derek de Solla Price in 1965 called science and technology "dancing partners," reflecting closely interactive but distinct activity. Arie Rip later cautioned that these dancers were like wooden puppets on a music box unless they considered their interactions as ongoing processes. Eda Kranakis saw a "hybridization" as the respective traditions, practices, knowledge, and values of science and engineering overlapped and merged. She could detect but small difference between engineers and scientists except what they called themselves. Walter Vincenti identified differences in process and intent as more significant than differences in activity. While NSF's policy makers couched their discussion more prosaically and in terms of practical considerations, they were groping toward a similar understanding.[37]

Meanwhile, Robert White, president-elect of the National Academy of Engineering, convened an ad hoc group of senior engineers from industry and academe to consider the NSF engineering mission. White's report, *Strengthening Engineering in the National Science Foundation,* saw reinvigorated engineering education and research as the keys to maintaining or restoring vital industrial productivity in the face of "severe economic competition" from abroad. The NAE study called on the Foundation to foster more multidisciplinary research, such as in manufacturing systems or the systems aspects of "megaconstruction" projects, which, it said, belonged in a class midway between traditional academic research and engineering practice (the latter not being NSF's concern). Although NSF was but one governmental player in engineering, and a minor one in terms of total budget, its role was vital, "given its broad charter for the health of science and engineering," its close relationship with universities, and the programmatic leverage of its funding choices.[38]

At the same time that it praised the Foundation for its increased attention to engineering, the academy committee chided NSF for various insufficiencies. Neither the Foundation nor the universities were "faring well" in their engineering missions; neither engineering education nor research were keeping pace with national needs. The engineers called for more generous grants for large-scale equipment and traditional individual research and more balance between analysis and experimentation. They championed enhanced opportunities for nontraditional students and a strengthened organizational "substructure" to serve fully the multiple disciplines of engineering. In particular, the NAE engineers recommended broadened engineering leadership at NSF, which was now, they said, dominated by science interests. Without "deep understanding and sympathy" for the "health of engineering" in the upper echelons, it would "remain as it is today, a secondary activity."

Having said all this, the NAE committee emphasized the Foundation's need for a greatly expanded *total* budget: "It is our strongly held view that an increase in the funding for engineering should not be made at the expense of the vital basic science activities of the NSF on which engineering so directly depends." Finally, as if unable to resist the temptation, the committee asked NSF to consider, "at an appropriate time," the "redesignation of the NSF as the National Science and Engineering (or Technology) Foundation to reflect its mission in both science and engineering."[39]

Branscomb thanked White for the academy's "very thoughtful remarks," which were "generally consistent" with the NSB's own. He did not call attention to their differing views on renaming the Foundation. As he had stated repeatedly, Branscomb thought such a move would be unnecessarily divisive within the political and scientific communities, where NSF most needed harmony. Keeping the idea of a separate technology foundation quiet would do more to enhance engineering than supporting it to the certain wrath of the basic-science interests. But, as noted at the June meeting, if NSF were to continue to oppose the establishment of a technology foundation on the grounds that science and engineering should not be separated, then "it must follow through with a viable program which addresses current engineering needs."[40]

Finally, on 18 August 1983, after modest revisions to a new version, which was considerably altered in style if not substance from the June working document, the National Science Board unanimously adopted the statement, *The Engineering Mission of the NSF over the Next Decade.* Because of its own difficulties over terminology, the board carefully included definitions so that it could state with clarity that the Foundation should not fund the "clinical practice" of engineering, for example, but should support the links between engineering principles applied in industry and those pertinent to academic research and teaching. Indeed, closer ties among the university community and industry, the professional societies, and other federal agencies were given as essential for identifying priorities and pursuing appropriate programs. To implement its broader conceptualization of engineering, the board charged the director to increase attention on four areas, a mix of cutting-edge and tradition: investigator-initiated research in the engineering sciences; computer capability and accessibility; engineering processes, materials processing, and manufacturing engineering; and engineering faculty and research facilities.[41]

NSF's new mission excited a flurry of interest throughout the engineering community. Headlines in the professional press proclaimed "more emphasis and new direction" for engineering at the National Science Foundation. The *Chronicle of Higher Education,* for one, suggested its endorsement of the shift by noting that although engineering had been "generally neglected" at NSF, it was now seen as critical to renewed U.S. competitiveness and would enjoy increased weight and visibility in future years. It quoted the understandably ebullient engineering director, Jack Sanderson, who declared that engineering "is clearly the expanding part of the foundation now." *Machine Design* approvingly declared that the new policy "enhances status, sharpens focus of engineering."[42] Both *Science* and *Engineering Education News* picked up on White's recommendation (not incorporated by the board) that NSF leadership needed to be stronger in numerical force and philosophical affinity for engineering. *Science* also observed how much active assistance NSF had garnered within the engineering community, especially the National Academy of Engineering, in the search for the Foundation's proper mission. Moreover, "having been invited to participate, these organizations now seem disposed to remain part of the process."[43]

Devoting two pages to NSF's expanded engineering mission, John Walsh, long-time *Science* reporter for Foundation matters, concentrated on the reaction of Brown, Congress's science and technology watchdog, who apparently remained "underwhelmed." "Exasperated" with decades of NSF concentration on basic research in the natural sciences, Brown told Walsh he intended once again to reopen hearings on his bill to establish a parallel technology foundation. Although pressure from engineers for increased clout had abated somewhat following the formation of the Engineering Directorate in 1981, the severe budget cuts of the early Reagan years and administration plans to reorganize the technical functions of the Department of Commerce had raised new concerns within the profession.[44]

To Walsh there was "no great mystery" why engineering had had such a difficult time in the National Science Foundation. The passive formula of awarding grants for individual research upon favorable peer review, which worked so well for basic science, poorly served the bulk of engineering, which went beyond engineering science. While everyone agreed that engineering practice should be funded elsewhere, Walsh wrote, the Foundation should encourage research on engineering methodologies or "generic technologies"—such as engineering design, materials processing, risk analysis, manufacturing engineering, and testing and quality control. He concluded that the NAE engineers' "enlightened" recognition of their "fundamental dependence" on a vigorous basic-science base could "forestall a head-on clash with science over federal funding." But with a "tough budget year in prospect," it was likely that "engineering's new phalanx of friends" would have to "exercise further patience."[45]

NSF's Advisory Committee for Engineering, meeting at the Foundation in late 1983, celebrated and endorsed the new engineering policies, but tempered their satisfaction upon hearing the fiscal realities. The acting assistant director for Engineering, Carl Hall, told the group that Engineering's total budget for FY 1984 was $121.8 million, an increase of 18 percent. While that sounded like a handsome increase— more than most disciplines got—much of the difference was earmarked for special programs, international activities, and equipment support. Further, as Hall's successor, Nam Suh, later emphasized, a high-percentage increase from a small base amounted to little in actual dollars compared to the smaller percentages for the more generously endowed directorates. Among the committee's varying recommendations for maximum stretching of Engineering's still-thin funds was the now-recurring one that NSF create "Centers of Excellence," modeled after the Materials Research Laboratory concept, to enhance academic engineering research, education, and links with industry all at once. This was an idea whose time had come.[46]

A SEARCH FOR NEW STRATEGIES IN ENGINEERING EDUCATION

To carry out its new mission, the Foundation's engineering leaders saw the need to develop new educational approaches. Some innovations, like the Presidential Young Investigator program, could be adopted quickly.

The more perplexing issues, such as the shape of academic training, required extensive studies that continued through the decade.

Although NSF had been offering fellowships, research grants, and training programs since the 1950s, the faculty shortages of the 1980s required renewed efforts to encourage the academic careers of engineers, from graduate students to fledgling professors. After surveying engineering deans, the American Association of Engineering Societies developed a four-part plan in 1982. The AAES's proposed "White House" engineering fellowships and professorships would offer initially lighter teaching loads, equipment and summer research support, five-year appointments sweetened with federal stipends, and improved campus environments. Such a program, with institutional resources matching federal dollars, could cost $467 million over ten years—a cost the nation could not afford not to meet, the umbrella organization insisted.[47]

In response, the Foundation in 1984 initiated a program of Presidential Young Investigator (PYI) awards, aimed particularly at engineers and physical scientists. Coordinated by SEE but funded by the research directorates, the program was designed to improve the ability of universities to attract top young people to conduct academic and industrial research and to teach the next generation. Each of these two hundred of the best and brightest, one hundred of them engineers, won a base grant of $25,000, renewable for five years, which could be augmented up to an additional $37,500 by matching NSF and industrial contributions. Meanwhile, the university provided the PYI's regular academic-year salary. NSF made clear its expectation of full complementary support from both industry and the university to ensure success of the program.[48] With its generous, long-term funding and impressive research results, the prestigious PYI program makes almost everyone's short list of NSF's showcase efforts of recent years.

Mark Davis, a professor of chemical engineering at the Virginia Polytechnic Institute, became a Presidential Young Investigator in 1985. He used his five-year research grant to collaborate with Dow Chemical Company's Juan Garces to develop a molecular sieve, a crystal of clay-like material laced with holes to separate large molecules from small ones. Their design offered the first improvement in the size of molecular sieve pores in 150 years and promised important applications in the petroleum industry, as it could filter a significantly higher yield of gasoline from a barrel of oil or catalyze new cracking processes. The sieve could also purify certain drug compounds previously too difficult or costly to filter. PYI Doreen Weinberger, an electrical engineer at the University of Michigan, conducted experimental research combining optics and semiconductors for ultrafast switching in communication systems and a new generation of computers called optical parallel processors.[49]

Despite ongoing attempts to improve engineering education, it was clear by mid-decade that further changes would be necessary to meet increasing competition from abroad. An NSF-requested study by the National Research Council in 1985 painted a gloomy picture. The NRC committee especially fretted about shortages of engineering faculty. And although it found the competence of recently graduated engineers—70 percent of whom ended their formal education with the bachelor's degree—"higher than ever before," these new practitioners, with their strong analytical skills and excellent theoretical grounding in engineering science, still needed six months or more of on-the-job training to become productive. The NRC laid the blame on inadequate academic instruction in design and engineering practice, vindicating Eugene Ferguson's argument that budding engineers were likely to be more creative and analytical *after* they understood the "art and practice" of design. While some industry-specific work experience would always be necessary and desirable, the report warned educators to remember that smaller companies could not afford lengthy training periods for new hires. Thus, as earlier, there was an appreciable time lag between general agreement on the need for educational change and its broad implementation.[50]

The NRC study also found the four-year curriculum increasingly strained, especially when it had to incorporate nontechnical training in areas such as engineering management, planning, teamwork, and communication skills. Yet teaching new techniques—computer-aided design and manufacturing, for example—that made engineering more mathematical and abstract, was also crucial. Here was a new area that not only embraced the scientific approach but furthered it. Options like five-year and dual-degree programs as well as continuing education were becoming increasingly attractive, if not seen as essential.

Among the NRC's twenty-three recommendations for academic engineering were broad efforts toward enhancing the appeal of faculty careers, more generous fellowship support, postponement of disciplinary specialties until the graduate level, inclusion of essential nontechnical training, and the encouragement of cooperative industry-university-government programs. The "second-tier" institutions, often undergraduate-oriented, which produced about half the nation's engineering graduates, should receive more support for research and facilities. This call for a more democratic distribution of funds came concurrently with NSF's initiation of the elitist Presidential Young Investigator program. It may well have been a deliberate attempt to address both constituencies. Meanwhile, the NRC committee felt safe in predicting a growing need for engineers as new technologies bred new industries.[51]

An ASEE task force, chaired by Edward David, produced the "National Action Agenda for Engineering Education" in 1987; like many other recent studies, it advocated curricular emphasis on design and manufacturing, with an orientation more on engineering practice than research. It also promoted master's degrees, raising the voice of that not-yet-successful chorus, and higher stipends to attract engineers into graduate study. The needs of the 1980s, particularly the international competitiveness crisis, thus drove academic emphasis on design and reinvigorated ties with industry, just as earlier times and different conditions had spawned a university-based scientific, engineering-science model. Professional credentials mattered more than ever.[52]

A STRATEGIC PLAN FOR THE 1990s

In May 1985 the Foundation completed a five-year strategic plan for education that demanded NSF's national leadership in mathematics, science, and engineering. Describing the Foundation's role as "energizing" and "catalytic," the plan focused on "points of strategic entry" where NSF could most effectively set educational standards and lever its own funding into broader support for meritorious collaborative proposals involving universities, industry, government, and others. By this statement, NSF acknowledged it was not, by dollar-wielding power, a muscular player; its influence rested on its reputation for demanding and recognizing excellence. The great challenge would be that while global economic competitiveness required ever-higher national standards, the political climate still called for a federal hands-off policy—"local diversity and individual flexibility."[53]

The Foundation's strategic plan for education stimulated keen interest among NSF's engineering constituents. When SEE director Bassam Shakhashiri established a program for undergraduate curriculum development in engineering that emphasized course content, teaching strategies, and the use of computers and other specialized equipment, the response was overwhelming. For FY 1989, 205 proposers requested $62.3 million for undergraduate education programs. Unfortunately, only $3.6 million was available; only 22 proposals could be funded. One was at Drexel University in Philadelphia, which was attempting to recast its "fragmented curriculum into a course load that combine[d] content with context." Drexel was replacing a dozen disciplinary courses with three broad multiterm courses on math and science, engineering, and laboratory fundamentals, all of which emphasized problem-solving techniques, scientific principles, and experimental methods. *Photonics Spectra* reported in January 1990 that such action was critical now, lest it become "too late to engineer another fix."[54]

Yet, for all the rhetoric, studies, new programs, and measurable progress, the late 1980s still saw engineering education described in crisis terms, a word NSF assistant director for Engineering John White used repeatedly. At the November 1988 International Forum on Engineering Education, he called on the entire engineering community to give "prompt and forceful attention" to issues like getting more students and faculty into the pipeline and encouraging leadership development among engineers to combat the paucity of engineers among American corporate executives, university presidents, and elected officials. He urged engineers to "think big," choose priorities to stretch limited resources, and work in strategic alliances to build and unify their support base.[55]

UNREALIZED RESOURCES: WOMEN, MINORITIES, AND DISABLED PERSONS

By the mid-1980s an overriding issue at NSF, particularly in engineering, had become concern over the homogeneity of the profession's practitioners. If social equity was insufficient reason to encourage underrepresented groups to pursue engineering, then national necessity was, said Engineering director White. Changing demographics were warning of a near future when the engineering profession's traditional pool of smart young white males could no longer supply its needs. Yet, according to White, in 1987 "not a single black American" received a doctorate in electrical engineering. Engineering had "a lower percentage of women graduates than any of the sciences" at any degree level. Only nine black women had won engineering doctorates in the previous decade.[56]

Encouraging and educating adequate numbers of able engineers was an NSF problem as old as the Foundation; discouraging some people from becoming these able engineers was a societal reality of much longer duration. In 1954 Office of Defense Mobilization director Arthur Flemming, who chaired Eisenhower's special committee on training scientists and engineers, acknowledged that "despite current shortages," which were then seen as urgent, there were "women, Negroes and handicapped persons who [were] sometimes not permitted to qualify for positions commensurate with their abilities and training." Further, "artificial barriers" prevented them from gaining access to necessary intellectual development. But Flemming's committee made no specific recommendations addressing any of these underrepresented— virtually unrepresented—groups except to draw "qualified married women teachers" back into high-school science classrooms and to provide college scholarships for "all highly qualified young men and women."[57]

Little was done for many years by the Foundation or any other entity to become more inclusive in the name of either national interest or ethics, although numerous reports lamented the thin ranks of various groups. In 1961, for example, NSF noted that only 4 percent of federally employed scientists and engineers were women. It was not until 1972 that Howard University became the first predominantly black school to offer degrees in chemical engineering. The professions themselves were slow to change attitudes and practices. In 1964 the *IEEE Spectrum* welcomed women "because they improve[d] an otherwise drab scene," and only secondarily, condescendingly, because "decreasing emphasis on hardware and testing" meant they could do the work.[58] Though the American Society for Engineering Education was seeking to promote the study of engineering, in its heavily statistical 1968 study, *Goals of Engineering Education,* it specifically limited itself to consideration of male students, whether addressing their aptitudes, aspirations, achievements—or the nation's needs. Would-be female engineers, deleted even from the discussion, could hardly expect to be included in the real-life fraternity.[59]

In 1968, still identifying a shortage of engineers, *Spectrum* carried a lengthy article by Irene Peden of the University of Washington. She emphasized that women constituted "the missing half of our technical potential" and that it was time to recognize and utilize women's "capabilities and creative talents." By Peden's reckoning, based on analyzing high-school aptitude tests, if all girls with "natural talents" in engineering became engineers, "we would have two women for every three men in the field." Instead, women then constituted less then 1 percent of all American engineers—a statistic still valid in 1974. In the latter year, Sol Cooper of the American Society of Civil Engineers conceded the existence of barriers beyond the formal ones: "We have a legal and practical obligation," he wrote, "not only to hire qualified people, but to create qualification."[60]

University of Illinois economics professor John Parrish noted in 1971 that although the 1960s had looked favorable for more women in engineering, "very little" had changed in absolute numbers or percentages. Women's rising rates of participation in education, the space race, equal-opportunity legislation, and increased activism by women's professional organizations, coupled with unprecedented shortages of technical personnel, had "forced and persuaded" industry, government, and academe to "pull down the barriers," however, so change in the new decade was more likely.[61]

In 1977, during the period of Carter's visible leadership for broadened opportunities, NSF conducted a major study, *Women and Minorities in Science and Engineering.* The data unsurprisingly confirmed the existence of "relatively few women scientists and even fewer engineers, and considerably fewer minority scientists and engineers of either sex." Women still represented less than 1 percent of the 1.1 million U.S. engineers—the largest proportion of these (22 percent), specialists in the fast-growing, high-demand field of computer science. Traditional barriers to employment did "tend to fall in the face of a skill shortage." Among racial minorities, including minority women, Asians constituted the vast majority. All these groups earned about 80 percent of the salaries of white males, which was better than in most professions. The report further stated that in engineering nearly three-fourths of the men but less than one-fourth of the women who planned in 1965 to enter the field were employed in it six years later. Women appeared to "drift into, choose, or [be] forced to enter 'less prestigious' and probably lower paying occupations relative to their original career choice." While wary of drawing conclusions, the NSF investigators projected greater participation by underrepresented groups only if societal attitudes, and, in the case of certain minorities, precollege educational opportunities, improved.[62]

National Academy of Sciences statistics showed twelve women earning doctorates in engineering in 1968 (0.4 percent of the total number of Ph.D.s awarded), with a growth to seventy-four (2.8 percent) in 1977.[63] The *IEEE Spectrum* reported encouragingly in 1983 that the number of undergraduate women studying engineering full-time had increased "well over 1000 percent in the last decade," to about 15 percent of all undergraduate engineering students.[64] But in 1985 *Mechanical Engineering* noted that the number of women graduating in engineering had peaked in 1983 and was on the decline. It also said that although the number of technically employed women had nearly doubled between 1976 and 1981, by NSF figures nearly 75 percent of women engineers had entered the professional workforce within the decade. In 1986, when women constituted almost one-half of the U.S. labor force, they represented only 4 percent of American engineers. They were twice as likely as men to be unemployed or underemployed.[65]

The Science and Technology Equal Opportunities Act of 1981, signed by Carter in the waning days of his presidency, authorized NSF to initiate a program of National Science Foundation Visiting Professorships for Women in Science and Engineering. In addition to conducting research and teaching at academic institutions, such visiting professors would counsel and mentor women students at all levels as well as provide greater visibility for technical women generally.[66] The Research Opportunities for Women (ROW) Program, begun in FY 1985 but based on a Carter legislative model, was designed particularly to assist women beginning or resuming interrupted research careers by providing an "alternative entry point" to NSF support. It also boosted academic women in general, since about two-thirds of them held the lowest faculty positions, which precluded them from becoming principal investigators in the normal grant process. Even though inadequate resources kept the initial funding rate at a disappointing 10 percent, the program was successful if measured by the fact that after three years, almost one-third of the initial ROW awardees, engineers in about the same proportion, subsequently won another, non-ROW NSF grant.[67]

The stories of other underrepresented groups were similar. The National Advisory Council on Minorities in Engineering reported with pleasure in 1977 that for the third successive year, the number of minority engineering freshmen had exceeded the most optimistic projections. More important, and more difficult, however, would be to increase their retention and graduation rates. And more compelling than statistics was personal reality. NSF director John Slaughter, who became in 1980 the first black to head a major federal science agency, said simply, "The first black engineer I ever saw was myself."[68]

In 1978 the National Science Foundation established the multimillion-dollar Resource Center for Science and Engineering (RCSE) program to promote increased participation by minority and low-income persons in science and engineering. By 1981 four centers had been created. That year NSF began the Minority Research Initiation (MRI) program to help minority faculty obtain their first research grants with such aid as proposal-writing assistance, funding of equipment, and time released from teaching. Indirectly, an important MRI outcome was the linking of these researchers with the broader NSF community and the universe of its programs. The Foundation also provided grants under the Research Improvement in Minority Institutions (RIMI) program to assist minority colleges in organizing and managing research programs, purchasing equipment, and publishing data.[69]

Although without organizational emphasis until the late 1970s and kept flat or cut from the budget during the early Reagan years, a variety of engineering research projects to aid disabled persons had received NSF funding from at least the early 1960s. In 1972, when Edward David, Nixon's science advisor, asked several federal agencies what they were doing to "use the 'skills that took us to the moon and back' to develop devices 'to help the blind to see, the deaf hear, and the crippled move,'" NSF admitted it had no targeted program and considered it beyond its purview to "solve the problems of the handicapped, as serious and important as we know them to be." But, responding to investigator interests, the Foundation was funding about $3 million in "active projects." In engineering, then and later, designing prostheses that used computerized gait analysis and minute electrical signals from remnant limbs was one promising effort. Developing reading machines based on optical character recognition devices coupled with computer-based synthetic speech generators was another direction pursued with significant results. But engineering advances to help physically challenged people function in the workplace, including an engineering laboratory, were but part of the problem. At least as great an obstacle was attitudinal barriers that prevented their full participation in the profession. On this front, the Foundation was slower to get actively involved than it was with other underrepresented groups, although today, at least in the Engineering Directorate, the three categories—women, minorities, and persons with disabilities—are generally considered under the same umbrella and given attention in the Division of Engineering Infrastructure Development.[70]

Thus, deliberate attempts to redress imbalances and inequities in the profession of engineering have increased in recent years. Indeed, in 1987 the National Science Foundation listed developing human resources and broadening participation in science and engineering first among its three top goals for the coming fiscal year. Beyond altruistic reasons, the nation needed people—rigorously, relevantly educated scientists and engineers—to restore the U.S. knowledge base and comparative standing in international productivity. But it was still true, as John White charged in 1988, that "engineering is losing market share with bright Americans. And, we do a poorer job in reaching bright women and bright minorities."[71] The issue remains both central and sore.

The NSF engineering community had thus at least identified the challenges of finding, encouraging, and properly preparing sufficient numbers of engineers to tackle current and emerging problems. It was gathering the necessary political, technical, and human resources to make a difference in the world of the mid-1980s. And it was armed at last with a mission that reflected recognition of its own distinctive role, although the respective boundaries of engineering and science would probably remain fluid and indistinct. But as civil engineer Samuel Florman stated in 1986, most engineers lived in the concrete. They were usually not very interested in theories, especially semantic ones. Glad for expanded resources, they were eager to get on with their work.[72]

CHAPTER 8

ACTING ON THE MANDATE

> The main way in which engineering research and education
> can contribute to the international competitive position of the
> United States is by bridging and shortening the gap between
> the generation of knowledge and its application in the mar-
> ketplace.
>
> —Roland W. Schmitt, "Engineering Research and
> International Competitiveness"

NSF engineering leaders continued to refine their planning and personnel policies, but the tools needed for dramatic new technical pushes
were falling into place. The necessity for bold action was evident from
America's deteriorating competitive posture; the prod as always was
political pressure. Engineering defined and introduced in 1984 a new
concept—Engineering Research Centers (ERCs)—to address in one
program the critical issues of promoting innovative basic engineering
research, rigorous and relevant engineering education, and mutually
supportive links with industry to ease and shorten the path between
technical discovery and application. Though never a dominant component of the directorate's budget, the ERCs became a showcase of interdisciplinary achievement responsive to national need.

Congress solidified engineering's place by specifically including engineering in the words of the NSF organic act. The Engineering Directorate reorganized itself in efforts to expedite effective action.

DEVELOPING A NEW CONCEPT: ENGINEERING RESEARCH CENTERS

NSF's Engineering Research Centers evolved from many separate threads that were woven together finally, rather quickly in the end, in 1983. That July, while the National Science Board was considering the final version of a new mission for engineering, the National Academy of Engineering urged NSF to "chart a new and more vigorous course" for engineering and engineering education.

In early August Jack Sanderson's Engineering staff generated an initiative that called for "strengthening the coupling" of academic research, education, and industry—terms increasingly stressed, and linked. Given that by the late twentieth century around 75 percent of engineers worked in business and industry, and most of the rest directly or indirectly (through contract) in government, NSF's pursuit of industry as an academic partner was sensible, if not overdue. Originally proposing $173 million for five engineering research centers and other multidisciplinary research focused on industrial processes, equipment and curriculum improvement, and human resources development, Sanderson's initiative had been pared by the mid-August board meeting to $77 million. Syl McNinch later observed that this proposal had "many of the right ingredients" but was "still somewhat fragmented" and "lacked an integrating mechanism." Perceived as "more of the same," it failed to win approval. The NSB did, at that meeting, adopt the new mission/policy statement for engineering.[1]

But an idea had begun to jell, and other initiatives soon followed, including the October 1983 "breakthrough" of Carl Hall, then acting assistant director for Engineering, who took the best of various previous proposals and integrated them into a single program he called EQUIP, or Engineering Quality Improvement Program. EQUIP would have provided generous long-range grants, also focused on institutional enhancements in instrumentation, equipment, and curriculum. It, too, "underwent some surgery" but took the Foundation farther along the path of innovation. On 27 October 1983 NSF submitted a revised budget to the Office of Management and Budget, which included $10 million for an engineering-grants initiative to improve the research quality and capability in Ph.D.-granting engineering departments that were "near the threshold of excellence."[2]

Two days later, on 29 October 1983, a Saturday, presidential science advisor George Keyworth convened a meeting of NSF officials and academic and industrial leaders to discuss a forthcoming COSEPUP (NAS Committee on Science, Engineering, and Public Policy) briefing paper on computers in design and manufacturing. The report made clear two critical problems: American industry, which increasingly relied on automated manufacturing, could not release its production equipment for training new engineers, and universities lacked both the interdisciplinary knowledge and costly instrumentation to provide such training to their students. Struck by these findings, Keyworth's group expanded their discussion to explore ways to strengthen NSF's engineering efforts vis-à-vis manufacturing. George M. Low, the chair of COSEPUP and president of Rensselaer Polytechnic Institute, opened the right door by relating how his students and faculty were profitably working jointly to design and build a glider from new composite materials. Low's pioneering idea was to provide engineering students with more experience in teamwork, hands-on experimentation, and cross-disciplinary and systems research in order to prepare them more appropriately for real-world technical careers. The real world also intervened when students were required to test-fly the glider as their final exam.[3] Thus was the Engineering Research Center concept born.

At that point it became important politically, as Lewis Branscomb remembered, for the "engineering world" to buy in to the concept—to want and ask for such a program. Accordingly, on 13 December 1983, NSF director Edward Knapp, former researcher and director of the Accelerator Technology division at Los Alamos National Laboratory, wrote again to Robert White, president of the National Academy of Engineering, to request NAE's help in developing and structuring a specific program in "what might be called 'engineering centers.'" Knapp reminded White that Keyworth had spoken favorably about engineering research centers to the NAE at its recent annual meeting. (Indeed, Keyworth would later pronounce the ERC program "among the two or three most important initiatives in the entire federal non-defense R&D budget." Looking back, Hall considered Keyworth's strong commitment crucial for the ERCs' success, in view of considerable skepticism, if not opposition, within the administration and elsewhere.[4])

Knapp pushed the academy to respond by the end of January 1984—scarcely six holiday-strewn weeks hence—in order to include its recommendations in NSF's FY 1985 budget submission to Congress. As it happened, the NAE report was not ready until mid-February, but the Office of Management and Budget had meanwhile allowed NSF to put in $10 million for some as yet undefined "Centers for Cross-Disciplinary Research in Engineering." Knapp outlined on-campus centers to house cross-disciplinary, experimental research activities and educational programs in much the same way that many others by now had done. Besides organizational and procedural considerations, he asked the NAE committee to study strategic questions, such as: How many centers should NSF support? For how long? How could the centers encourage cross-disciplinary research? Was it realistic to expect them to make a serious impact on undergraduate education? How should the relationship with industry be fostered? Should "smokestack" industries and small manufacturers be included? Branscomb later emphasized that although the Engineering Research Centers are popularly credited to the National Academy of Engineering, the real leadership came from the NSF director, guided by the NSB and staff studies.[5]

The NAE report, *Guidelines for Engineering Research Centers,* not surprisingly called the ERC concept ambitious and innovative. If "done well," such centers could significantly enhance future American competitiveness, although only if support for normal disciplinary activities were maintained as well. Chaired by W. Dale Compton, Ford vice president for research, the academy panel insisted further that Engineering Research Centers must complement and enhance, not replace, existing relationships between research universities and state governments, national laboratories, and the professional societies. The relationships with industry must be real and perceived by both sides as mutually beneficial. Perhaps remembering earlier, headlong innovations, the NAE also cautioned patience and realistic expectations; alterations should be expected over time in an effort that was essentially experimental.[6]

The National Academy of Engineering panel—whose membership included both Erich Bloch and Nam Suh, who were soon to take their places as NSF director and assistant director for Engineering, respectively—saw two purposes underlying the ERC concept. One was to enhance the ability of academic research institutions to conduct cross-disciplinary, systems-oriented engineering research on problems critical to industry. The other was to provide engineering students with a broader understanding of engineering practice by including industrial personnel and equipment in their training. For several reasons, inadequate funding being primary, too much of engineering education centered on small-scale, formally and analytically studied problems in traditional subdisciplines, whereas in industry the focus was on systems, teamwork, hands-on research, synthesis, and integration of effort.

In terms of details, the NAE panel specified that each center must have full-time leadership, even though its faculty would share a disciplinary and departmental base. The equivalent of at least three full-time faculty members would be needed to run a viable program, more at a larger institution. While ERC sizes would vary, each center must be sufficiently large to involve at least 10 percent of its home institution's graduate engineering students and substantially affect undergraduate education. Industrial involvement in the ERCs must be of sufficient scale and duration to make a significant and noticeable impact on the program.

As for how many Engineering Research Centers there should be, the NAE recommended a small number of outstanding, prestigious, and well-funded centers in preference to many inadequately supported ones. About twenty-five, developed in stages, were as many centers as the panel thought could provide disciplinary breadth and absorb the level of funding envisaged without distorting their overall research programs. Compared with total annual expenditures for academic engineering education and research, NSF's proposed target of $100 million for ERCs represented only about 2 percent.[7] The first year, funding for ERCs was $10 million. Traditionalists argued that this money came from their programs for individual grants, but this was not true; the ERC funds were new money.

The National Science Foundation, persuaded that research and education linked in academe and industry were interlocking keys to productivity and international competitiveness, accepted the NAE's recommendations almost entirely. Each innovative engineering center would be unique, defining its own subject area and tailoring its management structure to its own objectives. NSF further envisioned that the ERCs would reach out to undergraduates from other schools, to women and minorities, to existing university/industry projects, and to consortia incorporating the overall technological areas. Each center would receive $2.5 million to $5 million per year for five years; support for an additional five years would depend on the outcome of a thorough critique, including a site review to be conducted during the ERC's third year. Most ERCs would be expected to double their money with funds from industry, the state government, and the university. Ongoing NSF oversight would be the responsibility of Engineering's Office (later Division) of Cross-Disciplinary Research (and still later, Division of Engineering Centers), assisted by an advisory team for each center.

Accordingly, NSF issued an ERC program announcement in April 1984, which promoted Engineering Research Centers as a focal point of the Foundation's new commitment to an expanded engineering mission. Confirming scholarly and pragmatic interest in the concept—and overwhelming the Foundation's meager $10 million budget allotment—100 academic institutions submitted 142 proposals, altogether requesting some $2 billion over five years. To select the worthiest from these generally impressive ideas, NSF engaged experts from both the academy and industry, emphasizing the latter, as well as its own engineering staff in a complex, multistaged review process that culminated in visits and "final orals" for the finalists. After the initial screening by teams of staff-selected engineers, an ERC evaluation panel, co-chaired by Eric Walker and C. Lester Hogan, former president of Fairchild Camera, reviewed some forty semifinalists. Walker, who grumbled privately about Suh's heavy, interfering hand, had only positive comments in public about both the ERC concept and the selection process. He credited Suh's "ingredients for success" with guiding the panel's deliberations. Stated his own way, these ingredients were the proposed center's leadership; proper focus on important problems; bona fide industrial participation; appropriate infrastructure, including the university's internal organization and commitment to cross-disciplinary research goals; intellectual challenge that would establish new mental frontiers, contribute to the knowledge base, and provide graduate research topics; and enhanced educational opportunities for graduates and undergraduates. Further, a candidate institution had to demonstrate the advantage of a center approach over traditional modes of support.[8]

By February 1985 the meticulous, if not strictly harmonious, review process had yielded six Engineering Research Centers for initial funding. Seeking to provide each center with sufficient support to maximize the likelihood of success, NSF held the number at six, even though the ERC panel had identified three additional centers as deserving if money were available. Reasons for not funding specific cases ranged from inadequate management skills to vague industrial involvement, from weaknesses in conceptualizing a synthesis of the component parts (a collection of researchers doing business as usual would not pass) to an unlikelihood of significant breakthroughs.[9]

The National Science Board authorized the director to proceed with funding these six on 22 March 1985, and Bloch announced them to the public on 3 April in the context of their potential toward enhancing U.S. international competitiveness. As Bloch had recently put it, "If we can push technology hard enough, we can overcome disadvantages in labor, capital cost, and currency exchange. If we don't push technology hard enough, we don't compete. *It's that simple.*"[10] Together, the successful ERCs were slated to receive a total of $9.4 million in FY 1985.

Over five years the total for these and new ERCs would be $94.5 million. This premiere "class of 1985" varied in size, scope, and intellectual focus:

- The University of California, Santa Barbara, planned to establish a Center for Robotics Systems in Microelectronics to analyze, design, and build flexible, completely automated systems capable of generic tasks in microelectronics technology for the fabrication of advanced semiconductor devices.

- Columbia University's Engineering Research Center for Telecommunications would explore the integration—within a highly flexible telecommunications network test bed (MAGNET) and at the interface with users—of data, facsimile, graphics, and voice and video transmissions.

- The University of Delaware, in conjunction with a ceramics program at Rutgers University, expected to create an ERC for Composites Manufacturing Science and Engineering, an extension of a smaller, existing program on composite materials. Using cross-disciplinary training and research, the center hoped to develop crucial new materials for the commercial aircraft, automobile, trucking, and electronics industries, as well as consumer products.

- The University of Maryland's ERC on Systems Research stressed basic research to exploit recent advances in computer science and engineering, such as very large-scale integrated (VLSI) circuits, computer-aided design and manufacturing, and artificial intelligence, in the design of interactive automatic control and communication systems. Harvard University, in collaboration with Maryland, would add to the theoretical and applied systems-engineering aspects and conduct additional research, especially in robotics.

- The Massachusetts Institute of Technology Biotechnology Process Engineering Center anticipated joining biological and engineering perspectives in manipulating organisms or biological agents oriented toward the commercially viable manufacture of therapeutic products.

- Purdue University's Engineering Research Center for Intelligent Manufacturing Systems focused on automation for batch manufacturing of discrete products, where the "intelligent" system could provide at least semiautonomous reasoning to reduce production time, costs, and errors.[11]

Strictly speaking, the concept was not new. The Materials Research Laboratories bore many resemblances, especially in their interdisciplinary approach. Nor was the industrial component entirely innovative, as evidenced by the existing cooperation at such institutions as the Cornell submicron facility and the Stanford Center for Integrated Systems. The smaller-scale, much older Industry/University Cooperative Research Centers (IUCR) program, funded jointly by NSF and industry, was also similar, although the technical focus was more limited, education was not a major component, and Foundation support was limited to five years. After that the IUCR had to be self-sustaining.[12]

Yet in a real sense, the Engineering Research Center program represented a fundamental rethinking of traditional NSF engineering activity. It was itself an integrated systems approach, a synthesizing response that addressed immediate concerns in both engineering research and engineering education—concerns articulated by both academe and industry. It responded to a clear national need, according to NSF's charter.

Economically, ERCs would prove their worth by harnessing fundamental academic engineering knowledge in the solution of generic problems of national productivity. By concentrating on emerging technologies and improving traditional methodologies, they would help restore American competitiveness in world markets. A primary ERC thrust was to improve the academic infrastructure—such as facilities and instrumentation and technical and operations support for team efforts that would be too costly for most academic institutions, much less an individual investigator, to maintain.

Programmatically, the ERC endeavor was an innovative umbrella under which multiple NSF agendas could be gathered and advanced. In cooperation with industry and with significant funding from states and private sources, the Foundation would be a much more active partner in supporting engineering research and education across a broad range of objectives. ERCs would enable engineering schools to produce graduates with diverse skills in multidisciplinary areas and hands-on familiarity with the industrial environment—especially the application of computers and automation, which were revolutionizing engineering design analysis and manufacturing. Industries that anticipated employing these graduates would need to provide direct and sustained involvement—with people and ideas, not just money.

Politically, the program was particularly significant, because recognizing and encouraging Engineering Research Centers signaled the administration's decision that the National Science Foundation, not some new agency, would lead U.S. engineering and technology efforts. By addressing urgent technological issues, NSF also deflected once again congressional demands for a national technology foundation. Indeed, according to *Science & Government Report,* a Congress "more impatient than ever with NSF's treatment of engineering" had introduced a "record number of bills invoking the magic words 'research,' 'technology,' 'engineering,' and so forth." When the House Science and Technology Committee accepted NSF as "the chosen instrument for rebuilding academic engineering education and research," suggested Branscomb, it represented a "turning point in NSF history."[13]

The affected professional communities lost no time in registering their responses to Engineering Research Centers. Engineers, as exemplified by those of the Institute of Electrical and Electronics Engineers (IEEE), were predictably ecstatic. The IEEE's Russell Drew applauded the ERC initiative as an "eye-opener for the Administration and the NSF as a mandate for increased engineering funds" to "redress" the existing imbalance between scientific and engineering support. "This will enable us to thump on the [congressional budget] table with a little more credibility." While no one was suggesting that NSF's engineering support should come at the expense of science, Drew insisted that engineering's historical 10 percent must increase: "Most of the growth in the NSF budget should be slotted for 'hot areas' such as many fields of engineering."[14]

Many scientists, however, were nervous. Engineers' "huge response" to the ERC program announcement, which was attributable both to "the smell of money" and the opportunity to "establish a ground-floor presence," once again "aroused anxiety" among NSF's science clientele, who believed NSF engineering efforts (currently $143 million of the Foundation's total $1.5 billion budget) "came out of their hides." That Bloch and Suh—both engineers, both members of the NAE's ERC panel, and both now power figures at NSF—would guarantee top-level backing for Engineering Research Centers increased basic scientists' angst. Lewis Mayfield, head of Engineering's Office of Cross-Disciplinary Research, admitted, "They're concerned, there's no question about it," but he took pains to point out that NSF's new budget was maintaining the "same strong pace of growth for basic research" that it had for three years.[15]

ENGINEERING RESEARCH CENTERS: LAUNCHED AND MONITORED

The Engineering Research Centers moved quickly into the spotlight of the technical community. When the National Research Council sponsored an ERC symposium in Washington at the end of April 1985, the two-day high-level event attracted more than four hundred attendees from academe, industry, and government. The steering committee's opening words about the infant program were enthusiastic and hopeful: Engineering Research Centers "are the right step at the right time; they will inject into engineering new values and new approaches that are sorely needed. It behooves all of those involved in the engineering enterprise in the United States to ensure that this gem is highly polished, and that the sparkle and promise of this new beginning are not permitted to fade." Former NSF director H. Guyford Stever called the ERCs' debut a "brave new venture in American technological enterprise." Keyworth pronounced them "long overdue" and predicted that ERCs could outgrow NSF and continue on their own.

After NSF and White House science leaders outlined the national significance of the program, the academics heading the first six ERCs spoke of their goals and accomplishments. Several participants offered prescriptions for the future of American engineering from the ERC perspective. For example, industry must avoid applying pressure for near-term results if it was to take full advantage of the ERCs' potential. Universities must change their "campus sociology," which evaluated and rewarded faculty on individual achievement and dismissed results accomplished in groups across departmental lines. Government, beyond being a catalyst, must provide adequate, reliable financial support.

What differences—in research, in engineering education, in the health of U.S. industry—would be seen if the ERCs were successful? Suh envisioned an intellectual and cultural "climate of discovery" that would produce ideas to "change the way we live, the way we function, and the way we produce goods." National Science Board chair Roland Schmitt saw an improvement in the nation's competitive position as the real test: "If the ERCs can provide a strong link between academe and industry, research and development, education and practice, they can vastly improve the effectiveness with which we apply our rich national resources of knowledge and talent. If they can bridge the traditional engineering disciplines, they can be the catalyst for a needed reshaping of research approaches and values, in universities as well as in industrial manufacturing practices." As Schmitt noted, in current practice, engineering was the only profession whose teachers were "not, by and large, experienced practitioners." ERCs would enable engineering students to acquire both scientific knowledge and a practical research "internship." Further, they would become effective advertisements for attracting the next generation of top students into the program.[16]

The ERCs enjoyed, or endured, constant scrutiny. After one year the National Research Council pronounced them "leaders in change." Suh observed with gratification that American engineering schools were responding eagerly and ably to the demands and potential of cross-disciplinary research, backed by strong support from their respective universities. Other agencies and state governments, recognizing the benefits, were coming forth with funding, as hoped. Industry's intellectual and financial commitment was keen; industry had thus far invested $1.24 in the ERCs for every federal dollar committed and was an active, participating partner in both research and educational program components. With a "combination of leadership and institutional strengths, plus first-rate research," the successful ERCs had become "models of the will to win in technology."[17]

Challenges remained, however. The proper balance had to be maintained among individual research, traditional approaches, and cross-disciplinary efforts in order to ensure the strongest overall engineering infrastructure. While the ERC program was a key element in international competitiveness, a far bigger, long-term investment—at least $500 million annually—had to support NSF's engineering program as a whole. The Engineering Research Centers had materialized after a gestation of only nineteen months, but other countries—rivals—could even more quickly emulate the established pattern. Because most ERCs sought no specific engineering product, such as a space shuttle, it was difficult to focus the team on goals at once broader and more basic. Operational and philosophical strategic planning would be crucial for keeping the competitive edge.[18]

Following the third-year reviews of the first ERCs, the National Science Board agreed with the evaluators that two should be phased out. (See appendix 3.) Later, a third center lost its support. Lynn Preston, deputy director of the Division of Engineering Centers, said afterward that the unsuccessful ERCs lacked leaders with vision and team-building ability. They were unable to attract high-quality faculty and administrative support at their institutions. Or they lacked the proper infrastructure to support the leaders' ideas. With "people problems" and no coherent focus, movement was too slow. Branscomb mused that the shutdowns were useful to show universities what could—and would—be done if they did not meet their programmatic or funding commitments, but Suh still recalled the process as painful: "I didn't win any friends over this one." (But he was also planning to leave the Foundation, and making that decision himself would ease the initiation of his successor.) Thinking

at first to close only one center, Suh soon learned that political considerations required that if one had to go, the Foundation must close two.[19]

The Engineering Research Centers winning initial and renewal
funding eventually grouped around six broad technical areas: manufacturing and design, materials processing for manufacturing, optoelectronics/
microelectronics, biotechnology/bioengineering, energy and resource
recovery, and, later, infrastructure (see appendix 3). Each one addressed an
urgent frontier-of-knowledge technological issue, responded to the particular demands of its own industrial and academic communities, boasted
strengths, and suffered weaknesses. Two examples illustrate the ERC
concept in action—one at a premier technology school in the East, the
other at two less well-known universities in the West; one topic rooted
in prehistory, the other futuristic. Both were successful. Together they
suggest the program's scope and variety of aim and accomplishment.

Advanced Combustion Engineering Research Center (ACERC). While
fire may not seem like a high-tech concept, efficient combustion technology is a major determinant in economic competition. Virtually all
industry, whether traditional or avant-garde, depends on an adequate and
affordable supply of high-quality energy, the production of most of
which ultimately rests on combustion technology. In the mid-1980s,
with the availability of oil either geophysically or geopolitically uncertain, the United States realized it had to be able to produce economically
and utilize cleanly and efficiently low-cost, low-grade fossil fuel resources, such as coal, shale oil, and heavy petroleum liquids. Unfortunately, burning such fuels also increases the pollution of smog and acid
rain and the "greenhouse effect." Disposing of hazardous wastes is an
equally serious problem.

Although the United States had long led the world in manufacturing energy-producing equipment—such as gas turbines, boilers, and
advanced gasifiers—that lead was evaporating. Foreign competitors
were encroaching in the world market, while at home outdated design
methods and technologies impeded critical advancements. Market pen-
etration of new combustion technologies took too long. Researchers

insufficiently understood combustion fundamentals, and they communicated and cooperated with each other far too little.[20]

To improve the competitive position of the U.S. fossil energy enterprise, the National Science Foundation in 1986 awarded Brigham Young University (BYU) and the University of Utah a five-year, $9.7 million grant to establish the Advanced Combustion Engineering Research Center (ACERC). The ACERC was to be an extension of work already in progress at these institutions, with additional research by investigators, some from as far away as the University of Leeds in England. The center also won generous support from the Department of Energy, other federal agencies and national laboratories, Utah's Centers of Excellence Program, and more than two dozen industrial collaborators.

The Utah ERC, directed by Douglas Smoot, BYU dean and professor of engineering and technology, sought to make three contributions, the first being a new understanding of combustion mechanisms and their relation to fuel properties, principally those of coal. To that end, over one hundred ACERC participants (faculty, professionals, and students) conducted coordinated research in areas such as fuel structure and its reaction mechanisms, pollutant formation and control, turbulent fluid mechanics, and heat transfer.

The center's second goal focused on next-generation computer modeling of combustion processes. Computer-generated simulations—which integrated kinetic and mechanistic data, physical and chemical fuels property data, process-performance characteristics, and other variables—could then be applied to varied combustion systems to record, control, and predict the performance of particular combustion chambers, say, coal-fired utility boilers. The resulting knowledge would then be transferred to industry and incorporated into textbooks, user manuals, and short courses. To pursue this work, the two universities acquired three mini-super-computers and eight computer workstations. Marketing the emerging software to industries then brought additional financial support to the center.

The third intended product of the ACERC was people—students educated in combustion engineering fundamentals and trained to solve a wide range of *real* industrial problems. The two universities developed and offered over twenty interdisciplinary courses related to combustion or fossil fuels taught in a systems context. ACERC's education program stressed laboratory and classroom experience that involved undergraduate, graduate, and postgraduate students.

Making it all worthwhile was the transfer to industry of these experienced graduates, combustion process innovations, fundamental and applied research results, and advanced computer-based combustion codes. That happened by involving industrial participants in center activities, perhaps as visiting fellows, software development managers, or joint experimenters. Every year an industrial advisory council reviewed research programs and provided technical and management direction. Industrial associates and, at a lesser level, industrial affiliates became "members" based on annual fees or research grants, which gave them the right of first access to technologies developed at the center, such as design software. To maintain communication and encourage technology transfer, ACERC sponsored monthly seminars with speakers of broad background and often international repute. Intercampus shuttle bus service, a periodic newsletter—*Burning Issues*—and journal publications also kept people and ideas in touch.[21]

Did it work? Having demonstrated to NSF substantial success during its first three years, ACERC in March 1989 won its five-year extension, with a "significant increase" in financial support. Adding together the center's assets from all sources, its total operating budget for the next five years approached $15 million. The ACERC could point to specific accomplishments: By the summer of 1987 doctoral candidate Larry Baxter had formulated the most accurate and efficient computer method yet for calculating turbulent particle dispersion.[22] In March 1988 the ACERC reported a new analytical theory for relating the structural properties of coal as measured by nuclear-magnetic-resonance to its rate of devolatilization—that is, how coal, when heated, initially liberates complex gaseous and tar products ("volatiles").[23] University of Utah professor David Pershing, one of the principal investigators of the ACERC, and his chemical engineering students were making important gains in understanding what happens to hazardous hydrocarbon and metallic contaminants during combustion. Their advanced computer-aided design methods evaluated probable emission levels and estimated the potential of various control options.[24] In general, *Burning Issues* boasted robust technical activity, lively intellectual exchanges among investigators at varying levels of formality, ongoing and positive relations with the National Science Foundation, and a sense of purpose and accomplishment.

What did the people think? Those involved with the Utah ERC were enthusiastic about its accomplishments and its potential. Having generous and assured funding for five years gave investigators the freedom to pursue their work independently of economic pressures. Pershing touted the results of the center's interdisciplinary research. For example, in hazardous waste incineration, mechanical engineers designed waste incineration centers, chemical engineers determined the effects of pollutants, and chemists measured the specific harmful trace substances emitted. Before the ACERC, these researchers would never have thought to work together. Now a sense of community was developing. Pershing emphasized that the ERC's high visibility and prosperity attracted *American* graduate students, who were interested in the hot environmental issues and wanted to address real-world problems. The cross-fertilization of ideas between academia and industry produced results. The list of industrial associates continued to grow.[25]

Those engineers at the Utah universities not involved in the ERC program tended to view it less favorably. At least one mechanical engineer thought the program hurt engineering funding at the University of Utah, because it "forced more eggs into fewer, more concentrated baskets." He also thought the ACERC's effect on engineering as a whole was less than NSF claimed.[26] Others argued that while intellectual interaction and research productivity could rise with large numbers of interdisciplinary researchers working jointly on a problem, those same large numbers could mask a lot of unproductive motion. Some second-rate researchers got funded within the group effort who would not have been supported as individuals. And independent-minded researchers, used to working alone, sometimes preferred to pursue their own goals, in their own way, on their own.[27]

Despite valid complaints, the ACERC's obvious plusses merited the applause. The proponents, who had every incentive to make the most of their achievements, surely erred on the side of rosiness in their reports, but their vigor, sense of purpose, and productivity were clear. Discussion across disciplinary lines that had become almost institutionalized necessarily produced new methodologies as well as new ideas. Newness and flattering attention are themselves invigorating. Of course, who could argue that there was not one sloth in the beaver lodge? Or deny that priorities by definition demote something else?

Biotechnology Process Engineering Center (BPEC). At the Massachusetts Institute of Technology's Biotechnology Process Engineering Center, established in May 1985, chemical engineers, electrical engineers, nuclear engineers, biologists, applied biologists, and computer scientists came together to reinforce efforts that had already made MIT a world leader in biotechnology process engineering.

The primary purpose of the BPEC was to make proteins of therapeutic value from genetically engineered mammalian cells. While much is known about the traditional fermentation processes that give us "products from bugs," such as antibiotics and vitamins, the skills needed to make cells to produce proteins like those occurring naturally in the human body are less understood. Although the processes have been known in the laboratory for a decade or more, the commercial bottleneck was knowing how to grow such proteins efficiently, economically, safely, and predictably so that industries could manufacture and market therapeutic products made from them. Sometimes these proteins are important for people whose bodies, because of genetic disease, lack them—for example, the human growth hormone that is used to treat dwarfism. Sometimes they provide an answer that nature did not think of. The "engineered" protein TPA, or tissue plasminogen activator, which dissolves blood clots, is a significant therapy following heart surgery.[28]

MIT researchers in the engineering center sought to understand the principles and mechanisms that govern how mammalian cells process and secrete proteins. Then they could design and operate large-scale animal cell bioreactors for optimum protein production and control. In downstream processing, they would pursue efficient isolation and recovery of these genetically engineered, complex therapeutic proteins in their desired structural, functional, and purified forms. Finally, they planned to develop advanced computer modeling and simulation to improve the center's overall capabilities in designing, analyzing, and operating biochemical processes.

Education was a centerpiece of the BPEC. Complementing a training grant from the National Institutes of Health for interdisciplinary predoctoral candidates, the center offered cross-disciplinary courses to graduate and undergraduate students and also provided wide-ranging instruction in biotechnology processes to the industrial sector. Through seminars and ongoing contact with researchers pursuing a broad range of concepts and projects, graduate students came to understand the research paradigm in a way not possible outside the center. The undergraduate education component, a "gem" of the center, capitalized on MIT's Undergraduate Research Opportunity Program, which gave students independent research opportunities under BPEC auspices. And everyone got to meet the numerous international and industrial visitors who streamed through the highly visible facility.

The industrial component was significant, even though a large percentage of the affiliated industries maintained a passive, if not "dormant," relationship. During its first year, four visiting engineers from industry pursued their own research in the ERC laboratory, each for three months or longer. Lecturers from industry helped teach two courses. Fifteen firms directly sponsored sixteen research projects at MIT, which were worth $1.5 million. Industry also donated $770,000 worth of state-of-the-art equipment and $2.4 million for constructing a fermentation and downstream pilot plant. MIT formalized the relationship with an industrial consortium, whose members paid annual subscription fees, but technology transfer was still the most difficult and weakest link for this ERC. Information exchange took place through symposia, on-site seminars, journal publications, and theses, but many companies were still reluctant to take up generic technological advances if more applied laboratory work remained to be done to make the process commercially viable.[29]

The combustion and biotechnology engineering research centers, pursuing widely differing technologies with different objectives in schools of differing complexions, shared the overarching goal of achieving breakthroughs relevant to American economic competitiveness. Both employed cross-disciplinary teams of academic and industrial researchers. Both prepared the next generation of researchers through comprehensive educational programs. Both depended on advanced computer techniques, which they themselves continued to develop. Both enjoyed the high visibility and prestige so important to their success, although these benefits could sometimes work against them. The "hype," as well as their relatively fat purses, created resentment among faculty in research areas not included in the center. Students could be intimidated by the charged atmosphere of extraordinary expectation.

These Engineering Research Centers, like the others, were instituted from the top down. NSF cultivated these communities, not waiting for them to get together on their own, although their general research agendas were already partially formed. Suh stressed that while both he and Bloch pushed hard for them, neither thought that ERCs should become a major part of the Foundation in terms of budget, although some people made it out that way. Engineering schools spend $3 billion per year, compared with NSF's $150 million engineering budget. With such a small amount of money, all that NSF could do, Suh concluded, was change the direction of the $3 billion enterprise, as a rudder steers a ship. NSF hoped that creating ERCs would help universities change their own atmosphere and infrastructure to support effective change. Walker particularly applauded the funding of excellent but second-tier institutions, since ERCs there would have a proportionally larger effect.[30]

In September 1988, as the number of Engineering Research Centers approached the projected total of twenty-five, Bloch asked the National Academy of Engineering once again to assess the impact and potential of the eighteen ERCs funded thus far and the program as a whole. The ERC mission, responded the warmly positive NAE, was as valid and timely to U.S. engineering schools and industry as it was when first implemented. Moreover, "today we see the challenge to U.S. global competitiveness more clearly than we did in 1984." Then, mature industries, such as steel, machine tools, and auto manufacturing, were identified among the chief problems. Five years later, the technology-intensiveness of newer fields, such as factory automation, consumer electronics, semiconductors, advanced materials, optoelectronics, telecommunications, and supercomputing, made an ERC program not only desirable, but imperative, the NAE unsurprisingly concluded.

If nothing else, the emulation (not always as successful) of the ERC model by other programs in NSF and by other agencies, such as the National Aeronautics and Space Administration, National Institutes of Health, Department of Transportation, and National Institute of Standards and Technology, testified to its appeal and merit. At industry request, ERC universities instituted new advanced-degree programs in fields such as interfacial engineering, systems engineering, and biotechnology process engineering. New, previously impossible and mutually reinforcing linkages were formed between academe and industry, although the NAE committee called NSF "overzealous" in seeking industry's financial support when it should be concentrating on its intellectual participation. Overall, the report pronounced the ERC program "outstanding," a "major asset to the nation."[31]

The NAE committee, including several who had served on the original Compton committee of 1983, did have concerns, however. The members underscored the idea that the ERC program should "not be diluted with new directions or eroded by being merged with other programs," and they emphasized that it was more important for engineering research and education to be linked to engineering practice than that it happen through centers. They recommended improvements to the proposal process, especially a preproposal mechanism to ease the costs to all parties, and warned against micromanagement.[32]

Addressing an oft-mentioned complaint, one seen in Utah, the academy reviewers observed with regret that in constant dollars, funding for individual engineering research had declined by 13.4 percent between 1984 and 1988, while the ERCs' support had grown—a problem of particular consequence at institutions with ERCs. While showing such favoritism was understandable, "this practice should stop. In effect, it penalizes faculty members, students, and institutions that support ERCs and exacerbates tensions between individual investigators and ERCs." Even within NSF there were tensions between staff involved with the ERC program and the rest of Engineering. The committee "urge[d] NSF management to strive to incorporate the ERC program into the long-range thinking of NSF program personnel."

A different NAE concern about money was that budget deficits had caused funding for Engineering Research Centers to fall "substantially below that originally envisioned with adverse effects of many kinds." As a result, fewer new ERCs were created each year, and worse, the average award size in succeeding years had dropped sharply after the initial grant. The academy deplored NSF's decision to underfund existing ERCs in order to set up new ones. The Foundation's failure to meet the financial expectations it had created forced ERC managers to sacrifice equipment and instrumentation needs in order to meet salary commitments, which could only threaten the viability of the ERC concept. Recognizing the reality of budgetary constraints and the difficulties of making choices, the NAE still insisted that prospering ERCs in fewer areas of technology would have greater impact than a wider range of undernourished ones.[33] (However, when Foundation Engineering Centers staff put the question of choosing deeper funding or more new centers to the engineering deans of ERC institutions in January 1991, they surprisingly chose the latter, citing the excellence of the ERC model and the importance of ERCs to the national interest.[34])

In conclusion, the National Academy of Engineering committee simply urged that the ERCs "stay on course. Fulfill the program objectives laid out five years ago. Ensure that the program reaches *at least* the size specified in the original report. Preserve the distinctiveness of the program, and resist attempts to dilute its mission by subsuming it within other programs." NSF should especially not try to reshape the program to NSF's more familiar traditional management style. For its part, Congress must meet the original ERC funding targets as a "cost-effective investment toward the long-term technological strength of U.S. industry." Finally, the academy thought the ERC program "precisely" the right tool for the federal government to assist industry in world competitiveness, for universities to build a knowledge base for industry, and for industry to help shape the policies and share the research that affected its long-range well-being.[35]

Most of the Engineering Research Centers have prospered and continue to make significant advances in their varied technical areas. That being so, what makes a good ERC good? Preston credited the vision and focus of their leaders. Agreeing, Walker said "the man who runs it" was the determining factor—an engineer with bright ideas, who could find the right people and the money. Walker compared successful ERC leaders to "oldtime entrepreneurial engineers," those who built something the public could and would buy. Bruce Woodson, MIT graduate student and technical coordinator of its Biotechnology Process Engineering Center, thought it was the teamwork and the motivation that came from a shared vision. Whether faculty or students, those who understood and accepted the research center concept were those who benefited the most from it and contributed the most to its success. Branscomb stressed the importance of the university's ability to attract good people, resources, and attention. Suh said simply that in the final analysis, none of the quantitative scales designed to measure success, such as the numbers of students graduated or papers published, were relevant. If the ERC was good, "everyone will know. Everyone will want to go there for that kind of work."[36]

It is, of course, too early for analysis based on long-range perspective. Interestingly, all of the technically oriented participants who offered evaluative commentary judged the ERCs by relatively unmeasurable qualities. Suh is convincing, although counting the graduates or papers or outside dollars attracted is still the proof of "everyone" knowing and wanting to be there that justifies budgets. Engineering Research Centers, and their appraisal, continue.

CONGRESS INCLUDES ENGINEERING IN THE NSF ACT

While the Foundation was working to solidify the new engineering mission and launch Engineering Research Centers, Congress and the professional community continued to press for other ways to strengthen NSF's engineering program—especially that fundamental engineering research defined as "academic, nonproprietary, and generic." In June 1983 Penn State's Richard Cunningham added the collective voice of the National Society of Professional Engineers (NSPE) to those who argued, yet again, that for the sake of America's technological future, engineering must have factual as well as titular coequal status with science within NSF. That would mean doubling, at least, engineering's budget in two years. In fact, in 1983 the directorate controlled only $100 million out of NSF's $1.1 billion budget. At a minimum, Cunningham said, NSF should be retitled the National Science and Technology Foundation, with two coequal deputy directors. As an alternative, the NSPE reiterated the now familiar recommendation that the feasibility of an independent engineering foundation be "thoroughly studied."[37]

California representative George Brown, a former aerospace engineer and the perceptive chair of the House Committee on Science and Technology, could be counted on to listen to such arguments, for they were his own. In his view, the nation needed a far greater engineering presence due to the increasingly competitive world economy, and if the administration opposed creating a technology policy on ideological grounds and funded accordingly, he would force the issue piecemeal through the legislative process. So no one was too surprised when Brown reintroduced his National Technology Foundation bill in early 1984 as H.R. 4245.[38]

This bill quietly disappeared. (Many NSF observers think Brown never intended to pass an NTF bill but rather to apply its leverage to more immediate objectives.) However, on 9 February 1984 Brown, for himself and New Mexico Republican Joseph Skeen, introduced a bill to amend the National Science Foundation's organic act of 1950. H.R. 4822 would rename NSF the National Science *and Engineering* Foundation, add "engineering" to the name of an enlarged National Science Board (to ensure broader participation by engineers), and insert "and engineering" or "and engineers" as appropriate wherever "science" or "scientists" appeared. Thus, in explicitly providing for NSF's initiation and support of fundamental engineering research and education, the legislation would solidify in law the Foundation's broadened mission. As the Senate Labor and Human Resources Committee later put it, Engineering's organizational instability of recent years "demonstrates that the mere inference of legitimacy is insufficient." The proposed legislation would "cement the status of engineering as an equal NSF partner."[39]

Brown's bill engendered a vigorous response from the engineering community. The Engineering Education Task Force, which represented the NSPE, the American Society for Engineering Education (ASEE), the ASEE Engineering Deans' Council, and the National Association of State Universities and Land Grant Colleges, had already been meeting for six months to "forge a long-term legislative solution" to the problem of engineering's inadequate recognition. This ad hoc consortium supported changing NSF's mission, though not its name. While the Foundation *could,* under its current charter, adequately support engineering, Robert Kersten, engineering dean at the University of Central Florida, voiced engineers' long-held fear that without formal pressure to the contrary, engineering would remain "subordinate" to science. NSF's enabling legislation had to be modified to "build engineering an institutional home in the federal government." Donald Glower, the group's chair and the engineering dean at Ohio State, went further: "Some engineering deans feel the bias against engineers is so entrenched" that it would take legislation or a new agency to give remedy. Yet the engineers' task force thought it politically and economically unlikely that the government would create a new agency, given the current context of rising deficits and receding government. In the end that group preferred that science and engineering be served from one agency to emphasize their linkages in the interests of American competitiveness.[40]

Russell Drew of the IEEE, representing through the American Association of Engineering Societies over half a million engineers, testified before the House science subcommittee in late February 1984. Although most engineers worked in industry, he said, they championed the need for cutting-edge academic research and education. Drew supported H.R. 4822 in principle but took a dim view of NSF's proposed budget increase of $26.4 million for FY 1985. At 14.6 percent of the Foundation's overall increase, it simply was "not an adequate reflection of the priority that is needed for engineering research, which has been consistently underfunded by the NSF." Worse, the demands of several targeted programs rendered the funding curve virtually flat. Drew did favor the funding, separately, of new multidisciplinary programs along the lines of the Engineering Research Centers.[41]

For Branscomb, the engineering academy, the White House science office, Knapp, and others, Brown's ostensibly friendly proposal threatened what Branscomb called in an internal memorandum a "very important, but very fragile consensus" on NSF's mission in engineering. He wanted time to make the new policy work, not new legislation, which could provoke a "major public controversy" with the "battle lines drawn between science and engineering." He feared that scientists worried about their future budgets would "accuse us of moving the NSF into 'industrial work,' which in turn would excite the conservative business community to conclude we were implementing an 'industrial policy' that duplicates or intrudes on the prerogatives of private industry." A "Science and Engineering Foundation" was "a good description of what we want to be," Branscomb wrote, but he hoped to see NSF "earn the title before we wear it." Ever the negotiator, he also reminded his NSF colleagues that "the shift of political pressure from the National Technology Foundation to the renaming of NSF is REAL PROGRESS. Let us not be too hostile to people who can help us build that consensus."[42]

Nothing came of H.R. 4822 either, although, as was often the case, its substance became incorporated into NSF's annual authorization legislation. But when the House Subcommittee on Science, Research and Technology marked up the NSF budget for FY 1985 in mid-March, it adopted what was sometimes called the Skeen Substitute (for its nominal proposer) for Brown's engineering amendments. (Brown, his many admirers at NSF noted, seemed more interested in passing favorable legislation than having his name on it; at NSF, Brown's name was used.) The substitute explicitly included engineering in the Foundation's mission by adding references to "engineering" or "engineers" throughout the text of the 1950 NSF act, as Brown's amendment would have done. It did not expand the size of the National Science Board. It did not change the name of the Foundation.

Knapp agreed with the more modest wording changes but took pains to oppose a "major restatement" of NSF's mission, name, or structure "without a full debate and consensus in the scientific and engineering communities." He cited the Foundation's overall stability under its existing legislative authority and, in any case, assured Congress that with or without amendments, NSF would "continue to develop and strengthen— on the firm basis of merit—our engineering program." By his figures, engineering had enjoyed a 170 percent budget increase, "in current dollars," since the formation of the engineering directorate. Knapp's protests seem overmuch, since the substitute amendment contained no sweeping restatement; perhaps he was seeking to scotch any future action in that direction. He clearly reflected a conservative institution still reluctant to shift its time-honored basic-science focus. NSF was doing what it must, but not more.[43]

Threading his way carefully, Branscomb expressed to subcommittee chair Doug Walgren the National Science Board's belief that support of research and education in science and engineering at NSF should "go hand in hand." The NSB appreciated how advocates of a separate technology foundation had focused national attention on engineering, which needed "strengthening and redirection," he said. But NSF was "already embarked on this task" and could "achieve it more quickly at less cost and with less danger of raising barriers—already too high—between science and engineering." Branscomb hoped Skeen's substitute amendment would establish Congress's intent that NSF proceed on its course to ensure U.S. "preeminence in engineering" without "crippling" its crucial scientific leadership and that it would "help replace a sometimes divisive debate in the technical community with a sense of common purpose." He said the board considered the amendment a "helpful clarification" of mission and direction rather than a fundamental, disruptive change in authority. The legislative language, in fact, reflected the board's own new policy of respecting the differences in intellectual tradition and practice between engineering and science. (In May, NSF's Advisory Committee for Engineering unanimously applauded this new charter language, which recognized engineering's unique goals, professional structure, and approaches to problem solving.) Branscomb suggested some relatively minor word changes and repeated Knapp's warning that the board would oppose more sweeping changes in the NSF name or structure without a full airing of views.[44]

Branscomb's leadership was important. Sympathetic with engineering, he also appreciated basic science and NSF's traditions—substantively and politically—and took a broad, long-term, inclusive view. Energetic, prolific, perceptive, and politically sensitive, Branscomb understood and articulated the science/engineering relationship in thoughtful working papers and conciliatory discussion points that helped bring the board and other leaders, however reluctantly, to consensus on incremental change while at the same time preventing catastrophic change from being politically imposed.

A long parade of other partisans also offered testimony on the engineering amendments of 1984. While Robert Rosenzweig, president of the Association of American Universities, was wary and lukewarm (the bill would be "unobjectionable" if the new language would cause no change but merely reflect changes already made), most witnesses applauded the legislative compromise. J. Thomas Ratchford, of the American Association for the Advancement of Science, called the amendment not a license to do anything that NSF was not already doing but rather well-deserved, explicit recognition of the importance of engineering, whose ever new and broadening fundamental research could "not be forced into the mold of traditional science classification schemes." Glower, speaking for the Engineering Education Task Force, urged the changes as "essential ingredients of innovation and technological progress," which efforts NSF should lead. "We all support science," Glower insisted, "but we must remember that while science wins Nobel prizes, it takes engineering to win markets. We need both! . . . The time has come to institutionalize our commitment to engineering." As MIT's Myron Tribus, representing the National Society of Professional Engineers, put it, "While excellence in science is clearly necessary, it is not sufficient if we are to maintain and extend our technological leadership."[45] Robert White, in voicing his approval, was careful to include the firm statement that "support for basic scientific research on which advances in engineering are so dependent must be sustained"—an interesting reversion to language of an earlier era.[46]

Conspicuous in his contrary view stood Frank Press, president of the National Academy of Sciences. He considered tinkering with legislative wording unnecessary, since positive steps to address engineering's weaknesses—a directorate formed, a mission articulated, a research-centers initiative launched, a budget appreciably increased—had occurred under the existing charter. Press feared that change would "dilute" the National Science Foundation's fundamental mission, that it could ultimately lead to a "discipline of the month" series of amendments. Engineering would prosper at the expense of science, he wrote.[47]

When Press's statement appeared as an editorial in *Science* in April 1984, Branscomb wrote one to refute it, which was published two weeks later. He called the amendments "reasonable and constructive." The language would stop defining engineering as a scientific discipline, which pitted science against engineering in a "fixed-pie scenario." Engineering was more than science, and the work of engineers should not, he wrote, have to compete for support as science. "Such pressures in the past have hurt U.S. engineering, have hurt the economy, and have not helped science." If the engineering community became so frustrated with NSF that it turned to a national technology foundation, Branscomb warned, much of NSF's traditional political and budgetary support for science would be peeled off to the new institution promising faster results. Lost as well would be the "fruitful interchange" between science and engineering that was possible in a single agency.[48]

This exchange typified the opposing attitudes. Press's worries that growth in engineering would bring dilution and financial penalties for science had appeared every time that NSF considered adding to its scope. Branscomb, like other engineering champions, tried to show by defining the terms that science and engineering, though different, could and should prosper together. Indeed, while science proponents still rarely included engineering in their rhetoric, as if to diminish the subject by ignoring it, engineers continued to recite their grounding in the knowledge and methods of science. But now, celebrating their own interests and approaches, they claimed their own identity, their own prestige. In the challenge at hand, while Press had the strong force of NSF tradition on his side, Branscomb used the danger of far worse consequences to that tradition if engineering got left out. Press did not acknowledge that engineering's recent gains, like earlier ones, had come in response to threatened outside action, but both sides were keenly aware of the power of politics.

Electronics colorfully weighed in on the side of engineering, which it called "no threat to science" in its long struggle for financial and operational equity. "Without [engineers'] work, science would have no value that would justify the Government in spending so much money on it," but the "poohbahs of science have opposed such parity as a threat to their most important source of funding." According to the writer, many sciences depended on engineering, "pure" science being "no more useful an idea than phlogiston." Indeed, science would have much to gain if NSF gave equal weight to engineering, "for it is engineers, with their can-do bent and fine-tuned commercial antennas, who perform the cross-pollination that prevents science from withering away into irrelevance."[49] More moderate, and probably more helpful, was Walgren's view: "I do think engineers need more visibility. I don't look forward to a fight between engineers and scientists, but engineers do provide most of the base for jobs and economic growth."[50] They also provided a base for science. Even a physics professor, Clemson's John McKelvey, carrying *Electronics'* point and recalling Project Hindsight of the 1960s, admitted that although "the history of science and technology has by and large been written by scientists," it was true "now, as in the past," that "technology is driving science as fast as it is being driven by science." Engineers' contributions in complex instrumentation, in particular, often preceded and enabled scientific advance.[51]

The House of Representatives passed the substitute engineering amendments to the National Science Foundation Organic Act as a piece of the FY 1985 Authorization Bill on 25 April 1984, but the battle was not yet won. The Senate, whose internal jurisdictional squabbles had interfered with NSF legislation for several years, attached the NSF engineering amendments to the authorization bill for the National Bureau of Standards and added a Title III to establish centers for manufacturing sciences and robotics research. Reagan vetoed H.R. 5172, which included these three disparate provisions, on 30 October 1984. He objected to both the philosophical and fiscal implications of manufacturing research centers, which, he said, represented industrial policy and an "unwarranted role" for the federal government.[52]

Finally, both houses passed H.R. 1210, NSF's authorization for FY 1986, which included the engineering amendments of the year before. They settled their differences in conference by late October 1985, and with the president's signature on 22 November 1985, the bill became Public Law 99-159, with appropriation legislation following three days later. After thirty-five years, engineering at NSF had explicit statutory recognition; it had a name in the NSF Act.[53]

ENGINEERS REORGANIZE ENGINEERING

Senior editor Wil Lepkowski of *Chemical and Engineering News,* a longtime NSF watchdog, gave a sense of the charged atmosphere in July 1984. He observed a new spirit pervading the National Science Foundation—

the "spirit of industrial relevance." Noting that over the years NSF had swayed with the political winds to incorporate such programs as applied research, social science, and minority education, he wrote, "Again, the winds are whipping through Washington, howling words like 'engineering,' 'technology,' 'competition,' and 'Japan.'" Lepkowski saw NSF responding to the challenges of industrial competition with actions he credited to the "bouncy, idea-spewing" Branscomb. The Engineering Directorate, the Science and Engineering Education Directorate, and Engineering's new mission were early results. The fledgling network of Engineering Research Centers promised to be another effective thrust, while Congress underscored the new direction by writing engineering into NSF's organic act.

During that summer of 1984 the Foundation was virtually leaderless, with Knapp recently gone and several assistant directorships vacant, but now, wrote Lepkowski, engineers were "beside themselves with joy" that Erich Bloch, vice president of IBM and an electrical engineer whose experience was exclusively industrial, had been nominated as director of the Foundation. Not the administration's first choice, or even among its first dozen, Bloch would prove outspoken, effective, and enduring, even though the academic and scientific communities initially eyed him warily. Roland Schmitt, General Electric vice president for research and another engineer, had just become chair of the National Science Board, and MIT's Nam Suh was Keyworth's choice to lead the NSF Engineering Directorate—the first engineer named to that position. In Lepkowski's words, engineering at NSF had the "momentum of a freight train." Not since the RANN days of the early 1970s had NSF seen "such intense change in direction."[54]

Bloch and Suh, both activist by nature and leadership style, intended to stir further the already dynamic NSF engineering scene. They arrived at NSF on the same day in October 1984 and, in Bloch's recollection, "went around together to see what the Foundation was all about." Bloch thought that NSF, "shielded from reality," was "pursuing a research objective that didn't mesh with the agenda of society," which was to improve U.S. economic competitiveness. He and Suh "unanimously concluded that engineering was about the worst." Bloch's major goal was to "bring the Foundation closer to some of the mainstream issues the country was facing."[55]

Suh, who said he was "kind of naive" when he accepted the job, agreed with Bloch's assessment. He determined to reorganize the directorate—not, he insisted, because there is any inherent value in reshuffling, but because the administrative pattern must reflect the goals of the institution if those goals are to be met. The current engineering organization, mirroring the classical discipline-bound university system, could not, he thought, meet current national needs for cross-disciplinary work in emerging technologies. Although NSF engineers had long justified the existence of engineering by calling it engineering science, Suh thought that interpretation was damaging to engineering. Engineering design, for example, whether for mechanical devices or computer software, required thought processes different from, and beyond the understanding of, scientific principles. More attention to design was critical.[56]

Intending to stay at NSF for just one year, Suh set about the task of effecting change at once. After systematically talking over his proposals with every person and organization he could think of, in-house and outside, including the national science and engineering academies, engineering deans' councils, and professional societies, he developed specific goals for engineering:

- to continue to excel in areas where NSF already had a strong science base;
- to create a science base where one was lacking, such as in design and aspects of computer technology;
- to create an infrastructure within the universities for emerging technologies, such as biotechnology;
- to support research in critical technologies needed by industry, such as robotics for manufacturing;
- to implement the Engineering Research Centers concept and the systems approach in general; and
- to promote improved engineering education.

Within the space of about two months, the new assistant director for Engineering established separate divisions to accomplish each of these goals, with each division designed in its own way to best meet its particular needs and objectives. Even unsympathetic observers admired Suh for "caring about engineering education," the express responsibility of another directorate, "before it was fashionable." The reorganization withstood the test of time, Suh later thought. But as goals change, the organization must change, too, and should continue to evolve.[57]

Bloch thought the engineering reorganization succeeded in exerting its desired influence beyond the Foundation, especially on academic engineering. "Money talks," he said, "even for academics, so I hear." In another context, speaking on the issue of economic competitiveness, he called American universities "a major national resource, and the reason why our research and technology are the envy of the world. But the very strength of the universities makes them resistant to change—probably more so than any other sector. Industries change when the market says they must, and governments must answer to the electorate—at least once in a while. But universities are remarkably autonomous. They have to be talked into changing. I would not have it otherwise, but it is time to do some talking." Reconfiguring NSF's engineering program was one way to direct the dialogue.[58]

That was the power figures' version of what happened. Some careerists of the NSF engineering staff, who had never seen such a tempest as Suh created and Bloch endorsed, had another. "He came in like a bull in a china shop," said one veteran senior staffer. Suh gave the impression that only he, the outsider, knew what needed to be done. He would "clean the place up," just as satisfied traditionalists were getting the directorate stabilized and moving forward. Instead, he "tore it up." With "arrogance" and whirlwind activity, Suh "needlessly trod on corns." He did talk to everybody, but privately and in disregard of hierarchical lines; it was "an open door in the worst sense." Morale among some of the professional staff plummeted as people felt mishandled and programs were jerked about. For example, Suh made all the engineering division directors "acting," giving them each three months to prove their worth. While the reorganization accomplished a small awakening within the research community, the price paid for it was too high, thought those who were unhappy. Still, even those with sore corns warmly respected Suh's tireless hard work, strength of conviction, and solid reasoning.[59]

For his part, Suh both valued the friendships he made at NSF and acknowledged that "some people are still mad at me," some understandably, because they had money taken away. He continued to believe that "we did the right thing." "Somebody had to say it, somebody had to do it." He listed as achievements the large number of professors and researchers now in the fields of design and manufacturing engineering: "We literally created a community. We created a philosophy of professors working with other professors." Students and faculty were learning how real engineering was done. But there was still a long way to go, Suh concluded.[60]

Indeed, the rate of change had been dizzying. By mid-December 1984 Bloch was ready to justify a new engineering organization and program structure to members of Congress on the Appropriations Committee. He said the new plan would help expand the knowledge base in engineering, create a new knowledge base for engineering systems, promote cross-disciplinary research and education, and strengthen fundamental research in areas important to national security and economic competitiveness. In January 1985 the Foundation went public with the reorganization, implemented to sustain U.S. leadership in engineering research and to "insure that our engineering schools produce the best engineers." Bloch said the new design, following the mission statement adopted by the National Science Board in August 1983, would provide a better balance between established engineering disciplines and those just emerging; it would create stronger links between academic engineering and industry.

Three of Engineering's new divisions continued to serve major engineering science fields along traditional disciplinary lines: Chemical, Biochemical, and Thermal Engineering; Mechanics, Structures, and

Materials Engineering; and Electrical, Communications, and Systems Engineering. The Division for Science Base Development in Design, Manufacturing, and Computer Engineering addressed areas of great need that Suh and others had identified, while the Division for Fundamental Research in Emerging and Critical Engineering Systems took up research in subjects from earthquake hazard mitigation to optoelectronic computing. The primary responsibility of the Office of Cross-Disciplinary Research was oversight of the new Engineering Research Centers.[61]

In addition, during his tenure, which in the end exceeded three years, Suh inaugurated creative programs to reward high-payoff risk and innovation, as the National Academy of Engineering had recommended. He established a pilot project encouraging engineering program directors to select for direct funding a few small, long-shot "gems" among proposals declined by peer reviewers, who "tended to endorse orthodoxy." Suh's view was that if every funded project succeeded, the Foundation's funding guidelines were too conservative. This general idea became institutionalized as the Small Grants for Exploratory Research program, which provided a "fast track" for small, short-term creative proposals. Program directors kept in reserve 5 percent of their budgets to fund such meritorious proposals on their own, bypassing the cumbersome external peer review process. Suh won credit for the successes gained from a willingness to stake good hunches.[62]

The National Academy of Engineering, asked in early 1985 to evaluate the new engineering structure, reported being "impressed with the seriousness" of the Foundation's self-examination and the "boldness of its plans for improvement." The NAE committee unanimously endorsed the premises underlying the reorganization of engineering, especially its innovative new programs, such as Engineering Research Centers, but also emphasized that NSF "must seek with special sensitivity to achieve a good balance between the twin motivations of potential utility and intellectual endeavor." The members further stressed the need to balance the quest for breaking new ground (those "moments of extraordinary progress") with "systematic deepening and strengthening of the bases of established disciplines."

At this time of Foundation transition and economic urgency, the NAE reviewers thought it imperative that NSF's Engineering Directorate "receive a clear charge and the resources to realize its objectives." Its budget, and thereby its leverage, required substantial expansion. They proposed separate operating and capital budgets for engineering, the first to grow from $142 million in FY 1985 to between $350 million and $410 million in five years. NSF should also "take a leading role in meeting what we assess as a billion-dollar national problem of capital needs for academic engineering research equipment and instrumentation"— providing more than $30 million per year over five years. They chided the Engineering Directorate for responding to the "pressures of small budgets and large responsibilities by growing cautious, and even conservative, in its commitments." Suh, who had persistently advocated a $500 million budget goal for engineering to force a focus for discussion, welcomed that kind of support. The engineering academy concluded that while NSF alone could not ensure the country's world leadership in industry, its recognized prestige, integrity, and visibility were pivotal in meeting the challenge. NSF was "now positioned to become truly a national foundation for science and engineering, by whatever name."[63]

In 1986 the National Academy of Engineering formed a committee to identify critical issues meriting federal support for engineering and technology, including education, to assist the work of a House science policy task force. Mindful of the administration's opposition to technology policy, the committee members agreed that it was "time for national policy to emphasize the critical role of engineering—alongside scientific discovery—in building U.S. technological advantage." They urged the federal government to support an "appropriate mix of research opportunities"—both individual (especially Presidential Young Investigator awards) and collaborative (especially Engineering Research Centers). The government should encourage the professional development of engineers with inducements to pursue the doctoral degree and provide university faculty with sufficient research opportunities, equipment, and facilities. Among the committee's many familiar recommendations stood out two that they considered most important: better communication and long-term cooperation among researchers across all sectors (academic, industrial, governmental) and stability over time of federal support.[64]

Although in 1986 the Engineering Directorate developed a strategic plan for the 1988–92 period along the same lines that figured in the reorganization of 1985, change within Engineering continued. In May of that year the Foundation formed the Directorate for Computer and Information Science and Engineering, which left a renamed engineering division, Design, Manufacturing, and Computer-*Integrated* Engineering. Established within Engineering in June, the Office of Engineering Infrastructure Development gave greater attention to the human resources problems. But all this was fine tuning.

Suh's successor, John White, an engineer from the Georgia Institute of Technology, moved programs around again in February 1989—to consolidate and reduce overlap, he said, and to encourage interdisciplinary and cooperative engineering, stimulate the generation of emerging technologies, and distribute the directorate's workload more equitably. He promised that no research topic supported under the previous structure would suffer in any way. White's divisions were Chemical and Thermal Systems, Electrical and Communications Systems, Mechanical and Structural Systems, Design and Manufacturing Systems, Biological and Critical Systems, and Engineering Centers. The changes were subtle, but these names, even with the clear emphasis on systems—a broad, integrative approach—would have a more familiar ring for traditionalists.[65] No one assumed, however, that a steady state had been reached.

So on the eve of the fortieth anniversary of the National Science Foundation, with forceful leadership and a clear mandate, Engineering was more prominent than it had ever been, if still a disappointingly small "equal." Engineering Research Centers, NSF's most visible, dramatic, systematic engineering program, focused on problems key to regaining global economic leadership. Congress had ensured the statutory legitimacy of engineering within the Foundation. And the directorate had survived, and perhaps was stronger for, the controversial, ambitious leadership of its first engineers.

Throughout, a current of wariness and dissent continued to run through NSF and its academic constituencies as engineering, perceived by the basic-science old guard as an intruding, alien culture, gained in status, visibility, and fortune. Forced by political pressures, which were in turn driven by economic and social imperatives, to accept engineering or lose it, the Foundation prudently, if not happily, chose the former course.

For their part, engineers had succeeded in proclaiming and validating their own culture and traditions. First they had to reshape their own identity, an ongoing process since the nineteenth century, and then win acceptance of that image from others. It seemed that by the late 1980s, engineering and science were coexisting in NSF relatively peaceably and cooperatively as academic scientific and engineering research had become increasingly difficult to distinguish.[66] Future change was certain, of course.

CHAPTER 9

NSF ENGINEERING OVER TIME

> Our vision is a U.S. society whose quality and vitality are
> recognized as preeminent in the world because of our effec-
> tive application of technology through engineering.
> —Directorate for Engineering Vision Statement, 1990

Policies and priorities as well as technical challenges continue to change—
although for engineering at the National Science Foundation, the internal turmoil of the late 1970s and early 1980s stabilized during the next
decade. Engineering emerged stronger, more assertive of its intellectual
base, its methods, and its contributions, past and potential. Themes endure
and evolve, always in the context of the reality that shifting political
needs determine, through funding and legislative fiat, much of the agenda.

TECHNICAL CHALLENGES OVER TIME

As ever, the changing political, philosophical, and economic climate
nudged research directions and relative levels of support for various
programs at the National Science Foundation. Sometimes innovative,
even revolutionary, efforts emerged from the organizational and conceptual jockeying. RANN and the Engineering Research Centers were
thus born. But a great deal of established and valuable engineering re-

250

search, such as the seeking of fundamental understanding that was easily identified as engineering science, proceeded as it had for years, sometimes under varied jurisdictions. Some current challenges involved work once in the limelight but now less noticed, less glamorous. Sometimes new breakthroughs reintroduced interests lying fallow for want of a critical link. Sometimes old engineering problems reappeared after periods of quiescence, when their solutions proved impermanent. Improved computer capabilities made the solution of previously unsolvable problems possible.

The following examples illustrate these situations. Beyond their intrinsic appeal as intriguing technologies or applications of technology, they all represent long-standing NSF concerns that have experienced a resurgence of interest. They show both continuity and change in NSF engineering support as well as increasing levels of engineering sophistication as new achievements have built on the old. They all touch basic-science research as well, perhaps as good an argument as any for maintaining one, inclusive Foundation. And they all have felt the impact of fickle federal policy and funding support.

Deep Ocean Drilling. Engineering challenges at the bottom of the sea experienced quiet progress continuing years after an embarrassing, noisy beginning in the harsh glare of unfavorable publicity. It was true that the costly, mismanaged Project Mohole of the 1960s had besmirched the reputations of those who touched it while failing in its primary goal of penetrating the earth's mantle. Still, Mohole's pioneering work proved the feasibility of deep-ocean drilling. Even at the time, engineers developed a dynamic positioning system that kept a semisubmersible drilling platform in place in the open sea and succeeded in retrieving test cores that yielded important geological information heretofore only indirectly or speculatively understood. Other engineering feats for this pursuit in basic science included the design, fabrication, and/or testing of a coring turbodrill, a prototype "revolutionary retractable diamond coring bit," a deep-ocean untended digital data acquisition system, and a sonar-based method of hole reentry.

After Congress specifically forbade the further funding of Project Mohole in 1966, a consortium of marine research centers formed, with NSF encouragement, the Ocean Sediment Coring Program. Often called JOIDES, for the Joint Oceanographic Institutions for Deep Earth Sampling, or sometimes just the Deep Sea Drilling Project (DSDP), the program was organized to drill beneath the sea floor in the earth's upper crust. It became international in the 1970s with the inclusion of the Soviet Union, the Federal Republic of Germany, France, Japan, and the United Kingdom. Drillers discovered oil deep under the Gulf of Mexico and, as early as 1969, successfully deployed a ship-based drill string more than 19,000 feet long. Using the rig *Glomar Challenger* and later the government-owned *Glomar Explorer,* JOIDES made over six hundred drillings in the Atlantic, Pacific, and Indian Oceans by the late 1980s.[1]

The NSF-supported Offshore Technology Research Center at Texas A&M University at College Station and the University of Texas at Austin, started in 1988, utilizes and extends such drilling technologies. Researchers at this Engineering Research Center pursue analytical and experimental research to improve the structural integrity of deep-water systems, including new concepts for sea-floor foundations, which operate in continuous tension rather than the normal compression. Investigators are also working to understand hydrodynamic forces in deep water and to develop higher performance materials, such as fiber-reinforced polymers, for use in the hostile saltwater environment. Most of the major U.S. energy and offshore engineering companies support and benefit from the ERC. Thus, an extravagantly ambitious, esoteric goal has spawned broad participation in related programs of more immediate application, and this practical research in deep-ocean drilling may yet provide the keys to unlocking the mysteries of the Mohorovicic discontinuity between the earth's crust and its underlying mantle.[2]

Superconductivity. Dutch physicist Heike Kamerlingh Onnes discovered the intriguing physical principle of superconductivity in 1911 when he observed that the electrical resistance of a mercury rod dropped precipitously to zero when cooled to near absolute zero. It remained an astonishing, little-known theoretical oddity for several decades, although a few researchers sought to find superconducting materials ever better at carrying large currents per unit area while remaining superconducting under intense magnetic fields. After high-field superconducting magnets were developed in the early 1960s, the National Science Foundation began funding superconductivity research under the rubric of low-temperature physics, a field that electrical engineers quickly adopted. Investigators made important progress by the mid-1970s toward developing superconducting cables for power transmission and superconducting magnets for particle accelerators, levitated trains, energy storage, electric generators, and electron microscopy. But federal funding for many of these expensive and futuristic technologies shriveled in the face of waning energy demand following the 1973 Arab oil embargo, despite their long-term commercial and environmental potential. Most of what practical research continued was pursued abroad, primarily in Japan and Germany.[3]

From the beginning, the research prize was seen to be new materials or alloys that were superconducting at higher temperatures, thus reducing the prohibitive costs of refrigerating to extreme cold. When IBM researchers in Zurich reported their discovery in 1986 of a ceramic compound that was superconducting a few degrees higher than any previously observed, interest in the quest for "room-temperature" superconductors surged. Early the next year NSF-supported researchers at the Universities of Houston and Alabama produced their own ceramic compound, which was superconducting at more than one hundred degrees Fahrenheit higher than before. Readily available, cheap liquid nitrogen could cool these materials to their superconducting state, which would facilitate applications of superconductors at substantially less cost than with liquid helium cooling.

In 1987, with Reagan's blessing, NSF doubled its support for superconductivity research to "ensure U.S. readiness to take the lead in commercial applications of the new research." The new figure, $10 million, included $1 million for additional research at three materials research laboratories and for rapid start-up grants for engineers working on processing new superconducting materials. By 1990 that figure was $29 million, but researchers still complained that the Foundation was feeding a few large institutes and laboratories at the expense of growing numbers of individual investigators eager to pursue the "tremendous potential of the materials and the great mysteries that still surround how superconductors work." As usual, it was politically easier to pry money from Congress for big, showy projects than for lone researchers. The controversy over the appropriateness of such a policy goes on within the larger question of the wisdom of having allowed this particular well-established, promising research to languish for several years. Whether the reinvigorated activity in superconductivity will result in significant permanent applications or prove only another fad remains to be seen.[4]

The Nation's Infrastructure. Once-dramatic but now-aging engineering innovations sometimes demand revisiting. Such was the case when by the mid-1980s engineers observed with alarm the state of the nation's infrastructure, that "public capital" of highways, bridges, transit systems, water supply and waste management systems, communication and power systems, schools, and other public institutions that enable society to function. A significant portion of currently used engineered systems came on-line during the great building and investment era of the 1950s and 1960s. Urban freeways and the interstate highway system, for example, spread a network of roads over the country. With multiple lanes, limited access, and restricted grades and curves, they changed the nature of American overland transportation. Dams and bridges also proliferated.[5]

    A key contributor to this civil engineering heyday in big construction was prestressed concrete, pioneering work on which was begun in the 1920s by France's Eugene Freyssinet and other Europeans. The first American bridge using this material, which opened in 1951 over Walnut Lane in Philadelphia, predated NSF involvement, but the Foundation gave early and generous support to basic research in prestressed concrete. Like stone, concrete is strong in compression but weak in tension and will crack under comparatively small loads. Adding steel reinforcement to concrete members in areas subject to tension allowed widespread utilization of concrete in the twentieth century. But if embedded passive reinforcement could create what amounts to a new construction material, the properties of this reinforced concrete could be greatly improved by prestressing the structural members, that is, by stretching the reinforcement before loads were applied to the structure. This could be accomplished by applying external loads through jacks or restraining the expansive concrete in molds while it set. Prestressing also eliminated problems of shrinkage and cracking and reduced creep. So a highway using prestressed concrete had greater durability and load-bearing capacity for a given thickness than conventionally reinforced slabs.[6]

One early NSF-supported project was basic research on the torsional strength of prestressed concrete at the University of Florida in 1957. Ralph Kluge directed experimental and analytical investigations of the effects of torsion, shear, and bending loads on increasingly slender prestressed concrete structures. In 1969 the Foundation funded Ned Burns of the University of Texas at Austin to develop a rational numerical method to reliably predict behavior for continuous precast-prestressed concrete structures at their joints. He also experimented on factors influencing the behavior of such joints, work he had begun under an NSF Research Initiation Grant. Using the results of such work over many years, the American Concrete Institute prepared new building codes that, with new highway codes of practice, encouraged the introduction of prestressed concrete into the building industry.[7]

When the National Academy of Engineering's Committee on Public Engineering Policy (COPEP) evaluated NSF priorities for applied research in 1969 (anticipating the RANN era), it noted the lack of established descriptive or quantitative techniques for characterizing the behavior of systems over time—such as when buildings settle, bridges crack, or highways crumble. COPEP emphasized the necessity for developing adequate methods for testing existing systems. It was not enough to know when new structures should be built; knowledge of the magnitude, frequency, and probability of forces that wear a system down could reduce risks to users and make possible more timely renovations or gradual phase-outs. But during the 1970s, a time of tighter budgets and greater public concern for social relevance, the "performance criteria" of the built environment received little attention. Rather, the money went to ancillary problems of earlier infrastructure development, such as noise, air pollution, and traffic flow, although classical basic work on such phenomena as corrosion, fatigue, and fracture of materials under stress continued—and increased as the computerized method of finite element analysis became more sophisticated.[8]

By the 1980s, decades of neglecting America's infrastructure had left it dangerously deteriorated. Engineers estimated that nearly half of all U.S. bridges might be unsafe. More than 9,000 small dams posed a serious threat to public safety. Someone claimed to have counted 56 million potholes, which would require 20 million tons of material to fill. By the year 2000, about 7.5 billion gallons of gasoline would be wasted because of traffic congestion alone on roads inadequate in capacity or wanting in maintenance. When it cost $1,700 per year for one truck driver forced to detour five miles a day because of one closed rural bridge, or, more dramatically, when commuters in Los Angeles County daily wasted 485,000 hours on clogged freeways, it was not difficult to extrapolate the toll on national productivity. From there it was a short leap to see the consequences to American competitiveness in the global market.[9]

But infrastructure maintenance had difficulty gaining a hold on the public purse, since problems were slow to develop or hidden within a structure in mundane daily use. Some other crisis would always appear more obvious, urgent, or expedient—even though "chronic underinvestment" and delaying attention to these systems would cost more money in the long run than was saved in the short. Nor was such inattention new: "Spending money unstintedly for construction, often under the supervision of the best engineers the country affords, and then being niggardly in maintenance and operation appropriations and leaving costly and perhaps complicated works to run themselves except for political heelers or lame ducks is the rule rather than the exception in many if not most American cities," complained *Engineering News* in 1917. In the 1960s the net national, state, and local government investment in public works averaged 2.3 percent of the gross national product. In 1984, when the Joint Economic Committee (JEC) of Congress took notice with hearings and *Hard Choices: A Report on the Increasing Gap between America's Infrastructure Needs and Our Ability to Pay for Them,* this figure was 0.4 percent. But to monitor, repair, retrofit, or reconstruct leaky water mains or unsafe airports could cost, by some estimates, $1 trillion by 2000, or $3 trillion over twenty years. The national infrastructure study, led by Henry Reuss, a former Democratic representative of Wisconsin and chair of the JEC, called the problem a "manageable crisis" if there were federal cooperation, policy-setting, and financial assistance to state and local governments, which bore the primary responsibility. It recommended a national infrastructure fund to supplement existing resources. Despite some strong support, funding such restorations was not popular as a budget item in the 1980s, even though federal spending for public works, at least new ones, enjoyed historical precedent from the earliest days of the republic.[10]

NSF engineers saw their clear and crucial role in the infrastructure challenge as ensuring that the research community, especially in academe, had the charge and the wherewithal to identify and solve the technological problems inherent in improving the design, construction, and cost-effective operation of public works. Engineers could "illuminate the tradeoff between performance and cost," which would help decision makers establish workable standards for building, maintenance, and, especially, rehabilitation. In 1983 funding of research by all sources amounted to only about 0.2 percent of total spending on the infrastructure, not more than one-sixth of which was for basic research. NSF's Engineering Directorate established in February 1984 the Construction Engineering and Building Research Program to concentrate specifically on problems of the infrastructure and productivity. Any part of this program might have received funding under existing criteria, but the new organizational unit guaranteed attention to these increasingly urgent problems and also tacitly acknowledged the need to upgrade U.S. efforts, especially when rivals such as Japan were pouring resources into construction and manufacturing engineering.[11]

In recent years, one specific engineering thrust to deal with infrastructure concerns has been the development of various techniques of nondestructive evaluation, again a field where Europeans were at least a decade ahead. Being able to predict with some assurance the remaining safe life of, say, a bridge—without pushing it to its limit and without being able to "see" slow internal deterioration taking place—has important economic as well as social consequences. Historically, bridge painters provided the most reliable observations of a bridge's structural health, as they saw crack patterns or other anomalies change over time. Now there are promising new techniques such as surface wave spectral analysis (to test a building material's compositional integrity), vibration analysis (to recognize its normal acoustic "signature"), and advanced finite element methods (to show how excessive deflections in a material over time could signal impending failure). The National Science Foundation keenly supports research in these and other methodologies for understanding a structure's soundness, within the budgetary constraints imposed by a continuing national reluctance to pump scarce money into unglamorous, sometimes invisible, problems.[12]

DRIVING FORCES

Over NSF's first forty years, engineering advances occurred within the ordinary patterns of day-to-day operations as well as under innovative new modes, such as the RANN program or Engineering Research Centers. But always present were the driving forces of external, societal demands as expressed through the political system. In the beginning it was the memory of the impact of science and technology on World War II and then the engulfing cold war. In 1957, *Sputnik* shook the Western world, not with a direct military threat but the technological power it symbolized. NSF's research budget skyrocketed in the name of national security, although engineering was a relatively minor player. In the volatile period of the late 1960s through the 1970s, Congress loosened the purse to ease domestic distress. NSF responded to "national needs," and the engineering community invested heavily in trying to save the environment, improve urban living, and provide more and cleaner energy under the controversial and short-lived RANN program.

By the 1980s the driving force was global economic competitiveness, and its impact was felt particularly in engineering. As National Science Board chair Roland Schmitt put it in 1986, "Engineering Research—the region where the leading edge of research meets the cutting edge of application—is becoming more than ever before the key battleground of international competition."[13] True, there were countervailing cooperative international programs in such areas as ocean drilling, earthquake hazard mitigation, and polar research, which, though comparatively minuscule, had generated significant results over many years. International conferences and bilateral exchange programs also contributed to knowledge and understanding in specific technological areas. But competition in the global marketplace was ever tougher and more prevalent, its consequences more serious, and NSF had no focal point from which to address it.

Recognizing the Foundation's need for a more systematic approach to international concerns in engineering (as it recently had addressed for science), the National Science Board's Committee on International Science asked the National Academy of Engineering in 1984 to identify relevant issues. H. Guyford Stever, now the NAE's foreign secretary, wrote the academy's response. Lauding NSF's successful collaborative efforts in earthquake engineering as a fine example, it suggested that the agency select a manageable number of similarly appropriate technical areas, such as automated manufacturing, "reengineering of the infrastructure," and telecommunications and information, for international cooperation and reciprocal exchange of information in various formats. Although there were "substantial differences" between and within academic and industrial engineers, the NAE considered one problem to be American engineers' general lack of knowledge about foreign engineering progress. Their historical disinterest seemed traceable, at least in part, to a lingering "strong, conservative impression of superiority."[14]

It had become obvious that circumstances no longer justified such attitudes. While there was a reluctance to acknowledge it, the U.S. economic dominance of the post–World War II period had passed. Other countries were growing in both technological strength and market share, adapting with alacrity the results of American basic science and engineering research to their commercial advantage, while U.S. policy restraints and engineers' inadequate grasp of other languages and cultures—especially Asian ones—further disadvantaged American industry. A follow-up National Academy of Engineering study in 1987 urged, with specific recommendations, that U.S. engineering education address international developments, that U.S. engineers become more involved with their global counterparts, and that NSF establish mechanisms and increase funding to support these goals.[15]

The story of engineering's contribution to U.S. economic competitiveness remains unfinished, but a key component requiring mention is the role of leadership. Just as Andrey Potter and Eric Walker had pushed the respective engineering agendas of their day, during the 1980s NSF director Erich Bloch, engineering head Nam Suh, and successive National Science Board chairs Lewis Branscomb and Roland Schmitt understood and acted on the overarching competitiveness issue. Engineering saw prosperity and visibility (as well as tension and conflict) under the purposeful direction of these leaders. The clearly defined and heavily promoted Engineering Research Centers were only their most dramatic effort to address the competitiveness crisis during these years. By contrast, the beleaguered RANN program never enjoyed the united support of NSF leaders, who in turn were never able to agree on its mission. As a result, RANN strayed, or got yanked, too fast and too far beyond NSF's institutional culture and could not be sustained. The ERCs' future appears more promising.

Congressional leadership also fundamentally moved the National Science Foundation's engineering efforts. Representative Emilio Daddario's amendments of 1968 permitted NSF to support applied research, which was broadly seen as engineering-dominated, although the resulting RANN program went beyond Daddario's imaginings. Representative George Brown successfully prodded NSF for years to grant engineering more clout with his oft-proposed National Technology Foundation bill, without ever making a concerted effort to pass it. When Representatives Don Fuqua, chair of the House Committee on Science and Technology, and Doug Walgren, who was in charge of the Subcommittee on Science, Research and Technology, led the move to include specific engineering language in the NSF Act in 1985, it was with the careful cooperation of NSF and other engineering leaders and only validated what NSF had already been doing. The subtler approach seems to have won more friends, promising a greater likelihood of long-term benefit for engineering within the Foundation family.

At the most elemental level, the external force driving NSF and its components, including engineering, was politics, which itself reflects societal demand. It had been so since Truman vetoed the first NSF bill, which would have put scientists in charge (where a different kind of politics would have operated). NSF's freedom to pursue basic-science research with relatively little attention lasted only during the smallness of its first fifteen years; as budgets rose, so did political pressures. Members of Congress and executive decision makers became adept at tying appropriations for NSF research to specific outcomes they or their tax-paying constituents identified as beneficial. For its part, NSF learned to couch its budget justifications in the political language of the hour—to "ensure national security," to "improve the environment," to "increase economic competitiveness." Today, a larger, wealthier Foundation feels politicians' muscle intensely. Engineering, which is inherently closer to the real world and its problems, sometimes profited more visibly from these outside pressures than basic scientists pursuing the acquisition of knowledge—just as those trying to conquer specific diseases did at the National Institutes of Health. It followed just as naturally that those representing basic science, NSF's first and primary mission, would fear and resent the perceived competition from engineers as well as political meddling with their institution and their research choices. But they have had their friends in Congress, too. Looking at the size and structure of the federal research budget over thirty years, NSF's 1990 study,

The State of Academic Science and Engineering, concluded that "basic research and academic research" were the only categories that had "increased appreciably (after eliminating the effects of inflation)" since 1967. Overall federal funding of engineering had "declined markedly."[16]

CHANGE WITHIN

Within the engineering profession, tools and terminology were ever changing. The computer, with its own phenomenal growth in capability and usability, has become *the* ubiquitous, essential tool for any kind of technical research. A trade-off noted by "one metallurgical engineer" was that "the half-life for knowledge seems to be about seven years. In other words, seven years from now, half of what you know is going to be worthless." (Others shrank that time estimate appreciably.) It was almost beyond reckoning how quickly and totally engineers had moved from reliance on slide rules and prototypes to finite element analysis and computer-aided design and manufacturing.[17]

The systems approach—drawing upon engineering's traditional core disciplines to view parts of a problem in terms of a broadly defined whole—has increasingly captured the attention of NSF and the engineering community. The excitement of electrical engineers in 1962 over the development of a large-scale integrated system for managing airline reservations to determine passenger space available at any given time, compute fares and itineraries, and issue tickets seems almost quaint today.[18] Over the years engineers increasingly appreciated the need for that broad view in both their work and their institutions. Systems thinking, already the dominant mode, became institutionalized in Engineering director John White's 1989 internal reorganization plan. The name of every division reminded its constituents of its focus on systems. Interdisciplinary research is likewise not new but remains in the forefront of engineers' thinking. Proficiency in just one traditional discipline is simply no longer sufficient for solving the complex problems of society. The Engineering Research Centers brought all these ideas together in the name of productivity and competitiveness, but even smaller efforts speak the now-common language. Group projects and "teaming," despite the mystique of the solitary investigator, have become more common and accepted in university laboratories, as they had earlier in industrial and government research settings.

NSF's engineering constituents have also evolved over the years. Academic researchers remain number one, but increasingly the NSF engineering community works in tandem with industry, state and local governments, and other federal agencies. Industry provides practical training to students along with expensive and specialized equipment and instrumentation that universities cannot afford. It is also a market for engineering graduates attuned to the needs of industry. These arrangements apply on a lesser scale to local governments.

The professional engineering labor force was also changing, and none too quickly, to respond to the urgent need for talent. After years of discouragement, the engineering community has begun to court and value the contributions of underrepresented populations, especially racial minorities and women. As John White emphasized in 1990, engineering's first priority had to be serious, sustained action to enlarge the human resource pool. Effectively educating these engineers must come next. Only then could engineers, minimizing and crossing artificial boundaries between disciplines and between science and engineering, begin to solve the problems before them.

If it is true, as financial officers convincingly argue, that the history of an institution is written in its budget numbers, NSF figures show that the agency has seen steady growth over time, as measured in current dollars. In real dollars, however, that growth has been uneven. NSF's tenfold budget increases during the decade following *Sputnik*—indeed the Foundation's golden age—were not matched for another twenty years, while inflation and frequent program reshuffling eroded the effects of apparent progress. Real growth did not begin again until 1983. In 1988 the Reagan administration promised to double NSF's budget in five years, a goal that remained in place even though behind in fulfillment.[19] Within the NSF framework, engineering fared better at times and worse at others, although analysis is difficult. Actual figures are uncertain because of changing definitions and organizational configurations. White complained in late 1988 that "despite all the rhetoric to the contrary, support for engineering at the National Science Foundation has not increased dramatically relative to the overall increases at NSF. For the past decade, support for engineering has ranged between nine and ten percent of the overall NSF budget."[20] That relative status was little changed from earlier years, which suggests that all the organizational shuffling had had minor practical effect.

Engineering, like science, received funding from several other federal sources besides the Foundation. Still, the Engineering Research Board of the National Research Council, which called engineering research "essential because all creative technological development in an intensely competitive world rests on it," charged in 1987 that it was "seriously undersupported in the United States." By the NRC's figures, federal funding for engineering research (compared with science) had remained nearly constant, at about 25 percent of the total federal research budget for twenty years, during which period the United States had suffered a "steady decline in productivity and competitiveness." Only about 5 percent of the federal total for engineering research went to basic engineering.[21] NSF funded only 4.7 percent of all federal engineering research in 1984 compared to its relative funding (19.6 percent) for academic research in general. That same year, among federal agencies, the Department of Defense supported 35.8 percent of all basic research in engineering, while NASA funded 26.5 percent to NSF's 18.7 percent.[22]

The NRC reinforced the National Academy of Engineering's recommendation that the Foundation's engineering budget be increased from $150 million in 1985 to $400 million in 1990. Suh at the time was pushing (without illusions) for $500 million. When FY 1990 arrived, NSF requested $211 million for the Engineering Directorate and got $200 million. So neither the Foundation nor Congress heeded the high-level outsiders, although NSF's FY 1988 request for $205 million (a 25.7 percent increase) suggests that responsibility for the shortfall lay more on Capitol Hill. At that time, the Department of Defense and NASA werc spending a little over and somewhat under $1.5 billion, respectively, on engineering research, the Department of Energy not quite $0.5 billion. Thus, while engineering at NSF was receiving more attention and resources in the recent decade, a commitment the NRC applauded, supporters insisted much more was needed.[23]

Engineering's organizational progression was easier to see, and over the long run its direction was clearly positive. The road to an independent, top-level directorate had been bumpy and full of detours, but the stability of engineering's current position over a decade suggests that recognition and some acceptance have been achieved. As NSB executive officer Margaret Windus noted, however, although the board's approval of creating an engineering directorate "clearly became our philosophy, at least at that time," it remained a question how much that philosophy "has really penetrated the Foundation."[24] The future will tell. Whether even nominal engineering equality would have happened without congressional nagging is not certain, but rather than lose engineering to a rival technology foundation, NSF leaders agreed to grant it more status, visibility, and support. Indeed, thoughtful engineers remain divided among themselves as to the wisdom of institutionally separating engineering from science. While weary of playing second fiddle to science at NSF, they disagree on whether the competition for money and attention would improve or worsen under an independent agency. The question lingers, although quietly of late.

A dominant and pervasive theme of philosophical differences accompanied the NSF organizational and funding parade over its lifetime. Often referred to as a debate between basic and applied research, the issue is more usefully addressed in terms of scientific and engineering research. NSF's mission, said Vannevar Bush, was to promote and support basic scientific research, and founders agreed that included research on fundamental engineering principles, called engineering science, but not the applications of engineering to real-world problem solving. Engineering leaders, then striving to ground their profession in long-advocated scientific and mathematical rigor, approved that arrangement but eventually found the rigid mold confining and uncomfortable when even the hot pursuit of practical leads from basic research was denied. The Daddario Amendments and the subsequent RANN era provided an opportunity for new directions and emphases, but the rewritten charter, with its inclusion of applied research and problem solving, was too radical a departure for much of the NSF community, which was, at best, inadequately prepared for it. Because so much of RANN was labeled as engineering, the program's eventual unpopularity sullied engineering by association. So it was not surprising that it then took several more false starts before NSF could accommodate engineering philosophically and structurally.

But engineers did find their own voice at last. They claimed the validity of their own intellectual culture and operational traditions, insisting on, and finally achieving, equal standing within the NSF circle.[25] Branscomb, focusing on the usage of terms, materially helped NSF leaders and then the broader community to reach consensus on engineering's proper role. He argued that the phrase "applied science" should be purged from NSF's lexicon. It implied a commitment to solve a problem, and that, in turn, attracted the "political attention that does the mischief, not the fact that the science we support may, in fact, be helpful to those who do have the job of solving it." RANN's great difficulty had been one of unrealistic expectations unmet. Rather than promise solutions that put too much responsibility on science, NSF should sponsor research to "develop the knowledge, the tools and the skills that make it possible for the problem solving institutions of the country to do their work imaginatively and efficiently"—*after* NSF had done its work, Branscomb insisted. That left engineering, not applied research, as the "perfectly good," proper word. It described a profession and the skills of its practitioners, not the "act of application" or a development process.[26]

Although it has not disappeared, the basic-versus-applied argument appears less frequently today. As Fuqua put it, "Neither persistent demands for 'relevance' nor a reactionary reverence for 'purity' should interfere with a sound program balance between the advancement of knowledge desired for its contribution to science and the vigorous pursuit of knowledge desired for an application in filling human needs."[27]

With refocused attention on the relationship of science and engineering, then, questions of differences between them or of one dominating or threatening the other remain. Historian Arthur L. Donovan, writing a summary report to the National Research Council in 1985, suggested that "it would be easier to distinguish between science and engineering if they did not have so much in common." Sharing a "common mathematical and methodological language," they differed "culturally in the meanings they attached to the uses of that language." Both views were valid and useful.[28]

Indeed, is the debate necessary? Do science and engineering have to compete? Bloch and Branscomb both argued that their relationship is intertwined and mutually beneficial. Bloch spoke of science and engineering as a continuum. Each can, and does, drive the other, he said. "The 'straight line' conceptual model—with progress passing from science through engineering to technology—is not only far too simple to describe the complex interactions, it is simply incorrect." History, he wrote, proves that science did not always lead the way: Only forty years after Volta invented the battery did Faraday explain how it worked. Artists, craftspeople, and other amateurs developed the technology of photography decades before physicists and chemists understood the underlying principles. Instead, Bloch proposed a triangular model, with science, engineering, and technology at the vertices. Vectors running in both directions from each point show the interactions of each with the other two.[29]

Branscomb, too, called "obsolete" the "old linear model, in which science begets applied research, which leads to development, invention and innovation." He saw "applications, tools, materials, experimental research and theoretical developments" all "running in parallel toward both economic and intellectual goals." Furthermore, the "blurring of distinctions between science and engineering" had "revolutionized both the tools of science and the creative power of engineering." He also acknowledged, though, that some scientists still could "not accept that the values they find in 'basic scientific research' can be found in engineering." Nor were some engineers over their historical distrust of NSF policies.[30]

So the tensions of fitting engineering research, whose ultimate goal is to create new things or processes in the service of society, into an agency devoted to research in basic science, whose goal is to advance knowledge, remain and undoubtedly will continue indefinitely. But dynamic tension can be healthy and useful. Most of those close to the heart of NSF and its mission, both scientists and engineers, seem to recognize the interwoven threads of their relationship. To pull out one would weaken not only the other but the fabric of the whole. Congress, too, has quieted its demands for separateness. But it is true, as historians and sociologists point out, and engineers know, that technological systems also include political and economic institutions as components. As Lewis Mumford put it, history shows that society and its values can shape, or "socially construct," technology as well be shaped by it. Further, how those factors interacted in the past bears on how they will mesh in the future.[31] As values, transmitted through the political system, and technological forces continue their interplay, NSF's engineering history will be continually rewritten.

APPENDIX 1

National Science Foundation Directors, 1951–90

Alan T. Waterman, 1951–63

Leland J. Haworth, 1963–69

William D. McElroy, 1969–72

H. Guyford Stever, 1972–76

Richard C. Atkinson, 1976 (acting director), 1977–80

Donald N. Langenberg, 1980 (acting director)

John B. Slaughter, 1980–83

Edward A. Knapp, 1983–84

Erich Bloch, 1984–90

APPENDIX 2

National Science Foundation Heads of Engineering in
Varying Organizational Configurations, 1951–90

1951–52   Paul E. Klopsteg—Mathematical, Physical and Engineering Sciences (MPE)

1952–57   Raymond J. Seeger, Acting MPE

1952–54   Ralph A. Morgen, Engineering Sciences Program

1954–56   George H. Hickox, Engineering Sciences

1956–57   E. E. Litkenhous, Engineering Sciences

1957–58   Englehardt A. Eckhardt, MPE

1957–58   Gene M. Nordby, Engineering Sciences

1958–69   Randal M. Robertson, MPE

1958–61   Arthur H. Waynick, Engineering Sciences

1961–63   Samual Seely, Engineering Section

1963–64   William E. Lear, Engineering Section

1964–72   John M. Ide, Engineering Division

1969–71   Joel A. Snow, Interdisciplinary Research Relevant to Problems of Society (IRRPOS)

1971–77   Alfred J. Eggers, Jr., Research Applications, Research Applied to National Needs (RANN)

1972–74   Frederick H. Abernathy, Engineering Division

1974   Israel Warshaw, Acting, Engineering Division

1974–75   Thomas P. Meloy, Engineering Division

1975–77   Edward C. Creutz, Mathematical and Physical Sciences and Engineering (MPE)

1976–77   Charles Polk, Acting, Engineering Division

1977–79   James Krumhansl, MPE

1977–79   Henry C. Bourne, Engineering Division

1977–83   Jack T. Sanderson, Research Applications (RANN 1977–78), Applied Research and Research Applications (ASRA 1978–79), Engineering and Applied Science (EAS 1979–81), Engineering Directorate (1981–83)

1983–84	Carl W. Hall, Acting, Engineering Directorate
1984–88	Nam P. Suh, Engineering Directorate
1988	Carl W. Hall, Acting, Engineering Directorate
1988–91	John A. White, Engineering Directorate

APPENDIX 3

Engineering Research Centers, to 1990

Center for Robotics Systems in Microelectronics, University of California, Santa Barbara, 1985–89

Center for Telecommunications Research, Columbia University, 1985–

Center for Composites Manufacturing Science and Engineering, University of Delaware, Newark, with an affiliated program in ceramics at Rutgers University, New Brunswick, N.J., 1985–89

Systems Research Center, University of Maryland, in collaboration with Harvard University, 1985–

Biotechnology Process Engineering Center, Massachusetts Institute of Technology, 1985–

Engineering Research Center for Intelligent Manufacturing Systems, Purdue University, 1985–

Advanced Combustion Engineering Research Center, Brigham Young University, in collaboration with the University of Utah, 1986–

Center for Compound Semiconductor Microelectronics, University of Illinois, Urbana-Champaign, 1986–

Advanced Technology for Large Structural Systems, Lehigh University, 1986–

Center for Net Shape Manufacturing, Ohio State University, 1986–

Engineering Design Research Center, Carnegie Mellon University, 1986–

Optoelectronic Computing Systems Center, University of Colorado, Boulder, in cooperation with Colorado State University, 1987–

Center for Emerging Cardiovascular Technologies, Duke University, with other North Carolina institutions, 1987–

Hazardous Substance Control Engineering Research Center, University of California, Los Angeles, 1987–91

Center for Interfacial Engineering, University of Minnesota, 1988–

Center for Advanced Electronic Materials Processing, North Carolina State University, in association with other North Carolina institutions, 1988–

Center for Plasma-Aided Manufacturing, University of Wisconsin, Madison, in cooperation with the University of Minnesota, 1988–

Offshore Technology Research Center, Texas A&M University, in cooperation with the University of Texas, Austin, 1988–

Engineering Research Center for Data Storage Systems, Carnegie Mellon University, 1990–

Engineering Research Center for Geometrically Complex Field Problems, Mississippi State University, Starkville, 1990–

Engineering Research Center for Interfacial Microbial Process Engineering, Montana State University, Bozeman, 1990–

APPENDIX 4

Engineers Serving on the National Science Board, 1951–90

Of a total of 178 National Science Board members to 1990, 27 were engineers. In addition to Stever, engineers John B. Slaughter and Erich Bloch served ex officio on the board as NSF directors.

A. A. Potter, 1950–58
Edward L. Moreland, 1950–51
C. E. Wilson, 1950–51
Keith Glennan, 1956–58
Julius Stratton, 1956–62, 1964–67
Kevin McCann, 1958–64
Morrough O'Brian, 1958–60
Eric A. Walker, 1960–66
William Hagerty, 1964–70
Charles F. Jones, 1966–72
Thomas F. Jones, Jr., 1966–72
H. Guyford Stever, 1970–72, then named NSF director
Hubert Heffner, 1972–75
Raymond Bisplinghoff, 1976–82
Herbert Doan, 1976–82
Joseph Pettit, 1976–82
David Ragone, 1978–84
Robert Gilkeson, 1982–88
Norman Rasmussen, 1982–88
Roland Schmitt, 1982–88, 1988–
Simon Ramo, 1984–86
James J. Duderstadt, 1984–90, 1990–
Warren J. Baker, 1986–
John C. Hancock, 1986–
Daniel C. Drucker, 1988–

Arden Bement, 1988–
Jaime Oaxaca, 1990–

APPENDIX 5
Statistical Figures, NSF Spending

Comparison of Actual Research Spending
(FY 1952 to FY 1959) for Engineering,
Compared with Total Spending for Mathematical,
Physical, and Engineering Sciences and Total NSF

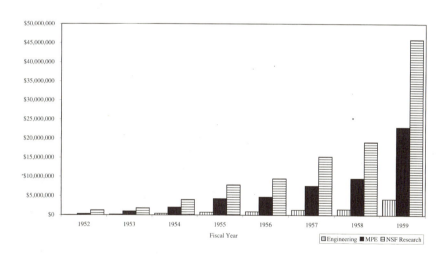

Engineering Research in the Second Decade
(FY 1961 to FY 1971) Actual Expenditures

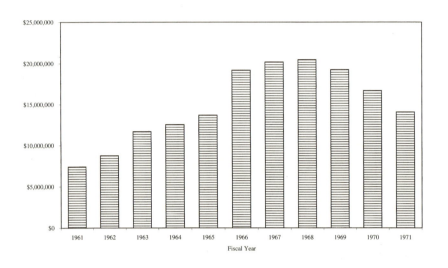

Engineering Research vs. Research Applications
(FY 1971 to FY 1979) Actual Expenditures

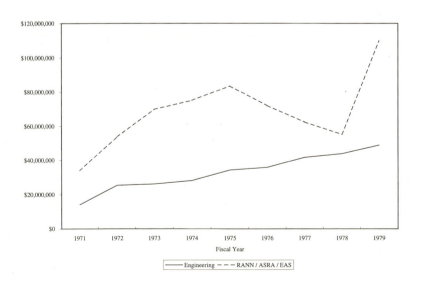

Comparison of Actual Research Spending
(FY 1971 to FY 1979) for Engineering,
RANN, ASRA, ESA, and Total NSF

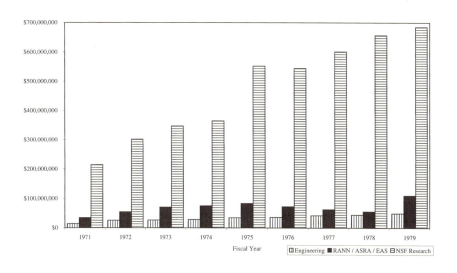

Engineering and Total NSF Research in the Fourth Decade
(FY 1980 to FY 1989) Actual Expenditures

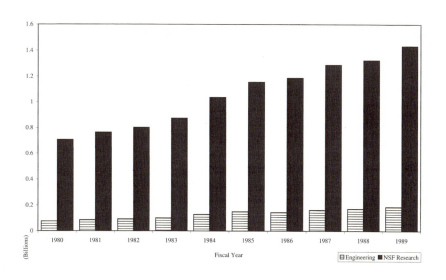

279

Engineering Research in Comparison to Total Research at NSF (FY 1952 to FY 1989) Actual Expenditures

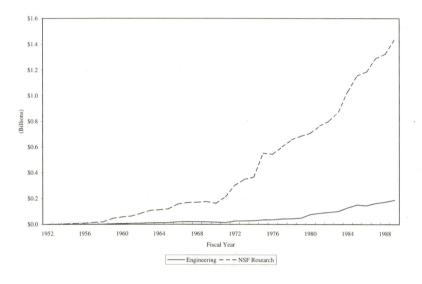

Percentage of Engineering Research to Total NSF Research Spending (FY 1952 to FY 1989)

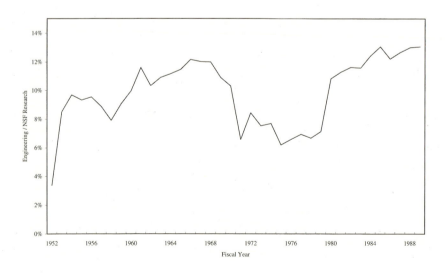

ABBREVIATIONS

AAAS	American Association for Advancement of Science
AAES	American Association of Engineering Societies
Acc.	Accession
ACERC	Advanced Combustion Engineering Research Center
AD/R	Assistant Director of Research
AD/Eng	Assistant Director of Engineering
AIME	American Institute of Mining Engineers
ASCE	American Society of Civil Engineers
ASEE	American Society for Engineering Education
ASRA	Applied Science and Research Applications
ATW	Alan T. Waterman
BYU	Brigham Young University
C&EN	Chemical & Engineering News
Chron	Director's Chron Files for Calendar Year, NSF History Office Files
COPEP	Committee on Public Engineering Policy
D/	Directorate for
EAK	Edward A. Knapp
EAS	Engineering and Applied Science
ECD	Division of Engineering Centers
ECS	Electrical and Communications Systems
EJC	Engineers Joint Council
ERC	Engineering Research Centers
FY	fiscal year
GPO	Government Printing Office
HGS	H. Guyford Stever
IBR	Integrated Basic Research
IEEE	Institute of Electrical and Electronics Engineers
IRPOS/ IRRPOS	Interdisciplinary Research Relevant to Problems of Our Society
JBS	John B. Slaughter

LJH	Leland J. Haworth
LMB	Lewis M. Branscomb
MIT	Massachusetts Institute of Technology
MPE	Mathematical, Physical, and Engineering Sciences (before 1975) Mathematical and Physical Sciences and Engineering (after 1975)
MPS	Mathematical and Physical Sciences
MRL	Materials Research Laboratories
NAE	National Academy of Engineering
NARA	National Archives and Records Administration
NAS	National Academy of Sciences
Notes	Director's Notes for Calendar Year, NSF History Office Files
NRC	National Research Council
NSB	National Science Board
NSB Book	National Science Board Agenda Book
NSF	National Science Foundation
NSF	Budget to Congress
NSF	Justification of Estimates on Appropriations, Office of the Controller Files
NSF	Legislative History Files
NSF	Legislative History Reports 1945–1950, NSF Controller's Office
OD	Office of the Director
OGC	Office of General Counsel
OMB	Office of Management and Budget
PL	Public Law
R&D	Research and Development
RA	Research Applications
RA/D	Research Applications Directorate
RANN	Research Applied to National Needs
RCA	Richard C. Atkinson
RG	Record Group

SEA	Directorate for Science and Engineering Applications
SEE	Directorate for Science and Engineering Education
Todd Chron 1965–1975	Ed Todd's Chron File, NSF History Office Files
UU	University of Utah
WDM	William D. McElroy
WNRC	Washington National Records Center (Suitland, MD)

NOTES

PROLOGUE: ENGINEERING: AN AMERICAN TRADITION

1. Silvio A. Bedini, *Thinkers and Tinkers: Early American Men of Science* (New York: Charles Scribner's Sons, 1975), xv–xviii, 195. Brooke Hindle and Steven Lubar, *Engines of Change: The American Industrial Revolution, 1790–1860* (Washington, D.C.: Smithsonian Institution Press, 1986), 45, 219. Edwin T. Layton, Jr., "Scientific Technology, 1845–1900: The Hydraulic Turbine and the Origin of American Industrial Research," *Technology and Culture* 20, no. 1 (January 1979): 88–89. Edwin T. Layton, Jr., "Science as a Form of Action: The Role of the Engineering Sciences," *Technology and Culture* 29, no. 1 (January 1988): 91.

2. Hindle and Lubar, *Engines of Change,* 102–4. Talbot Hamlin, *Benjamin Henry Latrobe* (New York: Oxford University Press, 1955), 545–46.

3. Terry S. Reynolds, "The Engineer in 19th-Century America," in *The Engineer in America: A Historical Anthology from "Technology and Culture,"* ed. Terry S. Reynolds (Chicago, Ill.: University of Chicago Press, 1991), 7–10, 25. Lawrence P. Grayson, "A Brief History of Engineering Education in the United States," *Engineering Education* 68 (December 1977): 246–48.

4. Reynolds, "The Engineer in 19th-Century America," 10–11, 16–17. Grayson, "A Brief History of Engineering Education," 247–48. Daniel Hovey Calhoun, *The American Civil Engineer: Origins and Conflict* (Cambridge, Mass.: MIT Press, 1960), 22, 42–43. American Public Works Association, *History of Public Works in the United States, 1776–1976,* ed. Ellis L. Armstrong (Chicago, Ill.: American Public Works Association, 1976), 666–69.

5. Darwin H. Stapleton, "Benjamin Henry Latrobe and the Transfer of Technology," in *Technology in America: A History of Individuals and Ideas,* ed. Carroll W. Pursell, Jr. (Cambridge, Mass.: MIT Press, 1981), 33–44. Todd Shallat, *Structures in the Stream: Water, Science, and the Rise of the U.S. Army Corps of Engineers* (Austin: University of Texas Press, 1994), 11–14. See also Hamlin, *Benjamin Henry Latrobe,* 547–61.

6. Elting E. Morison, *From Know-how to Nowhere: The Development of American Technology* (New York: Basic Books, 1974), 40–71. See also F. Daniel Larkin, *John B. Jervis: An American Engineering Pioneer* (Ames: Iowa State University Press, 1990), xi–xiv, xvii–xix, and passim.

7. Robert Rosenberg speaks of Americans' nineteenth-century passion for practicality in "American Physics and the Origins of Electrical Engineering," *Physics Today* 36 (October 1983): 53. Edwin T. Layton, Jr., *The Revolt of the Engineers: Social Responsibility and the American Engineering Profession* (Cleveland, Ohio: Press of Case Western Reserve University, 1971), viii–ix. Hindle and Lubar, *Engines of Change,* 15–21, 218.

8. Hindle and Lubar, *Engines of Change,* 249–68.

9. Raymond S. Merritt, *Engineering in American Society, 1850–1875* (Lexington: University Press of Kentucky, 1969), 22.

10. Ibid., 2–3, 7, 12, 13.

11. Layton, *The Revolt of the Engineers,* viii–ix.

12. See Monte A. Calvert, *The Mechanical Engineer in America, 1830–1910: Professional Cultures in Conflict* (Baltimore, Md.: Johns Hopkins University Press, 1967).

13. NRC, *Engineering in Society: Engineering Education and Practice in the United States* (Washington, D.C.: National Academy Press, 1985), 2–3. Rosenberg, "American Physics and the Origins of Electrical Engineering," 50. Terry S. Reynolds, "Defining Professional Boundaries: Chemical Engineering in the Early 20th Century," *Technology and Culture* 27 (October 1986): 694–95. Grayson, "A Brief History of Engineering Education," 252–53, 255, 257, 260.

14. Reynolds, "The Engineer in 19th-Century America," 24. Grayson, "A Brief History of Engineering Education," 253. See Terry S. Reynolds, *75 Years of Progress: A History of the American Institute of Chemical Engineers, 1908–1983* (New York: AIChE, 1983), 3–5.

15. Carroll W. Pursell, "'What the Senate Is to the American Commonwealth': A National Academy of Engineers," in *New Perspectives on Technology and Culture,* ed. Bruce Sinclair (Philadelphia, Pa.: American Philosophical Society, 1986), 19–29. W. Bernard Carlson, "Academic Entrepreneurship and Engineering Education: Dugald C. Jackson and the MIT-GE Cooperative Engineering Course, 1907–1932," *Technology and Culture* 29 (July 1988): 536–37.

16. A. Rupert Hall, "Engineering and the Scientific Revolution," *Technology and Culture* 2 (Fall 1961): 340.

17. Bruce Sinclair, "Inventing a Genteel Tradition: MIT Crosses the River," in *New Perspectives,* 1–18. Grayson notes that neither engineers nor scientists enjoyed the status of the more elite liberal-arts students at Harvard and Yale in the mid-nineteenth century. Grayson, "A Brief History of Engineering Education," 250.

18. Benjamin F. Thomas, "Technical Education in Colleges and Universities," *Proceedings of the American Association for the Advancement of Science,* 41st Meeting, Rochester, N.Y., August 1892 (Salem: Permanent Secretary, December 1892), 67, 70, 74–75.

19. Calhoun, *The American Civil Engineer,* 47.

20. Grayson, "A Brief History of Engineering Education," 248–49. Calhoun, *The American Civil Engineer,* 43–45. Calhoun gives the founding date of Partridge's school as 1820.

21. Merritt, *Engineering in American Society,* 2–3, 7, 12, 13. Grayson, "A Brief History of Engineering Education," 249–51. Robert M. Vogel, *Building Brooklyn Bridge: The Design and Construction, 1867–1883* (Washington, D.C.: Smithsonian Institution Press, 1983), 11. Reynolds, "The Engineer in 19th-Century America," 24.

22. Grayson, "A Brief History of Engineering Education," 251. Merritt, *Engineering in American Society,* 61.

23. Roger L. Williams, *The Origins of Federal Support for Higher Education: George W. Atherton and the Land-Grant College Movement* (University Park: Pennsylvania State University Press, 1991), 179.

24. Bruce Seely, "Research, Engineering, and Science in American Engineering Colleges, 1900–1960," *Technology and Culture* 33 (April 1993): 344–46, 358–67. Grayson, "A Brief History of Engineering Education," 246, 251. Lawrence P. Grayson, *The Making of an Engineer: An Illustrated History of Engineering Education in the United States and Canada* (New York: John Wiley & Sons, 1993), 44–45. Ronald Kline, "Construing 'Technology' as 'Applied Science': Public Rhetoric of Scientists and Engineers in the United States, 1880–1945," *Isis* 86 (June 1995): 201–4. W. E. Wickenden, "Research in the Engineering Colleges," *Mechanical Engineering* 51 (August 1929): 586–87. See also Rosenberg, "American Physics and the Origins of Electrical Engineering"; and Reynolds, "Defining Professional Boundaries," 694–716.

25. Kendall Birr, "Industrial Research Laboratories," in *The Sciences in the American Context: New Perspectives,* ed. Nathan Reingold (Washington, D.C.: Smithsonian Institution Press, 1979), 195–201. George Wise, *Willis R. Whitney, General Electric, and the Origins of U.S. Industrial Research* (New York: Columbia University Press, 1985), 2–3, 76–80. Daniel J. Kevles, *The Physicists: The History of a Scientific Community in Modern America* (New York: Alfred A. Knopf, 1978), 61. Carroll W. Pursell, Jr., "Introduction," *Technology in America,* 4–5. A. Hunter Dupree, *Science in the Federal Government: A History of Policies and Activities to 1940* (New York: Harper & Row, 1957), 297–99.

26. Dugald C. Jackson, *Present Status and Trends of Engineering Education in the United States* (New York: Engineers' Council for Professional Development, 1939), 17–18. Carlson, "Academic Entrepreneurship and Engineering Education," 537–41. Carlson discusses numerous interpretations of the effects of the relationship between industry and the academy on the evolution of the engineering profession, especially the work of David Noble and Edwin Layton. Grayson, "A Brief History of Engineering Education," 246.

27. See David F. Channell, "The Harmony of Theory and Practice: The Engineering Science of W. J. M. Rankine," *Technology and Culture* 22 (January 1982): 44–52.

28. David F. Channell, "Engineering Science as Theory and Practice," *Technology and Culture* 29 (January 1988): 98–103.

29. Layton, "Scientific Technology, 1845–1900," 64.

30. Seely, "Research, Engineering, and Science," 358–67. For a lively discussion of Thurston's efforts at Cornell, see Calvert, *The Mechanical Engineer in America,* chapter 5. See also Kline, "Construing 'Technology' as 'Applied Science,'" 201–4, esp. 202. Grayson, *The Making of an Engineer,* 44–45. Thorndike Saville,

"Achievements in Engineering Education," *Journal of Engineering Education* 43 (December 1952): 224.

31. Grayson, "A Brief History of Engineering Education," 254, 258. Terry S. Reynolds and Bruce E. Seely, "Striving for Balance: A Hundred Years of the American Society for Engineering Education," *Journal of Engineering Education* 82 (July 1993): 136.

32. Grayson, "A Brief History of Engineering Education," 258. Reynolds and Seely, "Striving for Balance," 139.

33. Reynolds and Seely, "Striving for Balance," 139.

34. Seely, "Research, Engineering, and Science," 344–46. Wickenden, "Research in the Engineering Colleges," 586–87. Reynolds and Seely, "Striving for Balance," 140.

35. Seely, "Research, Engineering, and Science," 346–54.

36. Wickenden, "Research in the Engineering Colleges," 588.

37. Seely, "Research, Engineering, and Science," 346–54.

38. Ibid. Jackson, *Present Status and Trends of Engineering Education,* 140–41. Jackson placed the importance of research seventeenth in his list of twenty-one important trends in engineering education.

39. Hindle and Lubar, *Engines of Change,* 78–81. Jeffrey K. Stine, *A History of Science Policy in the United States, 1940–1985,* Report No. 1 for the Task Force on Science Policy, House Committee on Science and Technology, 99th Cong., 2d sess., September 1986 (Washington, D.C.: GPO, 1986), 6. NSF, Directorate for Scientific, Technological, and International Affairs, *The State of Academic Science and Engineering,* January 1990, IB 1.

40. Dupree, *Science in the Federal Government,* 289–91. The Department of Agriculture was raised to cabinet status in 1889. Alex Roland, *Model Research: The National Advisory Committee for Aeronautics, 1915–1958,* vol. 1 (Washington, D.C.: National Aeronautics and Space Administration, 1985), 201–2.

41. Williams, *Origins of Federal Support for Higher Education,* 180–81. Daniel J. Kevles, "Federal Legislation for Engineering Experiment Stations: The Episode of World War I," *Technology and Culture* 12 (April 1971): 182–83, 188–89. Kevles, *The Physicists,* 150–52, 259–60.

42. Dupree, *Science in the Federal Government,* 296–301. Dupree did not specifically connect applied research with engineering schools.

43. Kevles, *The Physicists,* 150–52, 259–60. Stine, *A History of Science Policy,* 8–9, 12

44. Roger L. Geiger, *To Advance Knowledge: The Growth of American Research Universities, 1900–1940* (New York: Oxford University Press, 1986), 257.

45. Ibid., 256–57. Dupree, *Science in the Federal Government,* 340–41, 348–49.

46. Geiger, *To Advance Knowledge,* 255–67.

47. Ibid., 260–61. Carroll W. Pursell, Jr., "A Preface to Government Support of Research and Development: Research Legislation and the National Bureau of Standards, 1935–41," *Technology and Culture* 9 (April 1968): 161–64.

48. Geiger, *To Advance Knowledge,* 262–64. Stine, *A History of Science Policy,* 13.

49. Geiger, *To Advance Knowledge,* 255–67. See Stanley Coben, "American Foundations as Patrons of Science: The Commitment to Individual Research"; and Birr, "Industrial Research Laboratories," both in *The Sciences in the American Context,* ed. Nathan Reingold. By 1940, there were more than 2,000 industrial laboratories, employing 70,000 people altogether and dominated professionally by chemists and engineers.

50. Stine, *A History of Science Policy,* 15–19.

51. Terry S. Reynolds, "The Engineer in 20th-Century America," in *The Engineer in America: A Historical Anthology from "Technology and Culture,"* ed. Terry S. Reynolds (Chicago, Ill.: University of Chicago Press, 1991), 172–80.

52. Layton, "Science as a Form of Action," 92.

53. Seely, "Research, Engineering, and Science," 358–67. Reynolds, "The Engineer in 20th-Century America," 180–82. See also Eugene S. Ferguson's interpretation of Bush's rueful acknowledgment that calling engineers scientists during the war to elevate their status in the end left engineers with less respect in *Engineering and the Mind's Eye* (Cambridge, Mass.: MIT Press, 1992), 157–59.

54. Edwin T. Layton, Jr., "American Ideologies of Science and Engineering," *Technology and Culture* 17 (October 1976): 689.

55. Ferguson, *Engineering and the Mind's Eye,* preface and passim. Layton, "Scientific Technology," 88–89. Layton, "Science as a Form of Action," 91.

56. John W. Miller, *Men and Volts at War* (New York: Bantam Press, June 1948), 72–75.

57. W. H. G. Armytage, *A Social History of Engineering* (London: Faber and Faber, 1976), 276. Geiger, *To Advance Knowledge,* 250–51, 260.

58. Armytage recalls Goethe's phrase in *A Social History of Engineering,* 324.

CHAPTER 1: ENGINEERING UNDER A MANDATE FOR BASIC SCIENCE

The statement in the epigraph originated with Oliver Buckley, president of Bell Telephone Laboratories, in Isaiah Bowman's Committee on Science and the Public Welfare. OSRD Report to the President—No. 3, Minutes of Meeting of Steering Committee, 26 December 1944, NARA, RG 227, OSRD, Entry 2, Box 3.

1. Vannevar Bush, *Science—The Endless Frontier: A Report to the President on a Program for Postwar Scientific Research* (Washington, D.C.: NSF, 1945; reprint, 1980).

2. J. Merton England, *A Patron for Pure Science: The National Science Foundation's Formative Years, 1945–57* (Washington, D.C.: NSF, 1982), 3, 6–7, 10. Daniel S. Greenberg, *The Politics of American Science* (Harmondsworth: Penguin Books, 1969), 142–43 (this book is more widely known as *The Politics of Pure Science* [New York: New American Library, 1967]. My quotations are from the former edition, which was the only one available to me).

3. Earle B. Norris, "Research as Applied to Engineering: Its Importance in Industry and Economics Stressed," *Civil Engineering* 5 (May 1935): 408–9.

4. Daniel J. Kevles, "The National Science Foundation and the Debate over Postwar Research Policy, 1942–1945," *Isis* 68 (March 1977): 8, 15–17. Robert Franklin Maddox, "The Politics of World War II Science: Senator Harley M. Kilgore and the Legislative Origins of the National Science Foundation," *West Virginia History* 41 (Fall 1979): 31–32. Lyman Chalkley, "Prologue to the National Science Foundation" (Ann Arbor, Mich.: University Microfilms International, 1979), 28–31.

5. Bush, *Science—The Endless Frontier,* 78–79. Although Bush sometimes used the term "pure" science (or research), more often the favored words, by him and others, were "fundamental" or "basic." Kline writes that after the 1930s, these latter terms "began (slowly) to be substituted for *pure,* whose moral connotations, once an asset, were now objectionable to many researchers." Ronald Kline, "Construing 'Technology' as 'Applied Science': Public Rhetoric of Scientists and Engineers in the United States, 1880–1945," *Isis* 86 (June 1995): 196.

6. Bush to Isaiah Bowman, 10 January 1945; and Memorandum, Bush to Carroll Wilson, Agenda for Committee no. 3, 15 January 1945, both in NARA, RG 227, Office of Scientific Research and Development (hereafter OSRD), Entry 2, Box 3: "General: Report to Pres., Committee #3." Notes on Meeting of the Chairmen and Secretaries of the Four Committees in Dr. Bush's Office on March 8, 1945; and attached Agenda, both in NARA, RG 227, OSRD, Entry 2, Box 1. Eric A. Walker, *Now It's My Turn: Engineering My Way* (New York: Vantage Press, 1989), 143–45; and interview with author, 1 October 1990. Kline, "Construing "'Technology' as 'Applied Science,'" 219–21. Vannevar Bush, *Pieces of the Action* (New York: Morrow, 1970), 53–54.

7. Maddox, "Politics of World War II Science," 25, 35.

8. See, for example, England, "The Long Debate, 1945–50," part 1 of *A Patron for Pure Science;* or, for a briefer account, Milton Lomask, "Old Fears and New Ideas," chap. 3 of *A Minor Miracle: An Informal History of the National Science Foundation* (Washington, D.C.: NSF, 1976). See also Daniel J. Kevles, "Principles and Politics in Federal R&D Policy, 1945–1990: An Appreciation of the Bush Report," ix–xxxiii, preface to NSF 90-8, a fortieth-anniversary reprint of Bush, *Science—The Endless Frontier* (Washington, D.C.: NSF, 1990).

9. S. 825, Senate, 79th Cong., 1st sess., 4 April 1945 (legislative day 16 March). S. 1248, Senate, 79th Cong., 1st sess., 19 July 1945. See also S. Rept. 5, Senate Military Affairs Committee, Subcommittee on War Mobilization, "Findings and Recommendations," part 2 of *The Government's Wartime Research and Development, 1940–44,* 79th Cong., 2d sess., 23 July 1945. NSF maintains in bound volumes the complete legislative history of the Foundation's formation; see "NSF Legislative History: Reports, 1945–1950," NSF Controller's Office. Lomask, in *A Minor Miracle* (30), writes that between 1945 and 1950 no fewer than twenty-one national science agency bills were "introduced and heatedly discussed in Congress and in the press."

10. Harry S. Truman, Special Message to Congress Presenting a 21-Point Program for the Reconversion Period, 6 September 1945, *Public Papers of the Presidents of the United States, 1945,* 292–94.

11. S. Rept. 8, Senate Subcommittee, *National Science Foundation,* 79th Cong., 2d sess., 1946, 1, 4, 29. See also the section-by-section analysis of S. 1850, 79th Cong., 2d sess. A typescript (unfortunately undated and unattributed) in the bound NSF Legislative History, "Analysis of Legislation Now Pending in Congress with Regard to a Postwar Program for Scientific Research," provides a detailed comparison of numerous 1945 bills, especially those of Kilgore and Magnuson.

12. "Excerpt from Report on History of Science Foundation Legislation in Bulletin of the Engineering Section of the Association of Land-Grant Colleges and Universities—October 1947," typescript in bound NSF Legislative History, 2. See also PL 507, *The National Science Foundation Act of 1950.*

13. House Subcommittee of the Committee on Interstate and Foreign Commerce, Hearings, *The National Science Foundation Act,* 79th Cong., 2d sess., 28–29 May 1946, 31–33.

14. House Committee on Interstate and Foreign Commerce, Hearings, *National Science Foundation,* 80th Cong., 1st sess., 6–7 March 1947, 191–97.

15. The EJC, formed in 1941, was a joint committee of the American Society of Civil Engineers, the American Institute of Mining and Metallurgical Engineers, the American Society of Mechanical Engineers, the American Institute of Electrical Engineers, and the American Institute of Chemical Engineers. See T. A. Marshall, Jr., "Engineers Joint Council," *Mechanical Engineering* 76 (July 1954): 582–84; and House Committee on Interstate and Foreign Commerce, Hearings, *National Science Foundation,* 80th Cong., 1st sess., 6–7 March 1947, 272–77. Boris A. Bakhmeteff, "Science and Engineering: Engineer's Scientific Approach Needed to Guide Man's Destiny," *Civil Engineering* 16 (March 1946): 99–100. Kline, "Construing 'Technology' as 'Applied Science,'" 219.

16. House Committee on Interstate and Foreign Commerce, Hearings, *National Science Foundation,* 80th Cong., 1st sess., 6–7 March 1947, 76–78.

17. See Jeffrey K. Stine, *A History of Science Policy in the United States, 1940–1985,* Report No. 1 for the Task Force on Science Policy, House Committee on Science and Technology, 99th Cong., 2d sess., September 1986 (Washington, D.C.: GPO, 1986), 25–34; opening chapters in England, *A Patron for Pure Science;* and Lomask, *A Minor Miracle,* for general treatment of the origin of NSF. See also H. Rept. 796, House Committee on Interstate and Foreign Commerce, *National Science Foundation Act of 1949,* 81st Cong., 1st sess., 14 June 1949, 4.

18. The bill was S. 526, introduced by NSF supporter H. Alexander Smith (R-N.J.). Harry S. Truman, Memorandum of Disapproval of the National Science Foundation Bill, 6 August 1947, *Public Papers of the Presidents, 1947,* 368–69. Alex Roland, *Model Research: The National Advisory Committee for Aeronautics, 1915–1958* (Washington, D.C.: NASA, 1985), 201–2. Roland notes (p. 20) that the early NACA "skirted" the issue of "whether aeronautics was properly in the realm of science or of engineering," another NSF parallel. Lomask, *A Minor Miracle,* 54, 62–64.

19. Truman Speech, 13 September 1948, President's Secretary's file, Truman Library, cited in Maddox, "The Politics of World War II Science," 38. President's

Scientific Research Board, John R. Steelman, Chairman, *A Program for the Nation,* vol. 1 of *Science and Public Policy* (Washington, D.C.: GPO, 27 August 1947), 6–7.

20. For a detailed legislative history, see England, *A Patron for Pure Science,* chaps. 3–6, p. 109. Senator Elbert Thomas (D-Utah) introduced S. 247, which passed on 18 March 1948 without discussion; Rep. J. Percy Priest (D-N.J.) introduced "clean" bill H.R. 4846 on 24 May 1949. The conference committee reported out an amended S. 247, which the House passed on 27 April 1950, the Senate one day later. It became PL 507, *The National Science Foundation Act of 1950.*

21. Lawrence P. Grayson, *The Making of an Engineer: An Illustrated History of Engineering Education in the United States and Canada* (New York: John Wiley & Sons, 1993), 170–71. Bruce Seely, "Research, Engineering, and Science in American Engineering Colleges, 1900–1960," *Technology and Culture* 33 (April 1993): 353–54. *Who's Who in America,* 1958, s.v. "Potter, Andrey A." J. M. England, on the strength of the recollection of a Budget Bureau staff member who attended the NSB's first meeting, believes it was held in the Fish Room. See England, *A Patron for Pure Science,* 123.

22. A. A. Potter, "National Science Foundation Developments," *Journal of Engineering Education* 42 (October 1951): 81–85.

23. Lomask, *A Minor Miracle,* 67–73. England, *A Patron for Pure Science,* 133.

24. See, for example, NSF Annual Report, FY 1957, Director's Statement, ix–xvi; Director's Report to NSB, NSB Minutes, 14th Meeting, 13 June 1952, 6–7; NSF Staff Meeting Notes, 15 October 1952, 1; England, *A Patron for Pure Science,* 148–50, 181–202.

25. *Chemical & Engineering News* editorial, as quoted in NSF Annual Report, FY 1950–51, 17. NSF-1, press release, 15 February 1951.

26. NSF Annual Report, FY 1950–1951, 25–26. Press releases: NSF-10, Dr. Paul E. Klopsteg to Head National Science Foundation Division of Physical, Mathematical, and Engineering Sciences, 10 October 1951; and NSF-24, National Science Foundation Appoints Ralph A. Morgen to Direct Engineering Research Support Program, 22 May 1952.

27. NSF Annual Report, FY 1950–51, v-31, esp. 4, 10–12, 14.

28. NSF Budget to Congress, FY 1952, 21 May 1951, 18. Budget document, 25 April 1951, NSB Book, 6th Meeting, 11 May 1951. NSF Budget to Congress, FY 1954, 7 January 1953, 2:1–2:9. NSB Books, 8th Meeting, 7 September 1951, 6; 12th Meeting, 29 February 1952, Table 3. May 1951, NSF Agenda, 1952 Budget, ATW Notes 1951. A. A. Potter quoted in England, *A Patron for Pure Science,* 158. NSF Annual Report, FY 1952, 6, 19, 44.

29. NSB Books, 7th Meeting, 27–28 July 1951, agenda items 3–5; 8th Meeting, 7 September 1951, budget pp. 4–5 and appendix 1. NSF Budget to Congress, FY 1953, 1.

30. Alan T. Waterman, Statement to House Subcommittee of Committee on Appropriations, 20 January 1954, NSB Book, 25th Meeting, 29 January 1954. NSF

Annual Report, FY 1954, 4, 118–19. Executive Order [10521] Concerning Govern-ment Scientific Research, the National Science Foundation, and the Interdepartmental Committee for Scientific Research and Development, 17 March 1954, in *Federal Register,* 19 March 1954. See also White House press release, 17 March 1954, copy in NARA, RG 307, Box 23, File: Coordination of Basic Research.

31. Waterman to Warren Weaver, 30 January 1953, NARA RG 307, Office of the Director, Box 15: Basic Research Group, Research Development Board. Klopsteg to Roger Adams, 4 March 1958, NSF Historian's files: NSB, 1951–58. See Dian Olson Belanger, "Does Applied Research Drive Out Basic?" unpublished manuscript, NSF Historian's Files.

32. NSB Book, 15th Meeting, 7–8 August 1952, agenda item 2. According to Waterman's figures to the NSB, fourteen dollars were requested for every one dollar granted. NSF Budget to Congress, FY 1954, 7 January 1953, 2–7. See also NSF Annual Report, FY 1953, 34–35.

33. NSB Book, 14th Meeting, 13 June 1952, 8, 13–14. NSF Annual Report, FY 1952, 13, 46, 74.

34. NSB Book, 8th Meeting, 7 September 1951, appendix 1, esp. 19, 23. NSF Budget to Congress, FY 1953, 47.

35. NSB Book, 29th Meeting, 15 October 1954. NSF Budget to Congress, FY 1954, 2–21.

36. NSB Book, 31st Meeting, 6 December 1954, and 1950s board books, passim.

37. Seely, "Research, Engineering, and Science," 367–86. Edwin T. Layton, Jr., "American Ideologies of Science and Engineering," *Technology and Culture* 12 (October 1971): 689–95.

38. NSB Book, 29th Meeting, 15 October 1954. NSF Budget to Congress, FY 1954, 7 January 1953, 2:3–2:9. NSF Annual Report, FY 1954, 20. Staff Notes, MPE Divisional Committee, 3rd Meeting, 21 November 1952, 3, NSF History Office Files.

39. Terry S. Reynolds, "The Engineer in 20th-Century America," in *The Engineer in America: A Historical Anthology from "Technology and Culture,"* ed. Terry S. Reynolds (Chicago, Ill.: University of Chicago Press, 1991), 182.

40. Memorandum, Waterman to NSB, 11 October 1954; and attached draft "Report to the President by the Special Interdepartmental Committee on the Training of Scientists and Engineers, October 1954," both in NSB Book, 29th Meeting, 15 October 1954. NSB Book, 30th Meeting, 5 November 1954. L. E. Grinter, "Report on the Evaluation of Engineering Education," *Journal of Engineering Education* 46 (April 1956): 25–63. See also Terry S. Reynolds and Bruce E. Seely, "Striving for Balance: A Hundred Years of the American Society for Engineering Education," *Journal of Engineering Education* 82 (July 1993): 141–42.

41. Waterman, telephone conversation with Saville, 18 March 1954; and Waterman to Hauge, 23 April 1954, both in ATW Notes 1954. Staff Notes, MPE Divisional Committee, 6th Meeting, 4 February 1954, 1; 12th Meeting, 7–8 March 1957, 3.

42. Waterman to Saville, 5 November 1953, ATW Notes 1953. Saville to Waterman, 1 August 1955, attached to Waterman to Saville, 13 September 1955, ATW Notes 1955.

43. Seeger to Morgen, 3 January 1957; and Seeger diary notes, 6 and 13 May 1957, both in ATW Notes 1957. Litkenhous appears as a crotchety, difficult program director who was careless about his administrative duties.

44. Spangenberg to Waterman, 16 December 1954, ATW Notes 1954. Waterman had offered Spangenberg the job of heading engineering sciences at NSF, but he seems not to have accepted. ASEE, Engineering College Research Council, *A Survey of Research in the Nation's Engineering Colleges: Capabilities and Potentialities, July 1956 to June 1957* (Washington, D.C.: NSF, 1958), 45.

45. My emphasis. Randal Robertson, transcription of interview with J. M. England, 21 June 1985, 32–33, NSF History Office Files.

46. Minutes, MPE Divisional Committee, 13th Meeting, 24–25 October 1957, appendix B.

47. Ibid.

48. "Engineering Science and the National Science Foundation: A Report to the American Society for Engineering Education," undated but by internal evidence prepared between the summer of 1958 and September 1959, RANN Files, Office AD/Eng, File 25: RA/D. See also Waterman to DuBridge, 15 May 1959, statistics, ATW Notes 1959. NSF Annual Reports, FY 1957, 54; FY 1958, 47.

49. NSB Book, 48th Meeting, 6 September 1957, 2–4.

50. NSF Annual Reports, FY 1953, 35; FY 1957, 54.

51. Lomask, *A Minor Miracle,* 113.

52. Dwight D. Eisenhower, Radio and Television Address to the American People on Science in National Security, 7 November 1957, *Public Papers of the Presidents of the United States, 1957,* esp. 790, 792–93.

53. NSB-79-155, Memorandum, General Counsel to Director, Some Key Events in the Development of NSF Programs, 5 June 1972, 2–3, NSB Subject Files: RANN. NSF Budgets to Congress, FY 1960, FY 1961. Lomask, *A Minor Miracle,* 113. Minutes, MPE Divisional Committee, 15th Meeting, 20–21 October 1958, 2. NSF Annual Report, FY 1958, 3–13.

54. NSF Annual Report, FY 1958, 11. NSF-57-146, press release [speech on reaction to Sputnik], 7 November 1957.

55. NSF Annual Report, FY 1959, xi–xvii, 4–5.

56. NSF Annual Report, FY 1959, 8–10. The weather modification act became PL 85-510 of the 85th Congress, 1958.

57. NSF Budget to Congress, FY 1962, 58–60. MPE Annual Report, FY 1959, 17, NSF History Office Files.

58. MPE Annual Report, FY 1959, 16–17. Robert Schmidt, University of Detroit, Bending and Buckling of Multi-Layer Shells, WNRC, RG 307, Acc. #307-73-003, Box 51, File: University of Detroit.

59. Edwin T. Layton, Jr., "Mirror-Image Twins: The Communities of Science and Technology in Nineteenth-Century America," *Technology and Culture* 12 (October 1971): 579.

CHAPTER 2: TOWARD A LARGER ROLE

1. Gene M. Nordby and Robert N. Faiman, "NSF Promotes Basic Research in Civil Engineering," *Civil Engineering* 28 (April 1958): 64–67. MPE Annual Report, FY 1958, 1–2, 12, NSF History Office Files.

2. NSF Annual Report, FY 1962, 10, 39. University of Wisconsin, Automatic Optimization of Continuous Processes, 20 May 1960, WNRC, RG 307, Acc. #307-73-003, Box 21, File: Automatic Optimization.

3. NSF Annual Reports, FY 1959, 41; FY 1960, 49–50. Engineering Divisional Committee, 3rd Meeting, 22–23 July 1965, 6–7, WNRC, RG 307, Acc. #307-70A-3621, Box 14, File: Advisory Committee for Engineering FY 1965. NSF Annual Report, FY 1961, 33–34.

4. NSF Annual Report, FY 1961, 34.

5. MPE Annual Reports, FY 1957; FY 1959, 2–3; Staff Notes, MPE Divisional Committee, 23rd Meeting, 12–13 October 1962, 2–4, NSF History Office Files.

6. J. Merton England, untitled, unpublished, and unpaginated manuscript history of NSF, "Chapter 4: Little Science and Distribution Patterns"; see text accompanying nn. 20, 31. Daniel J. Kevles, "Principles and Politics in Federal R&D Policy, 1945–1990: An Appreciation of the Bush Report," preface to NSF's 1990 reprint of *Science—The Endless Frontier,* xiii.

7. England manuscript, chapter 4, text accompanying nn. 35–36, 43. Eric A. Walker, *Now It's My Turn: Engineering My Way* (New York: Vantage Press, 1989), 146. Walker wrote that he joined the National Science Board in 1962, but the October 1960 *Journal of Engineering Education,* in a biographical sketch on his becoming ASEE president for 1960–61, noted that he had "just this fall" been appointed to the NSB. Walker to Waterman, Resolution to the National Science Board, 23 November 1960, NARA, RG 307, OD, Box 48, File: MPE Division—Engineering. NSB-413, Memorandum, Waterman to NSB, Support of Engineering Sciences, 1 November 1961. See also ATW Notes 1961.

8. England manuscript, chapter 4, text accompanying nn. 37–39. Waterman, Notes for MPE Board Committee Meeting, 17 January 1961, ATW Notes 1961. Waterman to Morris, 9 February 1961, NARA, RG 307, OD, Box 48, File: MPE Division—Engineering. Robertson to Waterman, Position on Engineering Sciences for National Science Board Meeting, 10 March 1961; and Waterman, Comments on Robertson's memorandum dated 10 March 1961, 15 March 1961, both in ATW Notes 1961.

9. England manuscript, chapter 4, text accompanying nn. 3, 40–42. Randal Robertson, transcription of interview with J. M. England, 21 June 1985, 10–14, NSF History Office Files.

10. Notes, MPE Divisional Committee, 20th Meeting, 13 April 1961, 2–3. NSB, Approved Minutes, 69th Meeting, 19 January 1961, 13; 70th Meeting, 16–17 March 1961, 18.

11. Memorandum, Robertson to Waterman, Position on Engineering Sciences, 10 March 1961, ATW Notes 1961.

12. Memorandum, Robertson to Waterman, Reorganization of Division of MPE, 1 June 1961, NARA, RG 307, OD, Box 48, File: Division of MPE. Staff Notes, MPE Divisional Committee, 21st Meeting, 30–31 October 1961, 4; 22nd Meeting, 23–24 March 1962, 8.

13. England manuscript, chapter 4, text accompanying n. 43. NSF-61-153, press release, Basic Research in Engineering to Be Given Impetus by New Foundation Unit, 15 November 1961. Waterman to Lohmann, 6 October 1961, ATW Notes 1961.

14. NSB-491, Memorandum, Waterman to NSB, Policy Relating to the Support of Projects in the Engineering Sciences, 15 May 1962, NARA, RG 307, OD, Box 76, File: MPE Engineering Sciences Division. Walker, *Now It's My Turn,* 244.

15. "Engineering and the National Science Foundation," *Journal of Engineering Education* 53 (January 1963): 278.

16. Wisely to Haworth, 21 August 1963; and Van Antwerpen to Haworth, 18 September 1963, both in WNRC, RG 307, Acc. #307-70A-3621, Box 5, File: Engineering.

17. W. R. Marshall, "Engineering and Government," *Chemical Engineering Progress* 59 (December 1963): 51.

18. Lomask, *A Minor Miracle,* 113–18. Philip Handler, transcription of interview with Milton Lomask, 20 February 1974, esp. 2, 5–6. NSF Annual Report, FY 1964, x–xii.

19. Leland J. Haworth, "In Support of Engineering Excellence," *Proceedings of the Institute of Electrical and Electronics Engineers* 52 (January 1964). Lawrence P. Grayson, "A Brief History of Engineering Education in the United States," *Engineering Education* (December 1977): 261.

20. Eric Walker, interview with author, 1 October 1990. Walker to Anderson, 2 December 1980, NSB Subject Files: Engineering—General. Walker, *Now It's My Turn,* 244. NSB-638, Provisional Minutes, 81st Meeting, 15–17 November 1962, 22. Leland Haworth, interview [unattributed], 8 May 1973.

21. NSB-64-19, Minutes, Executive Session, 16–17 January 1964, ES:91:4. NSF 64-131, press release, Dr. Eric Walker Elected Chairman, National Science Board, 22 May 1964.

22. NSB-64-19, ES:91:9, attachment 1, 4–5. Walker, *Now It's My Turn,* 244.

23. NSB-64-19, ES: 91:9, attachment 1, 4–5.

24. Report of Approved Grants and Contracts and of Proposals Declined, NSB Book, 92nd Meeting, 20–21 February 1964. NSB-64-119, Report of Approved Grants and Contracts and of Proposals Declined, 11 June 1964.

25. Memorandum, Robertson to Haworth, Divisional Status for Engineering Section, 22 April 1964, WNRC, RG 307, Acc. #307-70A-3621, Box 23, File: Division

of Engineering 1965–66. NSB-64-87, Provisional Minutes, 93rd Meeting, 23–24 April 1964, 93:17–18. NSB-64-120, Memorandum, Haworth to NSB, Suggestions for the Creation of a Divisional Committee for Engineering, 11 June 1964.

26. NSB-64-168, Provisional Minutes, 96th Meeting, 10–11 September 1964, 96:3. NSB-64-134, Provisional Minutes, 95th Meeting, 18–19 June 1964. NSF-64-138, press release, NSF Establishes Division of Engineering; Dr. John Ide Appointed Division Director, 9 July 1964.

27. Eric Walker, interview with author, 1 October 1990. Walker, *Now It's My Turn,* 244. NSF-64-138. Robertson to Haworth, Divisional Status for Engineering Section, 22 April 1964. SACLANT stands for Supreme Allied Commander Atlantic.

28. Staff Notes, Engineering Divisional Committee, 1st Meeting, 17–18 December 1964, esp. 5–6, WNRC, RG 307, Acc. #307-70A-3621, Box 4, File: Divisional Committee for Engineering.

29. NSB-79-155, Memorandum, General Counsel to Director, Some Key Events in the Development of NSF Programs, 5 June 1972, 4, NSB Subject Files: RANN. NSF-64-100, press release, NSF Graduate Traineeship Program Announced at Engineers Joint Council Meeting, 14 January 1964. NSF Annual Report, FY 1964, xxiv.

30. Engineering Section Annual Report, FY 1964, 10–11, NSF History Office Files.

31. Engineering Division Annual Report, FY 1965, 1, 12–13, 16, NSF History Office Files.

32. Ibid., 3.

33. Ibid., 3–11. NSF Annual Report, FY 1965, 76.

34. Engineering Division Annual Report, FY 1965, 20–22.

35. Hollomon to Walker, 27 April 1965, WNRC, RG 307, Acc. #307-70A-3621, Box 23, File: Division of Engineering 1965–66. See also "Administration Witness Criticizes NSF," *C&EN* 43 (26 July 1965): 31. Ide to Haworth, Consensus of Divisional Committee for Engineering in Their Meeting 22–23 July 1965, 27 August 1965, WNRC, RG 307, Acc. #307-70A-3621, Box 14, File: Divisional Committee for Engineering 1965.

36. By Reorganization Plan No. 5 of 1965, NSF's statutory Divisional Committees were abolished and replaced with Advisory Committees. They functioned similarly. See O/D 65-19, Staff Memorandum, Establishment of Advisory Committees, 24 September 1965, WNRC, RG 307, Acc. #307-70A-3621, Box 14, File: Divisional Committee for Engineering 1965. Minutes, Advisory Committee for Engineering, 1st Meeting, 29–30 November 1965, 5–6; and Engineering Divisional Committee, 3rd Meeting, 22–23 July 1965, 8–10, and attachment, both in WNRC, RG 307, Acc. #307-70A-3621, Box 14, File: Advisory Committee for Engineering 1965. Walker, interview with author, 1 October 1990.

37. NSF Annual Report, FY 1965, vii–x, xiv–xvi.

38. See Lomask, *A Minor Miracle,* chap. 11. See also "Whatever Happened to . . . Project Mohole," *IEEE Spectrum* 27 (September 1990): 58–60. Daniel

Greenberg also writes, caustically, of Project Mohole in *The Politics of American Science.* NSF Annual Report, FY 1965, xiv–xvi, 94–97, 100–101.

39. NSF Annual Report, FY 1965, vii–xxxiii, esp. xxv–xxvi, xxxi–xxxiii.

40. Donald F. Hornig, "National Science Foundation," *C&EN* 43 (5 July 1965): 62–65.

41. "NSF May Be More Aggressive in the Future," *C&EN* 43 (5 July 1965): 26–27. Library of Congress, Legislative Reference Service, Science Policy Research Division, *The National Science Foundation: A General Review of Its First 15 Years,* Report to the House Committee on Science and Astronautics, Subcommittee on Science, R&D, 89th Cong., 1st sess., 1965, 1–12. See Lomask, *A Minor Miracle,* chap. 11, "Project Mohole and JOIDES."

42. "NSF May Be More Aggressive in the Future," 26–27. Alan T. Waterman, Statement, House Subcommittee on Science, R&D, Committee on Science and Astronautics, 30 June 1965, LJH Notes 1965.

43. House Subcommittee on Science, R&D, Committee on Science and Astronautics, *Hearings,* vol. 1, 89th Cong., 1st sess., June–August 1965, 572. Walker interview.

44. Edwin T. Layton, Jr., *The Revolt of the Engineers: Social Responsibility and the American Engineering Profession* (Cleveland, Ohio: Press of Case Western Reserve University, 1971), 251–52.

45. Ibid. Bruce Seely, "Research, Engineering, and Science in American Engineering Colleges, 1900–1960," *Technology and Culture* 33 (April 1993): 381–83.

CHAPTER 3: APPLIED SCIENCE AND RELEVANCE

1. Emilio Q. Daddario, interview with author, 16 August 1990. "New Overseers for Federal Science," *Science* 142 (11 October 1963): 210.

2. Emilio Q. Daddario, "A Revised Charter for the Science Foundation," *Science* 156 (1 April 1966). Daddario interview.

3. D. S. Greenberg, "Congress and Science: NSF Hearings Provide Some Illuminating Insights on the Deteriorating Relationship," *Science* 142 (18 October 1963): 368.

4. Daddario's H.R. 13696, lightly altered as H.R. 14838, passed by voice vote in the House in 1966, but, with no Senate action, the effort died at the close of the session. At the start of the 90th Congress in January 1967, Daddario introduced a new bill, presented "clean" in February as H.R. 5404. The House passed it in April; the Senate Special Subcommittee on NSF, chaired by Kennedy, took up a companion bill, S. 2598, in October and held hearings in November 1967. In February 1968 it favorably reported the House version, which was accepted by the Senate Committee on Labor and Public Welfare and the full Senate in May. The House agreed to Senate amendments in June, and in July H.R. 5404 became PL 90-407.

5. NSF Annual Report, FY 1968, xi–xv, 3–4. See NSB-68-186, Memorandum, Haworth to NSB, Revisions of NSF Act, 8 July 1968; and attachment, Explana-

tion of H.R. 5404: Major Changes Effected, 5 July 1968. NSB/C-III-66-10, Memorandum of Discussion, Committee 3, 13th Meeting, 7 February 1966.

6.   Kusinski to General Counsel, Brief Legislative History of Section 3(c) [Applied Research] of the NSF Act, 22 February 1971, NSF History Office Subject Files: NSF Act Amended (Daddario). See also NSB/C-III-66-10. To some, it seemed symbolic as well as sadly real that Waterman had died in late November 1967.

7.   H. Rept. 34, House Committee on Science and Astronautics, *Amending the National Science Foundation Act of 1950 to Make Improvements in the Organization and Operation of the Foundation: Report to Accompany H.R. 5404,* 90th Cong., 1st sess., 6 March 1967, 17.

8.   House Subcommittee on Science, R&D, Committee on Science and Astronautics, *Hearings on a Bill to Amend the National Science Foundation Act of 1950,* 89th Cong., 2d sess., 19–21 April 1966, 39.

9.   Senate Special Subcommittee on Science, Committee on Labor and Public Welfare, *Hearings on the National Science Foundation Act Amendments of 1968,* 90th Cong., 1st sess., 15–16 November 1967, 69–70. Haworth to Daddario, 18 April 1966, 2–3, LJH Notes 1966. Haworth to Hollomon, 22 September 1967, LJH Notes 1967. Leland Haworth, interview with Lomask and England, 17 April 1974. House, *Hearings on a Bill to Amend the National Science Foundation Act,* 19–21 April 1966, 10.

10.   Bruce Seely, "Research, Engineering, and Science in American Engineering Colleges, 1900–1960," *Technology and Culture* 33 (April 1993): 383, 385.

11.   Engineering advocate Eric Walker was not reappointed to the board when his term expired in 1966. By his account, he disturbed and embarrassed "the 'pure' scientists who considered the NSF to be their private property." He had also antagonized Haworth with his independence and Lyndon Johnson by refusing to help in his reelection campaign. Letter to author, 1 August 1991. (Walker evidently remembered the year and particular election incorrectly. He wrote it as 1968, but Johnson did not run that year. Handler assumed the chairmanship in 1966. He himself was appointed to the board in 1960.)

12.   NSB, Commentary Concerning H.R. 13696, *A Bill to Amend the National Science Foundation Act of 1950,* 15 April 1966, 4–6, NSF History Office Subject Files: NSF Act Amended (Daddario). Senate, *Hearings on the National Science Foundation Act Amendments,* 15–16 November 1967, 83–84. See also NSB/C-III-66-10. NSB/C-I-68-14, Committee I Paper on Applied Research, 18 March 1968, 1–2.

13.   Senate, *Hearings on the National Science Foundation Act Amendments,* 15–16 November 1967, 99. House, *Hearings on a Bill to Amend the National Science Foundation Act,* 19–21 April 1966, 98.

14.   House, *Hearings on a Bill to Amend the National Science Foundation Act,* 19–21 April 1966, 111. House, Subcommittee on Science, R&D, Committee on Science and Astronautics, *Report on the National Science Foundation: Its Present and Future,* 89th Cong., 1st sess., 1966, 50.

15. See Chalmers W. Sherwin and Raymond S. Isenson, *First Interim Report on Project Hindsight* (Washington, D.C.: Office of the Director of Defense Research and Engineering, 1966). Jeffrey K. Stine, *A History of Science Policy in the United States, 1940–1985,* Report No. 1 for the Task Force on Science Policy, House Committee on Science and Technology, 99th Cong., 2d sess., September 1986 (Washington, D.C.: GPO, 1986), 59–60. Daniel S. Greenberg, *The Politics of American Science* (Harmondsworth: Penguin Books, 1969), 58–61. D. S. Greenberg, "Hindsight: DOD Study Examines Return on Investment in Research," *Science* 154 (18 November 1966): 872–73. F. C., "Hindsight Study Adds Kind Words for Basic Research," *Science* 159 (26 January 1968): 413. James Kip Finch, *The Story of Engineering* (Garden City, N.Y.: Anchor Books, Doubleday, 1960), 526–27. Edwin Layton, "Mirror-Image Twins: The Communities of Science and Technology in Nineteenth-Century America," *Technology and Culture* 12 (October 1971): 563–64. "Projects Hindsight and TRACES: Quantitative Histories of Science-Technology Relationships," 28 December 1969, in "Organizational Notes," *Technology and Culture* 11 (April 1970): 221–22.

16. See Illinois Institute of Technology Research, *Technology in Retrospect and Critical Events in Science (TRACES),* 2 vols. (Washington, D.C.: NSF, 1968). Stine, *A History of Science Policy,* 60. Peter Thompson, "TRACES: Basic Research Links to Technology Appraised," *Science* 163 (24 January 1969): 374–75. "Projects Hindsight and TRACES."

17. See Mary Ellen Mogee, "Public Policy and Organizational Change: The Creation of the RANN Program in the National Science Foundation" (M.A. thesis, George Washington University, 1973), 107–8.

18. Ken Hechler, *Toward the Endless Frontier: History of the Committee on Science and Technology, 1959–79* (Washington, D.C.: GPO, 1980), 141. Daddario interview. Greenberg, *The Politics of American Science,* chap. 12.

19. Daddario, "A Revised Charter for the Science Foundation."

20. NSF Annual Report, FY 1969, 4.

21. Memorandum, Maechling to Director, Daddario Bill—New Factors, 27 October 1967, NSF History Office Subject Files: NSF Act Amended (Daddario).

22. NSB-69-16, Advisory Committee for Engineering, Annual Report, FY 1968. Robertson to Mogee, 6 March 1973, 2, NSF History Office Subject Files: RANN.

23. Lomask, *A Minor Miracle,* 215, 223–27. Philip Handler, transcription of interview with Milton Lomask, 20 February 1974. Bernard Sisco, transcription of interview with Milton Lomask and J. M. England, 23 May 1974, 6.

24. Lomask, *A Minor Miracle,* 251–53. See also H. Guyford Stever, transcription of interview with Arley T. Bever, 24 November 1976, 13–14.

25. Section 203 of PL 91-121, the FY 1970 military procurement bill. Stine, *A History of Science Policy,* 62–63. Mike Mansfield, "Too Many Research Eggs in Defense Baskets?" *Christian Science Monitor,* 12 September 1970, 9.

26. Cited in Thomas P. Hughes, *American Genesis: A Century of Invention and Technological Enthusiasm, 1870–1970* (New York: Viking, 1989), 444–50.

27. Bruce Sinclair, *A Centennial History of the American Society of Mechanical Engineers, 1880–1980* (Toronto: University of Toronto Press, 1980), 207. Sinclair limits his comment to mechanical engineers, but the literature suggests that this sentiment was common among engineers.

28. Larry Resen, "Opinion and Comment," *Chemical Engineering Progress* 64 (January 1968): 51. Louis N. Rowley, "In the Public Interest," *Mechanical Engineering* 90 (August 1968): 14–16. "Goals, a Proposed Statement," *Mechanical Engineering,* April 1970, 18–19. See Edwin T. Layton, Jr., *The Revolt of the Engineers: Social Responsibility and the American Engineering Profession* (Cleveland, Ohio: Press of Case Western Reserve University, 1971).

29. Engineering Division Annual Report, FY 1966, 24–26, NSF History Office Files.

30. NSB-70-6, Advisory Committee for Engineering, Annual Report, FY 1969, 2–3. Advisory Committee for Engineering, Annual Report, FY 1967, WNRC, RG 307, Acc. #307-74-038, Box 1, File: Advisory Committee for Engineering 1968.

31. NSB-70-6, 3–4. NSF Budget to Bureau of Budget, FY 1971, 50.

32. NSB-LRP-69-16, Draft Report, Long Range Planning Committee, esp. 3–4.

33. Lomask, *A Minor Miracle,* 215–16. Daddario interview. Joel Snow, transcription of interview with Milton Lomask and J. M. England, 17 July 1974, 28–29.

34. Randal M. Robertson to Mary Ellen Mogee, 6 March 1973. Robertson is quoting himself in a memorandum to the director on applied research, 7 February 1969, 10, WNRC, RG 307, Acc. #307-75-052, Box 10, File: Research, Applied Research.

35. Mogee, "Public Policy and Organizational Change," 41–42.

36. Robertson to Mogee, 6 March 1973, 3. Robertson to Director, Applied Research, and attachments, 7 February 1969, WNRC, RG 307, Acc. #307-75-052, Box 10. House Subcommittee on Science, R&D, Committee on Science and Astronautics, *Hearings on 1970 National Science Foundation Authorization,* 91st Cong., 1st sess., 25 March 1969, 231–40.

37. Robertson to McElroy, Initiation of New Program Interdisciplinary Research Relevant to Problems of Our Society (IRPOS), 3 October 1969, NSF History Office Subject Files: RANN. Memorandum, McElroy to Staff, Establishment of Office of Interdisciplinary Research, 27 October 1969, LJH-WDM Notes 1969.

38. Memorandum, McElroy to Staff, Establishment of Office of Interdisciplinary Research 27 October 1969; and McElroy to Russell et al., 19 December 1969, both in LJH-WDM Notes 1969. McElroy to Volpe et al., 24 February 1970, WDM Notes 1970. NSB-70-62, Memorandum, McElroy to NSB, 2 March 1970. NSF-69-168, press release, NSF Establishes Problem-Oriented Interdisciplinary Research Program, 22 December 1969. NSF-70-117, press release, Snow Describes Interdisciplinary Research Program in Authorization Testimony, 19 February 1970. Robertson to Mogee, 6 March 1973. Syl McNinch, "The Rise and Fall of RANN," 1984, 3, NSF History Office Subject Files: RANN.

39. Memorandum, Consolazio to McElroy, 5 December 1969; and William V. Consolazio et al., *The Foundation's Role in the Solution of Problems of Social Relevancy: A Report to the Director,* December 1969, unpaginated, both in WNRC, RG 307, Acc. #307-75-052, Box 10, File: Task Force on the Foundation's Role in the Solution of Problems of Social Relevance. This report was finalized in January 1970.

40. Memorandum, Todd to Planning Director, Problem-Oriented Research, 1 May 1970, Todd Chron 1965–75. W. R. Marshall, Jr., "Social Directions of Engineering," *Chemical Engineering Progress* 67 (January 1971): 11.

41. NSB-70-62, Memorandum, McElroy to NSB, 2 March 1970, Appendix E. Mogee, "Public Policy and Organizational Change," 50, 82. The Office of Management and Budget (OMB) superseded the Bureau of the Budget in 1970.

42. Minutes, Advisory Committee for Engineering, 11th Meeting, 26–27 February 1970, 5, WNRC, RG 307, Acc. #307-75-053, Box 2, File: Engineering—Advisory Committee 1970. McElroy to Tyler, 22 April 1970, with attached resolutions, WDM Notes 1970.

43. Advisory Committee for Engineering, Annual Report, FY 1970, 2–6, WNRC, RG 307, Acc. #307-77-056, Box 1, File: Engineering Advisory Committee.

44. Mogee, "Public Policy and Organizational Change," 49–56. John E. Marshall, "Social Science and Problems of Society: Some Issues, Challenges, and Opportunities for NSF," [March 1970], WNRC, RG 307, Acc. #307-75-053, Box 5, File: IRPOS 1970.

45. D. S. Greenberg, "Research Priorities: New Program at NSF Reflects Shift in Values," *Science* 170 (2 October 1970): 144.

46. Mogee, "Public Policy and Organizational Change," 64–65. Carlyle E. Hystad, transcription of interview with Lomask, 25 July 1974, 1–4. NSF Annual Report, FY 1971, Director's Statement and 57.

47. Lomask, *A Minor Miracle,* 239–41. NSB-79-155, Memorandum, General Counsel to Director, Some Key Events in the Development of NSF Programs, 5 June 1972, 14–16, NSB Subject File: RANN.

48. Lomask, *A Minor Miracle,* 241–50; NSF Budgets to Congress, FY 1968, 146, and FY 1969, 1.

49. NSB-79-155, 6. David quoted in Mogee, "Public Policy and Organizational Change," 81. Hystad interview with Lomask, 3, 9. Most of the engineering disciplines reacted to society's calls for confronting problems at home. As an example, *Chemical Engineering Progress* ran such articles as "Social Directions of Engineering" (January 1971), "Evolution or Revolution in a Professional Society" (August 1971), and "Utilization of Unemployed Engineers in Air Pollution Control" (September 1971).

50. COPEP, *Federal Support of Applied Research* (Washington, D.C.: NAE, 1970), esp. cover letter, Starr to Linder, 4, 6.

51. COPEP, *Priorities in Applied Research: An Initial Appraisal* (Washington, D.C.: NAE, 1970), esp. cover letter, Starr to Linder, 5, 11–31.

52. NSB-70-347, Approved Minutes, 134th Meeting, 17 December 1970, 134:2. Memorandum, Bisplinghoff to McElroy, Applied Research in NSF, 5 December 1970, Paul Herer RANN Files. Raymond L. Bisplinghoff, transcription of interview with Arley T. Bever, 28 September 1976, 13–14. Lomask, *A Minor Miracle,* 241–43.

53. Bisplinghoff interview with Bever, 15. Task force member Syl McNinch later credited Bisplinghoff for proposing the program title RANN. See McNinch, "The Rise and Fall of RANN," 2. Mogee, "Public Policy and Organizational Change," 67–75. Lomask, *A Minor Miracle,* 243–46. Herbert Carter, transcript of interview with Arley T. Bever, 4 November 1976, 6.

54. Joel Snow, transcription of interview with Milton Lomask and J. M. England, 17 July 1974, 10. Snow paraphrases Handler in this interview. Press releases include NSF-71-116, Research in Science, Engineering Offers Great Public Benefit, McElroy Says in Authorization Testimony, 25 February 1971; NSF-71-128, McElroy Says NSF to Respond to National Challenges, 23 March 1971; NSF-71-130, Knowledge of Societal Problems, Provision of Means to Solve Them Present Critical Challenge, Says Carter, 23 March 1971.

55. NSB-79-155, 15–16. McNinch, "The Rise and Fall of RANN," 3–5.

56. NSF-71-118, press release, Dr. Alfred Eggers Appointed to Head Major New NSF Unit, 2 March 1971. "A Conversation: Bisplinghoff and Eggers," *Mosaic* 2 (Spring 1971): 23–27.

57. Memorandum, Snow to Addressees, 10 February 1971, and attached Budget Justification, D:17-D:34, NSF History Office Subject Files: RANN Task Force. History and Issue Paper: Research Applied to National Needs, unattributed, undated, NSF History Office Files. McNinch, "The Rise and Fall of RANN," 1–2. Memorandum, Todd to Division Director, Engineering, Transfer of Research Activities to Research Applications, 19 July 1971, Todd Chron 1965–75.

58. William E. Cory, "EEs and Public Needs: A Recommendation," *IEEE Spectrum* 9 (February 1972): 33–35.

59. Alfred J. Eggers, Jr., Statement, Subcommittee on Science, R&D, Committee on Science and Astronautics, 24 February 1972, 1, 11–12, Office AD/Eng, RANN File 17: RA Advisory Committee.

60. "RANN: Growth at NSF Stirs Concern, but . . . ," *Science & Government Report* 2 (15 July 1972): 1–4.

61. Raymond L. Bisplinghoff, "NSF's Research Applied to National Needs Program—A Progress Report," Remarks at AAAS, 29 December 1972, esp. 2, 11, NSB Subject Files: RANN.

62. Research Applications Advisory Committee, Annual Report, FY 1971, Office AD/Eng, RANN File 17. White to Eggers, 28 December 1971; and Memorandum, Cowhig to AD/RA Senior Staff, University Views of the RANN Program, 3 December 1971, both in Herer RANN Files.

63. Memorandum, Hystad to Loweth and Young, Preliminary Results of Review of NSF Progress in Implementing RANN, 14 February 1972, Herer RANN Files.

64. Eggers to Wenk, 11 April 1972, Office AD/Eng, RANN File 23: COPEP. Memorandum, Consolazio to Director, NAE Study Report: *Priorities for Research Applicable to National Needs,* 8 November 1972; and attached memorandum, 3 November 1972, both in WNRC, RG 307, Acc. #307-77-080, Box 7, File: Priorities for Research Applicable to National Needs. See COPEP, *Priorities for Research Applicable to National Needs* (Washington, D.C.: NAE, 1973).

65. NSB-72-47, Advisory Committee for Engineering, Annual Report, FY 1971, 1.

66. Memorandum, Todd to AD/R, Bureaucratic Buccaneering, 29 January 1971, Todd Chron 1965–75.

67. NSB-72-47, 9. Advisory Committee for Engineering, 13th Meeting, 20–21 April 1971, WNRC, RG 307, Acc. #307-77-056, Box 1, File: Engineering Advisory Committee.

68. S. Rept. 92-1028, *National Science Policy and Priorities Act of 1972: Report to Accompany S. 32,* 92nd Cong., 2d sess., 9 August 1972, 1, 40, 41; S. Rept. 93-1254, *National Policy and Priorities for Science and Technology Act of 1974: Report to Accompany S. 32,* 93rd Cong., 2d sess., 9 October 1974, 16, both of Senate Committee on Labor and Public Welfare. S. Rept. 94-622, Senate Committee on Labor and Public Welfare and Committee on Aeronautical and Space Sciences, *National Policy, Organization, and Priorities for Science, Engineering, and Technology Act of 1976,* Joint Report to Accompany S. 32, 3 February 1976, 27. Stever, who was doing an almost impossible double duty as White House science advisor and NSF director, left NSF to head this presidential office in August 1976.

69. See Mogee, "Public Policy and Organizational Change," 87–93.

70. Congressman John W. Davis, the Georgia Democrat about to replace Daddario as head of the House science subcommittee, thought in 1970 that "the average representative or senator did not yet understand the difference between basic and applied research or the relation between teaching and research." John J. Beer, "Science and the Federal Government, 1970," *Technology and Culture* 12 (April 1971): 246 and passim.

CHAPTER 4: THE RANN YEARS

1. Syl McNinch, "The Rise and Fall of RANN," 1984, 6, NSF History Office Subject Files: RANN. Joel Snow, transcription of interview with Arley T. Bever, 4 October 1976, 14–15.

2. RA/D, NSF Terrestrial Solar Energy Program, 20 October 1972, Office AD/Eng, RANN File 25: Reorganization RA. NSB-71-216, Proposed Position Statement on National Science Policy and NSF RANN Project Selection and Coordination, 1 July 1971, 3. NSB-71-252, NSB Guidelines for RANN Project Selection and Coordination: NSF Policy Statement, 2 September 1971, 3.

3. Richard C. Jordan, "Solar Energy Powered Systems—History and Current Status," *ASTM Standardization News* 3 (August 1975): 13–14.

4.   Ibid. NSB-72-275, Memorandum, Stever to NSB, Energy Research and Technology Program, 15 November 1972, 4, 13–14. NSF Terrestrial Solar Energy Program, 1972, table 2.

5.   NSB-72-275, 1–6. Jordan, "Solar Energy Powered Systems," 14–16. NSB-73-228, Memorandum, Stever to NSB, Contracts for Studies of Solar Heating and Cooling of Buildings, 10 September 1973, 73-228-A—B. According to Jordan, by 1975 approximately one hundred solar houses had been built in the United States. [Jay Holmes], internal paper, appendix A: Background of RANN, 16 January 1974, 6, NSF History Office Subject Files: RANN. Holmes wrote that NSF was named the lead agency for solar research in early 1973. RA/D, NSF Terrestrial Solar Energy Program (Expanded), 27 June 1973, Office AD/Eng, RANN File 20: Terrestrial Solar Energy Program. Richard T. Duncan and Eugene R. Doering, "Solar Heating for Atlanta School," *ASHRAE Journal* 17 (July 1975): 35. ASHRAE is the American Society for Heating, Refrigeration, and Air Conditioning Engineers.

6.   NSB-74-263, Memorandum, Stever to NSB, Proposed Award from the RANN Program, 31 October 1974, 74-263-A—1-1. Richard T. Duncan, "Solar Energy Will Cool and Heat Atlanta School," *ASHRAE Journal* 16 (September 1974): 47–49. Duncan and Doering, "Solar Heating for Atlanta School," 35–39.

7.   Richard T. Duncan et al., "Lessons Learned," in *Solar Heating and Cooling of Buildings,* vol. 3 of *Sharing the Sun: Solar Technology in the Seventies,* ed. K. W. Boer, Joint Conference, American Section, International Solar Energy Society and Solar Energy Society of Canada, Inc., 15–20 August 1976, Winnipeg, 153–67. This article also appeared in the November 1976 issue of *ASHRAE Journal.* Garfield B. Harris, Assistant Superintendent, Facilities Services, Atlanta Public Schools, to Joan M. Zenzen, 5 November 1990.

8.   NSB-72-266, Memorandum, Stever to NSB, Proposed Grants from the RANN Program, 7 November 1972. NSF 74-52, NSF/RANN, Energy Research and Technology: Abstracts of NSF Research Results, August 1974, NSB Subject Files: RANN. B. O. Seraphin, *Optical Properties of Solids: New Developments* (Amsterdam: North-Holland Publishing Company, 1976), 929. Jordan, "Solar Energy Powered Systems," 13–14.

9.   NSF Organizational Charts, July 1, 1974–June 30 1975, 105–6.

10.   NSB-75-36, Provisional Minutes, 169th Meeting, 16–17 January 1975, ES-169:22. "Dateline Washington," *ASHRAE Journal* 18 (August 1976): 10–11. William Booth, "Solar Power: Costly Technology Whose Time Has Come?" *Washington Post,* 15 October 1990, A3. Marcel Belanger, Sunrayce participant, interview with author.

11.   [Unattributed], Program Plan, Environmental Aspects of Trace Contaminants, undated [folder labeled 1971], Office AD/Eng, RANN File 16: RA Advisory Committee.

12.   NSB-70-138, Memorandum, McElroy to NSB, Proposed Projects from the Office of Interdisciplinary Research, 11 May 1970, 70-138-5-1. NSB-72-132,

Memorandum, Stever to NSB, Proposed Grants from RANN Program, 8 May 1972, 72-132-5-1–5-6. NSB-74-196, Memorandum, Stever to NSB, Proposed Award from RANN Program, 29 August 1974, 74-196-1–13. Title 40, Code of Federal Regulations, section 80 (40 CFR 80). James Caldwell, Fuels Division, Environmental Protection Agency, telephone interview with Joan M. Zenzen, 16 November 1990.

13. NSB-72-205, Memorandum, Stever to NSB, Proposed Grants from RANN Program, 23 August 1972, 72-205-H. Bobby G. Wixson and J. Charles Jennett, "The New Lead Belt in the Forested Ozarks of Missouri," *Environmental Science & Technology* 9 (December 1975): 1128–33. NSF-RA-G75027, M. Simons, Jr., "Case Study No. 2: Environmental Pollution in the New Lead Belt," *RANN Utilization Experience: Final Report to the National Science Foundation,* 2-3–2-19, Richard Green Files.

14. NSB-73-4, Memorandum, Stever to NSB, Environmental Aspects of Trace Contaminants Program under Rabin, Carrigan, and Goor, 5 January 1973. TID-4500-R62, "Trace Contaminants in the Environment," *Proceedings of the Second Annual NSF-RANN Trace Contaminants Conference,* Asilomar, Pacific Grove, California, 29–31 August 1974, passim. Simons, "Case Study No. 2," 2-6, 2-17–2-19.

15. NSB-73-45, Memorandum, Stever to NSB, Urban Technology Program, 5 February 1973, 8–10.

16. NSB-72-7, Memorandum, Stever to NSB, Proposed Grants from RANN Program, 7 January 1972, 72-7-13-2. NSB-73-45, 8–15. John C. Rowley et al., "Rock Melting Subterrenes—Their Role in Future Excavation Technology," *Proceedings, 1974 Rapid Excavation and Tunneling Conference* 2 (New York: Society of Mining Engineers, 1974), 1777–95. NSF Budget to Congress, FY 1973, B-X-3.

17. ASCE, William H. Wisely, Planning and Initiating a Program for Improving the Effectiveness of Underground Construction, November 1970; Bloss to Wisely, 11 February 1971; Underground Construction Research Council of ASCE in cooperation with AIME, *Summary Report on the Use of Underground Space to Achieve National Goals* (New York: ASCE, 1973); ASCE news release, "Report of Underground Construction Study Available," 24 August 1973; Taylor to Long, Final Technical Letter Report, 6 June 1973; NSF-73-147, press release, Engineering Groups Propose Major R & D Program on Putting Freeways and Electric Lines Underground, 7 May 1973, all in WNRC, RG 307, Acc. #307-80-081, Box 3, File: ASCE, Wisely.

18. Charles Fairhurst, Rational Design of Tunnel Supports, and attached supporting papers, 1971, WNRC, RG 307, Acc. #307-80-081, Box 13, File: University of Minnesota, Fairhurst.

19. Elizabeth Maggio, "'Rock Burst' Detector Developed," *Arizona Daily Star,* 8 June 1976; Stuart A. Hoenig, Final Report, Detection of Pre-Rock Burst Phenomena, October 1980, both in WNRC, RG 307, Acc. #307-81-072, Box 4, File: University of Arizona, Hoenig.

20. H. Guyford Stever, "RANN and National Research," paper delivered at NSF-RANN Symposium, 18–20 November 1973, 5–6. H. Guyford Stever, "Whither

the NSF?—The Higher Derivatives," paper delivered at NAS Symposium Celebrating NSF's Twenty-fifth Anniversary, 21 April 1975, 10, and reprinted in *Science* 189 (25 July 1975): 264–67.

21. Quentin W. Lindsey and Judith T. Lessler, "Utilization of RANN Research Results: The Program and Its Effects," 30 March 1976, iii, Herer RANN Files.

22. Grant to Eggers, 24 February 1976, Office AD/Eng, RANN File 3.

23. NSB-77-188, Memorandum, Advisory Committee for Research Applications Policy to NSB, 21 April 1977. Research Applications Policy Advisory Committee, "Findings," 18–19 April 1977, Office AD/Eng, RANN File 3. NSB-77-228, Provisional Minutes, Open Session, 189th Meeting, 21–22 April 1977, 189:3.

24. Proxmire to Stever, 1 November 1974; and Stever to Proxmire, 3 January 1975, both in NSB Subject Files: RANN.

25. NSB-71-216. NSB-71-252. Eric Walker, interview with author, 1 October 1990. Walker's view on Handler is confirmed by an OMB review of RANN: OMB Memorandum, Hystad to Loweth and Young, Preliminary Results of Review of NSF Progress in Implementing RANN, 14 February 1972, 4, Herer RANN Files. Board member E. M. Piore also opposed the RANN concept, according to Hystad. Philip Handler, transcription of interview with Milton Lomask, 20 February 1974, 3, 5. NSB-71-278, Memorandum, McElroy to NSB, Proposed Projects from RA/D, 29 September 1971. NSB-71-314, Memorandum, McElroy to NSB, Proposed Projects from RA/D, 3 November 1971.

26. Handler interview, 5. NSF-PR76-69, Atkinson Named Acting Head of National Science Foundation, 19 August 1976. NSF-PR77-47, President Announces Intention to Nominate Atkinson as NSF Director, 21 April 1977. Atkinson served as acting director from August 1976 to April 1977, when the president nominated him for the director's post. NSF Annual Report, FY 1976, vii, xii. McNinch, "The Rise and Fall of RANN," 6–7.

27. Mary Ellen Mogee, "Public Policy and Organizational Change: The Creation of the RANN Program in the National Science Foundation" (M.A. thesis, George Washington University, 1973), 111–17, 142–45.

28. Detlev W. Bronk, "The National Science Foundation: Origins, Hopes, and Aspirations," *Science* 188 (2 May 1975): 414. H. Guyford Stever, transcription of interview with Milton Lomask and J. M. England, 15 July 1974, 16–17. Bernard Sisco, transcription of interview with Milton Lomask and J. M. England, 23 May 1974, 14. Handler interview.

29. RANN History and Issue Paper, undated, 10, NSF History Office Subject Files: RANN. MWD-75-84, Comptroller General of the United States, "Opportunities for Improved Management of the Research Applied to National Needs (RANN) Program, National Science Foundation," 5 November 1975, Office AD/Eng, RANN File 22: Background Material on RANN.

30. Mogee, "Public Policy and Organizational Change," 93–94.

31. McNinch, "The Rise and Fall of RANN," 4–5. Stever to Davis, 3 May 1974, HGS Chron 1974.

32. Draft Study of the Research Applied to National Needs Program, July 1975, i–iv, 24, 118, Office AD/Eng, RANN File 22.

33. Comptroller General, "Opportunities for Improved Management of RANN," i–ii. J. Ward Wright et al., "A Proposed RANN Program for Research Utilization," 4 February 1977, Office AD/Eng, RANN File 24: Memo of Understanding. House Subcommittee on Special Studies, Investigations and Oversight and Subcommittee on Science, Research and Technology, Committee on Science and Technology, *The National Science Foundation Program on Research Applied to National Needs (RANN): Goals and Objectives, an Investigative Report,* 94th Cong., 2d sess., March 1977.

34. National Research Council, *Social and Behavioral Science Programs in the National Science Foundation: Final Report* (Washington, D.C.: NAS, 1976), 71 and passim. Jorgenson to Simon, 4 May 1976; and Archibald to Walsh, 29 March 1976, both in Office AD/Eng, RANN File 30: NAS Report.

35. McNinch, "The Rise and Fall of RANN," 7–8. On the science applications task force, see chapter 5.

36. Ibid., 5, 8. Lepkowski, "NSF Mulls Reorganizing Applied Science," 19.

37. NSB-72-47, Memorandum, Stever to NSB, Advisory Committee for Engineering, Annual Report, FY 1971, 29 February 1972.

38. NSB-72-47, 1. Minutes, Advisory Committee for Engineering, April 1971, WNRC, RG 307, Acc. #307-77-056, Box 1, File: Engineering Advisory Committee.

39. NSB, *The Role of Engineers and Scientists in a National Policy for Technology* (Washington, D.C.: GPO, 1972), 2, 4.

40. H. Guyford Stever, "Looking Ahead at NSF," paper delivered at Eleventh Autumn Meeting, NAE, 10 November 1975, 8–9, NSF Library, Speeches of the Director 1975. Atkinson to Fox, 28 September 1975, RCA Chron 1976. Stever to Jericho, 1 October 1975, HGS Chron 1975.

CHAPTER 5: INTERREGNUM

1. NSF PR75-66, NSF Director Announces Agency Reorganization, 10 July 1975. H. Guyford Stever, "Looking Ahead at NSF," Remarks to National Academy of Engineering, Washington, D.C., 10 November 1975, 1–4, 6–7, NSF Historian's Office.

2. NSF PR75-66. Task Force on Engineering and Research Applications, 5 October 1976, 2, Office AD/Eng, RANN File 34: Establishment of ASRA.

3. NSF PR76-9, Applied Research on Problems Posed by Scarcities Is Outlined for House Subcommittee; NSF PR76-10, Stever Cites Importance of Basic Research Effort; and NSF PR76-8, Dr. Hackerman, NSB Chairman, Tells House Committee His Board Supports Strengthening of Basic Research, all 28 January 1976. NSB-76-353, Basic Research: Current Status and the Past Decade, 30 September 1976, 9, 11, NSB Subject Files: Basic Research—General.

4. NSB-76-301, Provisional Minutes, 183rd Meeting, 20 August 1976, 183:9.

5. Task Force on Engineering and Research Applications, 5 October 1976, 6–11.

6. Ibid. Memorandum, Averch to Director, Office of Planning and Resources Management, 21 January 1975, Comments on RANN-Related SRPS Activities (attached to Task Force paper), 2. Draft commentary on RANN, "Program Element Summaries," unsigned, n.d., attached to NSF FY 1976 Budget Request to OMB, see "Engineering" and "RANN."

7. Task Force on Engineering and Research Applications, 5 October 1976, 5.

8. NSF-PR77-87, Atkinson Announces Reorganization of RANN Directorate, 15 September 1977. Mary Ellen Mogee, *Reorganization of the Research Applications Directorate in the National Science Foundation: The Directorate for Applied Science and Research Applications* (Washington, D.C.: Library of Congress Congressional Research Service, 1978), 9–11, appendix 1: Charter. Jack Sanderson, interview with author, 28 November 1990.

9. NSB-77-191, Science Applications Task Force, Progress Report to NSB, 21 April 1977.

10. Mogee, *Reorganization of RA/D,* 12–14. Internal paper, "Part 2: General Recommendations and Comments on Organization," undated and unsigned draft, 1–3, Office AD/Eng, RANN File 34.

11. "Part 2: General Recommendations and Comments," 4–10. Mogee, *Reorganization of RA/D,* 14–15. See also NSB-77-345, Memorandum, Atkinson to NSB, Establishment of a Directorate for Science Applications, 17 August 1977. Report of the Science Applications Task Force to the Director of the National Science Foundation, July 1977, 9–14, Herer RANN Files.

12. John Whinnery, "General Recommendations and Comments on Organization," appendix 2 of final report: Comments on Organization, Office AD/Eng, RANN File 34. Mogee, *Reorganization of RA/D,* 15–18. Memorandum, Whinnery to Task Force on Science Applications, 13 September 1977, Office AD/Eng, RANN File 34.

13. NSB-77-345, Memorandum, Atkinson to NSB, Establishment of a Directorate for Science Applications, with attachments, 17 August 1977; and Memorandum, Atkinson to Science Applications Task Force, Science Application Activities, 23 August 1977, both in Office AD/Eng, RANN File 25: Reorganization of RA. Unattributed, Discussion Paper: Proposed Establishment of the Directorate for Applied Science and Research Applications, 27 December 1977, Office AD/Eng, RANN File 34. O/D 78-2, Atkinson to Organization, Establishment of the Directorate for Applied Science and Research Applications, 4 January 1978; and NSF-78-2, NSF Establishes New Directorate to Strengthen Links between Applied and Basic Research, 5 January 1978, both in NSB Subject Files: Engineering—General.

14. Mogee, *Reorganization of RA/D,* 22–24. Discussion Paper: Proposed Establishment of the Directorate for ASRA. O/D 78-2.

15. Bisplinghoff to Atkinson, 19 August 1977, Office AD/Eng, RANN File 34.

16. ASRA, Director's Briefing, Integrated Basic Research Division, 2 February 1978; and Memorandum, AD/ASRA to Management Council, Important Notice to Academic Institutions and Other Organizations Performing Basic Research, 18 December 1978, both in Office AD/Eng, RANN File 29: Office of IBR.

17. D/ASRA, Division of Applied Research, "Applied Research," *Program Report* 2 (November 1978): 1–8. Mogee, *Reorganization of RA/D,* 21. NSF-PR77-87. D/SEA, viewgraphs of presentation to NSB, 5 September 1977, Office AD/Eng, RANN File 34.

18. Bisplinghoff to Atkinson, 19 August 1977. Mogee, *Reorganization of RA/D,* 22. Sanderson interview.

19. Sanderson to Marriott, 1 February 1978, RCA Chron 1978.

20. D/MPE, Division of Engineering, "Engineering," *Program Report* 2 (May 1978): 1–3. NSF Budget to Congress, FY 1978, B-1.

21. D/MPE, "Engineering," 4–6.

22. Ibid., 6–7.

23. Ibid., 8, 53.

24. House Subcommittee on Science, Research and Technology, Committee on Science and Technology, *Hearings on 1979 National Science Foundation Authorization,* 94th Cong., 2d sess., January 1978, 352, 367–68.

25. NSF Annual Reports, FY 1978, vii; FY 1979, vii–viii.

26. Sanderson interview.

27. Wil Lepkowski, "NSF Mulls Reorganizing Applied Science," *C&EN* 58 (18 August 1980): 18.

28. "Proposed," [hand dated 30 May 1979], Office AD/Eng, RANN File 33: Establishment of EAS.

29. O/D 79-15, Atkinson to Organization, Establishment of the Directorate for Engineering and Applied Science (EAS), 29 May 1979, Office AD/Eng, RANN File 33: Establishment of EAS (comparing the draft document with the printed staff memorandum suggests that the date of the former was appended at a later time). NSF Annual Report, FY 1979, 85.

30. "Proposed," [30 May 1979]; and O/D 79-15.

31. Sanderson to Oden, 16 July 1979; Bourne to Naghdi, 13 August 1979; and others, all in Office AD/Eng, RANN File 33.

32. Weinschel to Sanderson, 25 June 1979, Office AD/Eng, RANN File 33.

33. House Subcommittee on Science, Research and Technology, Committee on Science and Technology, *Hearings on the 1981 National Science Foundation Authorization,* 96th Cong., 2d sess., February 1980, 387–89, 390–91, 396–98.

34. Ibid. On NSF-Cooperative Automotive Research Program (CARP), see NSB-80-528, Provisional Minutes, Open Session, 221st Meeting, 17 December 1980, 221:15–16.

35. House Subcommittee on Science, Research and Technology, Committee on Science and Technology, *Hearings on Government and Innovation: An Engineering Perspective,* 96th Cong., 1st sess., 5 June 1979, 1–2.

36. Sanderson to Glower, 8 November 1979; Sanderson to Brown, 9 November 1979; Atkinson, handwritten notes, Meeting with Engineering Deans 10 December 1979, n.d.; typed summary of draft bill—National Technology Foundation, all in WNRC, RG 307, Acc. #307-87-223, Director's Office Subject Files 1979: EAS.

37. House Subcommittee on Science, Research and Technology, Committee on Science and Technology, *Hearings on H.R. 6910: National Technology Foundation Act of 1980,* 96th Cong., 2d sess., 9–10 and 16–18 September 1980, 1, 38–39, 41, 66–67.

38. Ibid., 80–83.

39. Ibid., 260, 562, 582–83.

40. Ibid., 637, 665, 668, 841. The number of member societies in the AAES, successor to the Engineers Joint Council, apparently fluctuated. In the late 1990s, there were twenty-eight.

41. John Walsh, "NSF under Challenge from Congress, Engineers," *Science* 209 (26 September 1980): 1499. John Walsh, "NSF Boosts Engineering, Applied Research," *Science* 210 (5 December 1980): 1105.

42. See Stuart W. Leslie, *The Cold War and American Science: The Military-Industrial-Academic Complex and MIT and Stanford* (New York: Columbia University Press, 1993), 11–13 and passim; Lawrence P. Grayson, *The Making of an Engineer: An Illustrated History of Engineering Education in the United States and Canada* (New York: John Wiley & Sons, 1993), 178–79; Bruce Seely, "Research, Engineering, and Science in American Engineering Colleges, 1900–1960," *Technology and Culture* 33 (April 1993): 368–69.

43. NSB-80-289, Approved Minutes, Open Session, 217th Meeting, 18–20 June 1980, 217:15–17, 21. Memorandum, Atkinson to Director, OPRM, Task Force C Recommendations Regarding Engineering Research within NSF, 23 June 1980, NSB Subject Files: Engineering—General.

The construction of the phrase "engineering, or engineering science," which suggests the interchangeability of these terms, was the shorthand of M. Kent Wilson, the group's executive secretary. Pettit concluded that retaining both engineering sciences (traditionally supported) and engineering that shared operational paradigms with science would be advantageous to NSF.

44. NSB-80-289, appendix C, attachment 1, 217:22-23. See also Memorandum, Margaret L. Windus to Dick Morrison, Basic and Applied Research, and attachments discussing the evolution of policy, 16 November 1984, NSB Subject Files: Applied Research.

45. NSB-80-289, appendix C, attachment 2, 217:24-29.

46. Housner to Brown, National Science Foundation Program in Earthquake Hazards Mitigation, 27 June 1980, attached to NSB-80-289. Barash to Carter, 14 July 1980; and Sanderson to Barash, 9 October 1980, both in Office AD/Eng, RANN File 32: Reorganization EAS.

47. Lepkowski, "NSF Mulls Reorganizing Applied Science," 18. Walsh, "NSF under Challenge," 1499.

48. Walter G. Vincenti, *What Engineers Know and How They Know It: Analytical Studies from Aeronautical History* (Baltimore, Md.: Johns Hopkins University Press, 1990), 3–7, 13–14.

49. Walsh, "NSF under Challenge," 1499. Memorandum, Bourne to Director, Office of Planning and Resources Management, Further Comments on Basic and Applied Research, 20 August 1980, Office AD/Eng, RANN File 32.

50. NSB-80-289, appendix C, attachment 1, 217:23. Lepkowski, "NSF Mulls Reorganizing Applied Science," 19.

51. NSB-80-358, Memorandum, Langenberg to NSB, Organizational Philosophy and Rationale, 5 September 1980, 2, 4, Office AD/Eng, RANN File 32.

52. Memorandum, Senich to Langenberg, Some Considerations for the Proposed Reorganization of EAS, 29 October 1980, 3–4, Office AD/Eng, RANN File 32.

53. Memorandum, Wirths to Langenburg, Comments on Proposed Reorganization of Engineering & Applied Research at NSF, 6 November 1978, 1–3, Office AD/Eng, RANN File 32.

54. Havens to Branscomb and Langenberg, 1 October 1980, NSB Subject Files: Engineering—General.

55. Sanderson interview. Memorandum, Coakley to Assistant Director, EAS, Revised Recommendations for Reorganization of EAS resulting from the June 1980 NSB Meeting, 16 July 1980; Memorandum, Coakley to Slaughter, Submittal of Draft Plan for Office of Interdisciplinary Research, 5 January 1981; and O/D 81-32, Slaughter to Organization, Establishment of the Office of Interdisciplinary Research, 25 November 1981, all in Office AD/Eng, RANN File 32.

56. NSB-80-528, Provisional Minutes, Open Session, 221st Meeting, 20–21 November 1980, 221:7–10.

57. Walsh, "NSF Boosts Engineering, Applied Research," 1105. S. Rept. 97-131, Senate Committee on Labor and Human Resources, *Authorizing Appropriations for the National Science Foundation for FY 1982,* 97th Cong., 1st sess., 3 June 1981, 4.

58. NSF Annual Report, FY 1980, vii–viii.

59. O/D 81-6, Slaughter to Organization, Reorganization of the National Science Foundation, 19 February 1981, Office AD/Eng, RANN File 32. NSF-PR81-19, National Science Foundation Reorganization Announced, 4 March 1981. NSB-81-105, Provisional Minutes, Open Session, 223rd Meeting, 19–20 February 1981, 223:5.

60. Paul Herer, interview with Joan M. Zenzen, 14 November 1990. Sanderson interview. See also NSF Annual Reports, FY 1976–80; and Mary Ellen Mogee, "Public Policy and Organizational Change: The Creation of the RANN Program in the National Science Foundation" (M.A. thesis, George Washington University, 1973), analysis passim.

61. Sanderson interview.

CHAPTER 6: BY ANY NAME, ENGINEERING

1. "The Shake of Things to Come," *Mosaic* 2 (spring 1971): 11. Michael Gaus, interview with author, 27 November 1990. The section that follows is generally informed by this interview.

2. Gaus interview. NSB-70-251, Memorandum, McElroy to NSB, Proposed Grants for Research in Earthquake Engineering, 24 August 1970. NSF Annual Report, FY 1966, 44–45.

3. NSF Budget to Congress, FY 1968, 152. Engineering Division Briefing, 4 August 1969, 13–15, attached to Memorandum, Mayfield to Heads, Molecular Biology and Chemistry, and Deputy Director, BMS, Discussion of Enzyme Engineering, 14 August 1969, WNRC, RG 307, Acc. #307-75-052, Box 4, File: Division of Engineering 1969.

4. H. B. Seed, Soil and Foundation Response during Earthquakes, 15 April 1968, WNRC, RG 307, Acc. #307-77-002, Box 1, File: Seed, California. J. T. P. Yao, Adaptive Systems for Earthquake Resistant Structures, November 1967, WNRC, RG 307, Acc. #307-73-003, Box 65, File: Yao, New Mexico.

5. NSF Budget to OMB, FY 1972, D-1. NSF Annual Report, FY 1971, 61–62.

6. NSB-72-274, Memorandum, Stever to NSB, Disaster and Natural Hazards Program, 15 November 1972, 4 and passim. NSF Budget to Congress, FY 1973, I-I-5. Program Estimates FY 1978–82, Disasters and Natural Hazards Program Element, attached to Task Force on Engineering and Research Applications, Office AD/Eng, RANN File 34: Establishment ASRA. George Housner, "Earthquake Damage and Earthquake Protection," *Coping with Man-Made and Natural Hazards,* vol. 4 of *RANN 2: Realizing Knowledge as a Resource. Proceedings of the Second Symposium on Research Applied to National Needs,* Washington, D.C., 7–9 November 1976 (Washington, D.C.: NSF, 1977), 9–10.

7. NSF Annual Reports, FY 1976, 92; FY 1978, 102.

8. Draft Memorandum, Todd to Bisplinghoff, Proposal from Cal Tech on Prevention and Control of Natural Disasters, 28 September 1972, Todd Chron 1965–1975. Gaus interview.

9. Stever to Kennedy, 8 October 1975, HGS Notes 1975. NSB-72-274, 5–6.

10. Public Law 95-124, 7 October 1977, 91 Stat. 1098-1100.

11. Rachel Prud'homme, "Oral History, George W. Housner: How It Was," *Engineering & Science* 53 (Winter 1990): 35. This entire issue of the Caltech alumni magazine is devoted to the October 1989 Loma Prieta Earthquake and earthquake research in general. Stever to Ducander, 12 October 1972, WDM-HGS Notes 1972. Stever to Kennedy, 8 October 1975.

12. House Subcommittee on Science, Research and Technology, Committee on Science and Technology, *Earthquake* Hearings, 22–24 June 1976, 1–3. H. Rept. 95-286, part 2, House Committee on Interior and Insular Affairs, *Reducing the Hazards of Earthquakes, and for Other Purposes,* 95th Congress, 1st sess., 9 June 1977, 5–6.

13. Memorandum, Sanderson to Pimentel, Agreement on US-Japan Earthquake Research Program Involving Large-Scale Testing, 24 July 1979, WNRC, RG 307, Acc. #307-87-223, File: Director's Office Subject Files 1979: EAS. Slaughter directive attached to Implementing Arrangement between the Ministry of Construction of Japan and the National Science Foundation of the United States of America for Cooperation in the US-Japan Joint Earthquake Research Program, 11 May 1981, WNRC, RG 307, Acc. #307-87-220, Box 2, File: Engineering 1981. NSB-82-265, Memorandum, Slaughter to NSB, Status Report on US-Japan Cooperative Earthquake Research Program Utilizing Large-Scale Testing Facilities, 13 August 1982. Bye to Schmitt, supplemental testimony, 5 April 1982, WNRC, RG 307, Acc. #307-87-219, Box 2, File: Engineering 1982.

14. Memorandum, Senich to Atkinson, NSF Response to the El Centro-Calexico Earthquake, 18 October 1979, Office AD/Eng, RANN File 27: Division of Problem-Focused Research. NSF Annual Report, FY 1979, 97. See also Goldberger to Atkinson, 7 November 1979, WNRC, RG 307, Acc. #307-87-223, File: Director's Office Subject Files 1979: EAS; and Atkinson to Goldberger, 21 December 1979, RCA Chron 1979.

15. Sam Iker, "Earthquake Engineers Let It Slide," *Mosaic* 12 (Nov./Dec. 1981): 21–24.

16. Ibid., 24–26. NSF Annual Report, FY 1983, viii. Gaus interview and author's earthquake research laboratory visit, 27 November 1990.

17. Bye to Schmitt, 5 April 1982. Gaus interview.

18. Directorate of Engineering, Division of Civil and Environmental Engineering, "Civil & Environmental Engineering," *Program Report* 6 (February 1982): 9–10. Bye to Schmitt, 5 April 1982. Gaus interview.

19. NSF-PR85-72, NSF to Establish First Federally Funded Earthquake Engineering Research Center, 6 November 1985. NSB-85-154, Memorandum, Bloch to NSB, Earthquake Engineering Research Center, 10 June 1985; and attached Draft Program Announcement, Earthquake Engineering Research Center, 10 June 1985.

20. Gaus interview. Nam Suh, interview with author, 17 April 1991. John Scalzi, interview with author, 14 March 1990. PR 86:214, Wilson Calls for GAO to Investigate National Science Foundation's Decision to Place Earthquake Center in New York, 22 September 1986, Pete Wilson Press Release; and attached NSF Statement, 23 September 1986, both in NSB Subject Files: Engineering—General.

21. Gaus interview.

22. NAS, *Science and Technology: A Five-Year Outlook* (San Francisco, Calif.: W. H. Freeman, 1979), 301–2. NSF Annual Report, FY 1984, 62.

23. NSB-85-103, Memorandum, Bloch to NSB, Materials Research Laboratories, 18 April 1985; and attached MRL Program Policy Statement, February 1973, 1. NSF Annual Report, FY 1984, 62. See Graham to Krumhansl, 5 September 1978; Burnstein to Krumhansl, 13 December 1978; Cohen to Krumhansl, 18 December 1978, all in WNRC, RG 307, Acc. #307-87-220, NSF Director's Office Subject Files: MPE January–June 1979. NSF Budget to Congress, FY 1979, B-3.

24. NSB-85-103, MRL Program Policy Statement attachment, 1.

25. NSF Budgets to Congress, FY 1979, B-VI-2; FY 1980, A-VI-3.

26. D/MPE, Division of Materials Research, "Materials Research," *Program Report* 3 (March 1979): 23–26. NSF Annual Report, FY 1985, 26.

27. NSB-85-103.

28. NSB-75-370, Memorandum, Stever to NSB, Proposed Grant Renewals for the Materials Research Laboratories, 18 December 1975, 14-1–3.

29. NSF Budget to Congress, FY 1979, B-VI-2–3. NAS, *Science and Technology* (1979), 302–3.

30. NSB-85-103. See also NSF Annual Report, FY 1985, 24.

31. NSB-85-105, Memorandum, Bloch to NSB, The Center for the Study of Materials—Carnegie-Mellon University, 18 April 1985. NSB-85-106, Memorandum, Bloch to NSB, Materials Research Laboratory—Case Western Reserve University, 18 April 1985.

32. NSB-75-368, Provisional Minutes, 177th Meeting, 20–21 November 1975, 177:3–5, citing NSB-75-328, Memorandum, Stever to NSB, Proposal from the Engineering Division for Support of a National Research and Resource Center for Microfabrication, 30 October 1975. NSB-76-246, Memorandum, Stever to NSB, Progress Report on the Proposal from the Engineering Division for Support of National Research and Resource Center for Sub-Micron Structures (Microfabrication), 27 July 1976, 76-246-1. NSB-76-323, Memorandum, Atkinson to NSB, Project Announcement Requesting Proposals for Establishing a "National Research and Resource Facility for Sub-Micron Structures," 22 September 1976.

33. NSB-77-212, Memorandum, Atkinson to NSB, Proposed Project from Engineering, 25 April 1977. NSB-77-260, Provisional Minutes, Closed Session, 190th Meeting, 19–20 May 1977, CS:190:22–23. See also NSF Budgets to Congress, FY 1979, B-V-3; FY 1980, A-V-3–4. NSB-82-136, Memorandum, Slaughter to NSB, National Research and Resource Facility for Sub-Micron Structures, 19 April 1982.

34. NSF Annual Report, FY 1985, 3.

35. Ronald F. Balazik, "New Materials: How Can the U.S. Regain the Lead?" *Minerals Today* (January 1990): 13–17.

36. NSF Annual Report, FY 1977, 14.

37. NSB-66-103, appendix C: 1965 Annual Report to the National Science Board of the Advisory Committee for Engineering, 14 January 1966, 103:34. NSF Organization Charts 1975–86. NSF Annual Report, FY 1986, 72.

38. NSF Annual Report, FY 1968, 134–35. For a detailed NSF-supported study of computer modeling of police dispatching, see Richard C. Larson, *Urban Police Patrol Analysis* (Cambridge, Mass.: MIT Press, 1972).

39. Krumhansl to Press, 30 October 1978, and attached background paper on microelectronics, 1, WNRC, RG 307, Acc. #307-87-220, File: MPE. NSF Annual Report, FY 1982, 18.

40. Sanderson to Fuqua, 17 May 1978, RCA Chron 1978. NSF Annual Reports, FY 1982, 18; FY 1984, 38. NSF, Division of Electrical, Computer, and

Systems Engineering, "'An Engineering Seminar': Electrical, Computer, and Systems Engineering," August 1983, 9–11, NSB Subject Files: Engineering—Division of ECS. NSF Budget to Congress, FY 1985, Eng-I-1–4. George N. Saridis, "Intelligent Robotic Control," *IEEE Transactions on Automatic Control* 5 (May 1983): 547–48. Ronald K. Jurgen, "Detroit Bets on Electronics to Stymie Japan," *IEEE Spectrum* 18 (July 1981): 29–32. NSF Annual Report, FY 1983, 46. George DeFlorio, "Intelligent Automata and Man-Automaton Combinations: A Critique and Review," *Electrical Engineering* 82 (March 1963): 203.

41. NSF Annual Reports, FY 1982, 18; FY 1986, 5. NSF Budget to Congress, FY 1982, B-I-5–6.

42. NSF, *Recent Research Reports* (Washington, D.C.: NSF, 1981), 34–45.

43. Emory Kemp, project consultant, telephone interview with author, 16 January 1991. Ray W. Clough to Zenzen, 4 October 1991; and Clough, "Original Formulation of the Finite Element Method," manuscript of presentation at Robert Melosh Paper Completion, Duke University, February 1989. "Why Do Things Fail? FEA Has the Answer," *Design News* 44 (22 August 1988): 61–63.

44. Kemp interview. NSF Budget to Congress, FY 1968, 151. Seed, Soil and Foundation Response during Earthquakes, 15 April 1968; Arthur H. Nilson, Bond Stress-Slip Relations in Reinforced Concrete, December 1971, both in WNRC, RG 307, Acc. #307-77-002, Box 1. NSB-75-370, 17-2.

45. NSF Budget to Congress, FY 1968, 151. AFFDL-TR-68-150, Harold C. Martin, "Finite Element Analysis of Fluid Flows," *Proceedings of the Second Conference on Matrix Methods in Structural Mechanics* (December 1969): 517–19. C. Taylor, B. S. Patil, and O. C. Zienkiewicz, "Harbour Oscillation: A Numerical Treatment for Undamped Natural Modes," *Proceedings of the Institute of Civil Engineers* 43 (June 1969): 141–55. G. Wayne Clough and Abdulaziz Mana, "Lessons Learned in Finite Element Analyses of Temporary Excavations in Soft Clay," *Numerical Methods in Geomechanics,* vol. 1 of *Proceedings of the Second International Conference on Numerical Methods in Geomechanics,* ed. C. S. Desai (June 1976), 496. F. W. Smith, Theoretical and Experimental Modeling of Snow Avalanche Release Processes, 1977, WNRC, RG 307, Acc. #307-81-072, Box 4, File: Smith, Colorado.

46. Erich Bloch, interview with author, 23 August 1990; and numerous Director's Office documents.

CHAPTER 7: RESOURCES AND POLICIES FOR FULL PARTNERSHIP

1. NSB-82-81, Director's Statement on Long Range Planning, February 1982, 233:24–27. See also NSF-88-16, George T. Mazuzan, "The National Science Foundation: A Brief History," 26–29.

2. Lewis Branscomb, interview with author, 21 March 1991.

3. Memorandum, Waterman to NSB, Special Interdepartmental Committee on the Training of Scientists and Engineers, 11 October 1954, NSB Book, 29th Meeting, 15 October 1954. J. Merton England, *A Patron for Pure Science: The*

National Science Foundation's Formative Years, 1945–57 (Washington, D.C.: NSF, 1982), 248–54. H. Rept. 97-34, House Committee on Science and Technology, *Authorizing Appropriations to the National Science Foundation,* 97th Cong., 1st sess., 13 May 1981, 26.

4. House Subcommittee on Science, Research and Technology, Committee on Science and Technology, *1982 National Science Foundation Authorization,* 97th Cong., 1st sess., 10 March 1981, 361–62. Branscomb interview.

5. Frederic T. Mavis, "History of Engineering Education," *Journal of Engineering Education* 43 (December 1952): 219. Samuel Rezneck, "The Engineering Profession Considers Its Educational Problems: A Forgotten Episode of the Centennial Exposition of 1876," *Association of American Colleges Bulletin* 43 (October 1957): 410–18.

6. Lawrence P. Grayson, *The Making of an Engineer: An Illustrated History of Engineering Education in the United States and Canada* (New York: John Wiley & Sons, 1993), 48–49. Monte A. Calvert analyzes these two "cultures" in *The Mechanical Engineer in America, 1830–1910: Professional Cultures in Conflict* (Baltimore, Md.: Johns Hopkins University Press, 1967). See conclusions, 277–81.

7. Charles Riborg Mann, *A Study of Engineering Education Prepared for the Joint Committee on Engineering Education of the National Engineering Societies* (Carnegie Foundation for the Advancement of Teaching, Bulletin no. 11) (New York, 1918), vi, chapter 13. Society for the Promotion of Engineering Education, *Report of the Investigation of Engineering Education, 1923–1929 [Wickenden Report],* vol. 2 (Pittsburgh, Pa.: University of Pittsburgh, 1934), 1101–7. Terry S. Reynolds and Bruce E. Seely, "Striving for Balance: A Hundred Years of the American Society for Engineering Education," *Journal of Engineering Education* 82 (July 1993): 138–40. Grayson, *The Making of an Engineer,* 134–38, 186–87.

A second part of the Wickenden report, published in 1934, addressed engineering programs at the less-than-baccalaureate-degree level.

8. Arthur Bronwell, "Basic Conflicts in Engineering Education," *Mechanical Engineering* 76 (November 1954): 886–87, 890. Eugene S. Ferguson, *Engineering and the Mind's Eye* (Cambridge, Mass.: MIT Press, 1992), 159–61. See also William G. Shepherd, "Education for the Engineering Mission," *IEEE Spectrum* 5 (February 1968): 94–100; and Emmett M. Laursen, "Several Issues in Civil Engineering Education," *Engineering Issues,* Proceedings of the American Society of Civil Engineers, 103 (April 1977): 83–88. Reynolds and Seely, "Striving for Balance," 141–42.

The Society for the Promotion of Engineering Education (SPEE) became the American Society for Engineering Education (ASEE) in 1946.

9. Eric Walker, "Remarks on the Goals Study of the American Society for Engineering Education," paper delivered at NAE Symposium, New York, 13 October 1966, 25–29. Eric Walker, interview with author, 1 October 1990.

10. E. A. Walker, J. M. Pettit, and G. A. Hawkins, *Goals of Engineering Education: Final Report of the Goals Committee* (Washington, D.C.: ASEE, January

1968), 2, 4, 5, 7, 11, 47–49, 60. NSF Annual Report, FY 1967, xi–xii. Reynolds and Seely, "Striving for Balance," 143–44.

11.   American Society for Engineering Education, *The Goals of Engineering Education: The Preliminary Report,* E. A. Walker, Chairman (West Lafayette, Ind: Purdue University, October 1965), 26–30. Eric A. Walker and Benjamin Nead, "The Goals Study: An Interpretation by the Chairman, ASEE Goals Committee," *Engineering Education* 57 (September 1966): 13, 18, 19. See also "ASEE: The Sound and the Fury" and individual statements, *Chemical Engineering Progress* 62 (February 1966): 17–49; "ASCE Position on Goals in Engineering Education," *Civil Engineering—ASCE* 36 (June 1966): 87–88; "Goals of Engineering Education: ASME's Critical Analysis," *Mechanical Engineering* 88 (August 1966): 32–35.

12.   Reynolds and Seely, "Striving for Balance," 143–44.

13.   Thorndike Saville, "Achievements in Engineering Education," *Journal of Engineering Education* 43 (December 1952): 228. Charles A. Rowe and William H. Spaulding, "An Industry Viewpoint on Curricula," *Chemical Engineering Progress* 59 (March 1963): 20–21.

14.   McElroy to DuBridge, 1 April 1970, WDM Notes 1970. Krumhansl to Quittenton, 7 December 1978, and attached correspondence, WNRC, RG 307, Acc. #307-87-220, Director's Office Subject Files: MPE.

15.   See W. Bernard Carlson, "Academic Entrepreneurship and Engineering Education: Dugald C. Jackson and the MIT-GE Cooperative Engineering Course, 1907–1932," *Technology and Culture* 29 (July 1988): 536–67. Erich Bloch, interview with author, 23 August 1990. Nam Suh, interview with author, 17 April 1991.

16.   NAE Task Force on Engineering Education, *Issues in Engineering Education: A Framework for Analysis* (Washington, D.C.: NAE, April 1980), 6.

17.   ASEE/AAES, Memorandum on Engineering Education, 27 March 1980, 9, 11, NSB Subject Files: Engineering—General.

18.   ASEE/AAES, Memorandum on Engineering Education, 1, 3–6.

19.   NAE, *Federal Actions for Improving Engineering Research and Education* (Washington, D.C.: NAE, 1986), 15.

20.   Lewis M. Branscomb, Statement, House Committee on Science and Technology, 6 October 1981, 3, 4, NSB Subject Files: Engineering—General. ASEE/ AAES, Memorandum on Engineering Education, 1, 3–6. David to Slaughter, 28 May 1982; and attached speech, Edward E. David, Jr., "Propping Up the Tree of Knowledge in the Information Age," paper delivered at the Annual Dinner Meeting, MIT Club of Northern New Jersey, Convent Station, N.J., 19 May 1982, 4–5, both in WNRC, RG 307, Acc. #307-87-219, Box 2, File: Engineering 1982. NRC, Engineering Research Board, *Directions in Engineering Research: An Assessment of Opportunities and Needs* (Washington, D.C.: National Academy Press, 1987), 50. Betty M. Vetter, "Demographics of the Engineering Student Pipeline," *Engineering Education* 78 (May 1988): 737.

21.   NSB-82-398, Memorandum, Windus to NSB, Article on National Policy toward Engineering Education, 2 December 1982, and attachments.

22. ASEE/AAES, Memorandum on Engineering Education, 7–8, 10–11. Donald E. Marlow, ASEE executive director, who drafted this memorandum, reiterated these points in "Engineering Education: Issues and Answers," *Mechanical Engineering* 102 (December 1980): 26. See also NRC, *Directions in Engineering Research*, 60. The NAE's own *Issues in Engineering Education: A Framework for Analysis* appeared about the same time as the ASEE/AAES report, with virtually the same recommendations.

23. Memorandum, Sanderson to Branscomb, 3 November Meeting of EAS Advisory Committee and Subcommittees, 5 November 1980, and attachments, Office AD/Eng, RANN File 32: Reorganization of EAS.

24. NSB-81-335, Provisional Minutes, Closed Session, 227th Meeting, 17–19 June 1981, CS:227:11; and attached Revised Draft Statement on Education, 17 July 1981.

25. H.R. 5254, *A Bill to Provide a National Policy for Engineering, Technical and Scientific Manpower, to Create a National Coordinating Council on Engineering and Scientific Manpower, and for Other Purposes,* 97th Cong., 1st sess., 16 December 1981. Memorandum, Kemnitzer to Windus, Selected Materials from House Science and Technology Committee Hearing on H.R. 5254, 27 April 1982, NSB Subject Files: Engineering and Science Manpower Act of 1982. H. Rept. 97-34. NSF Budgets to Congress, FY 1983–1984, Budget Summaries. Slaughter to Gregg, 2 April 1982, JBS Chron 1982. OGC, Background Paper, H.R. 5254, National Engineering and Science Manpower Act of 1982, 15 January 1982, NSB Subject Files: Manpower Act 1982.

26. S. L. Glashow, "Science Education and U.S. Technology," Testimony, House Committee on Science and Technology, 29 April 1982, NSB Subject Files: Manpower Act 1982.

27. Memorandum, Kemnitzer to Windus, 27 April 1982; and Charles S. Robb, Testimony, House Subcommittee on Science, Research and Technology, 29 April 1982, both in NSB Subject Files: Manpower Act 1982.

28. George A. Keyworth, II, [untitled] paper delivered to National Engineering Action Conference, New York City, 7 April 1982, NSB Subject Files: Manpower Act 1982.

29. George A. Keyworth, II, Statement [presented by Douglas Pewitt of the Office of Science and Technology Policy], before the Subcommittee on Science, Research and Technology, 29 April 1982; and Staff Summary Paper, Hearings on H.R. 5254, 29 April 1982, both in NSB Subject Files: Manpower Act 1982.

30. Draft LMB Comments on H.R. 5254, 9 April 1982; H.R. 5254 Discussion Agenda with handwritten notes, 12 April 1982; and NSB-82-178, Memorandum, Branscomb to NSB, Comments on H.R. 5254, 10 May 1982, all in NSB Subject Files: Manpower Act 1982.

31. Fuqua's bill got combined with a general Department of Education bill, which included controversial language about school prayer. Meanwhile, the Senate passed its own Education for Economic Security Act prescribing specific activities for

NSF such as partnerships with state agencies. Rather than deal with the school prayer issue, the House passed the Senate version, cutting off NSF opportunities to press for Fuqua's initial intent in conference. NSF Legislative Summary Documents and Joel Widder, interview with author, 9 May 1991. H.R. 5842, *A Bill to Authorize Appropriations for Activities of the National Science Foundation for FY 1982–1983,* 16 March 1982, 97th Cong., 2d sess., 7–8. Knapp to Hatch, undated [after 1 October 1983], EAK Chron 1984.

32. NSB-83-250, Statement on the Engineering Mission of the NSF over the Next Decade, as Adopted by the National Science Board at Its 246th Meeting on August 18–19, 1983, 1.

33. NSB-83-127, *Issues and Analysis,* vol. 1 of *Discussion Issues 1983: The Engineering Mission of NSF over the Next Decade* (Washington, D.C.: NSF, June 1983), 18–19, 32, 42, 44. NSB-83-250. President's National Productivity Advisory Committee, Engineering Education for Higher Productivity, 14 December 1982, in NSB-83-190, Replies of Outside Experts: The Engineering Mission of NSF over the Next Decade, June 1983, 98.

34. Branscomb interview. NSB-83-127, *Issues and Analysis,* 2–4. NSB-83-250, 1. NSB-83-190, v–xvi.

35. NSB-83-250, 2. Internal Notes, Engineering Discussion, 17 June 1983, 4; undated draft notes, "Dr. Edward Knapp's Comments at PPC [NSB Planning and Policy Committee] Meeting of March 17, 1983"; and Memorandum, Carl Hall to AD/Eng, June Issues by NSB—What Is Foundation's Role in Engineering, 22 March 1983, all in NSB Subject Files: Engineering—General.

36. Internal notes, Engineering Discussion, 17 June 1983, 2–5. NSB-83-221, Minutes, Open Session, June 1983, 6:83-4–6.

37. Eric A. Walker, *Now It's My Turn: Einigneering My Way* (New York: Vantage Press, 1989), ix. Arie Rip, "Science and Technology as Dancing Partners," 231, 256–58; and Eda Kranakis, "Hybrid Careers and the Interaction of Science and Technology," 178–79, both in *Technological Development and Science in the Industrial Age: New Perspectives on the Science-Technology Relationship,* Peter Kroes and Martijn Bakker, eds. (Dordrecht: Kluwer Academic Publishers, 1992). Bruce Seely, "Research, Engineering, and Science in American Engineering Colleges, 1900–1960," *Technology and Culture* 33 (April 1993): 383–85.

38. NSB-83-229, Memorandum, Windus to Members of the Executive Committee, NAE Engineering Recommendations, 15 July 1983, and attachment, White to Branscomb, "Strengthening Engineering in the National Science Foundation: A View from the President of the National Academy of Engineering," 15 July 1983, NSB Subject Files: Engineering—General.

39. Ibid.

40. Branscomb to White, 26 July 1983, NSB Subject Files: Engineering—General. Branscomb interview. NSB-83-221, 6:83-5.

41. NSB/Res-83-50, as referenced in Excerpt from August 1983 Open Session Meeting, with attached NSB-83-250.

42. NSF-PR83-59, NSF Engineering Activities to Receive New Direction and Emphasis, 18 August 1983. Kim McDonald, "Engineering Research to Get More Emphasis in National Science Foundation Programs," *Chronicle of Higher Education* (20 September 1983); and "Board Revamps NSF's Mission," *Machine Design* (6 October 1983), both in NSB Subject Files: Engineering—General, 1977–83.

43. "Science Board Calls for New Emphasis on Engineering in NSF," *Engineering Education News* 10 (October 1983): 6. John Walsh, "NSF Seeks Expanded Role in Engineering," *Science* 222 (9 December 1983): 1101–2.

44. Walsh, "NSF Seeks Expanded Role in Engineering," 1101–2.

45. Ibid.

46. Summary Minutes, Advisory Committee for Engineering, 30 November–1 December 1983, 1–2, NSB Subject Files: Engineering—General. Suh interview.

47. AAES Special Task Force, *A Working Plan for Attacking the Engineering Faculty Shortage Problem* (Washington, D.C.: AAES, 1982). "Plan May Ease Faculty Shortage in Engineering," *C&EN* (11 October 1982), summarizes the AAES report. Memorandum, Berry to Fechter, Drafts of Notes on SRS [Division of Science Resources Studies] Projections of Scientists and Engineers, 1982–87, and attached survey questionnaires and results, 20 April 1983, NSB Subject Files: Engineering—General.

48. NSF, SEE, "Enhancing the Quality of Science Education in America: NSF's Role and Strategy, 1985–1990," 29 May 1985, 26–27, Yankwich Files. Erich Bloch, "Responsibilities in Science and Technology," paper delivered to Council of Graduate Schools Annual Meeting, Washington, D.C., 7 December 1984, in NSB Book, 258th Meeting, 17–18 January 1985.

49. NSF Annual Reports, FY 1988, 10; FY 1986, 38–39.

50. NRC, *Engineering Education and Practice in the United States: Foundations of Our Techno-Economic Future* (Washington, D.C.: National Academy Press, 1985), 1–17. Ferguson, *Engineering and the Mind's Eye,* 161.

51. NRC, *Engineering Education and Practice,* 1–17.

52. Reynolds and Seely, "Striving for Balance," 146.

53. NSF, "Enhancing the Quality of Science Education in America," 1–4.

54. NSB-89-118, Memorandum, Shakhashiri to NSB, Undergraduate Curriculum Development in Engineering, 3 August 1989. David Lytle, "The Government Factor: Engineering Education in Crisis," *Photonics Spectra* (January 1990): 64, 66.

55. John A. White, "Current and Emerging Issues Facing American Engineering Education," paper based on speech at International Forum on Engineering Education, Washington, DC, 30 November 1988, 2, 14, NSB Subject Files: Eng—General (3), 1986—. White was NSF assistant director for Engineering from 1988 to 1991.

56. White, "Current and Emerging Issues," 2–3. *Engineering Education* devoted its May 1988 issue to concerns over human resources. See Vetter, "Demographics of the Engineering Student Pipeline," 735–40; and "Findings & Recommendations from the Report of the Task Force on the Engineering Student Pipeline," 778–80.

57. Memorandum, Waterman to NSB, Special Interdepartmental Committee on the Training of Scientists and Engineers, 11 October 1954, and attached report, 8, 11, 13, in NSB Book, 29th Meeting, 15 October 1954.

58. See, for example, NSF-61-146, press release, Women in Scientific Careers Are Subject of NSF Report, 26 October 1961. Herbert M. Katz, "Howard U. to Graduate First Ch.E.'s," *Chemical Engineering Progress* 68 (February 1972): 22–24. J. D. Ryder, "Spectral Lines: The Ladies in Our Midst," *IEEE Spectrum* 1 (July 1964): 65.

59. Walker, Pettit, and Hawkins, *Goals of Engineering Education,* 11 and passim.

60. Irene C. Peden, "New Faces of Eve: Women in Electrical Engineering," *IEEE Spectrum* 5 (April 1968): 81–84. NSF-PR76-60, The Nation's Resources: Men and Women in Science and Engineering, 1974, 2 July 1976. Sol E. Cooper, "Minority Employment—a Job for ASCE," *Engineering Issues,* Proceedings of ASCE, 100 (July 1974): 194.

61. John B. Parrish, "Women in Engineering: Some Perspectives on the Past, Present and Future," in *Women in Engineering: Bridging the Gap between Society and Technology,* Proceedings of an Engineering Foundation Conference, 12–16 July 1971, New England College, Henniker, N.H., ed. George Bugliarello et al. (Chicago: University of Illinois Press, 1972), 38–39.

62. NSF-77-304, *Women and Minorities in Science and Engineering* (Washington, D.C.: NSF, 1977), vii, 2, 5, 12, 14, 17, 22.

63. NAS, *Science and Technology: A Five-Year Outlook* (San Francisco, Calif.: W. H. Freeman, 1979), 476–77.

64. Carol Truxal, "The Woman Engineer," *IEEE Spectrum* 20 (April 1983): 58. For a detailed historical analysis of the status of women engineers in employment, see Betty M. Vetter, "Changing Patterns of Recruitment and Employment," in *Women in Scientific and Engineering Professions,* ed. Violet B. Haas and Carolyn C. Perrucci (Ann Arbor: University of Michigan Press, 1984), 59–74.

65. "Engineering Not Attracting Women: Enrollment Down and Leveling Off," *Mechanical Engineering* 107 (May 1985): 10. NSB-87-1, *Science & Engineering Indicators—1987* (Washington, D.C.: NSF, 1987), 61.

66. NSB-81-25, Memorandum, Slaughter to NSB, NSF Visiting Professorships for Women in Science and Engineering Program, 9 January 1981. NSB-81-64, Provisional Minutes, Open Session, 222nd Meeting, 15–16 January 1981, 222:15-18. NSF Annual Report, FY 1983, 80–81.

67. NSF Report 90-13, *NSF's Research Opportunities for Women Program: An Assessment of the First Three Years* (Washington, D.C.: NSF, January 1990), 2, 4, 6.

68. Report of the Chairman, Committee on Minorities in Engineering, to the National Advisory Council on Minorities in Engineering, 16 February 1977, 1–2, WNRC, RG 307, Acc. #307-87-228, Box 2, File: NAE 1977. [J. Merton England],

"Biographical Sketch of Dr. John B. Slaughter, Director of the National Science Foundation," NSF History Office Files.

69. NSB-81-172, Memorandum, Slaughter to NSB, Recommendation for Establishment of a Resource Center for Science and Engineering at the City College of the City University of New York, 10 April 1981. NSF Annual Report, FY 1986, 36.

70. Stever to Leonard Laster, 22 June 1972, with attachments, WNRC, RG 307, Acc. #307-77-080, Box 2, File: Engineering, Division of. Mann to Kahne, 14 July 1981, with attachments: Robert W. Mann, "Technology for Human Rehabilitation," *Technology Review* 81 (November 1978); and excerpt from Annual Report Academic Year 1979–1980, Department of Mechanical Engineering, MIT, WNRC, RG 307, Acc. #307-87-220, Box 2, File: Engineering 1981. Slaughter to Austin, 23 July 1982, WNRC, RG 307, Acc. #307-87-219, Box 2, File: Engineering 1982. NSF Budget to Congress, FY 1981, F-VI-5.

71. NSF Budget to Congress, FY 1988, NSF Summary 2-5. White, "Current and Emerging Issues," 3.

72. Seely, "Research, Engineering, and Science," 383–86. Samuel C. Florman, "An Engineer's Comment," *Technology and Culture* 27 (October 1986): 680–82.

CHAPTER 8: ACTING ON THE MANDATE

1. Syl McNinch, "Engineering Research Centers (ERC): How They Happened, Their Purpose, and Comments on Related Programs," 14 September 1984, 3, NSF library. Terry S. Reynolds, "The Engineer in 20th-Century America," in *The Engineer in America: A Historical Anthology from "Technology and Culture,"* ed. Terry S. Reynolds (Chicago, Ill.: University of Chicago Press, 1991), 178–79. NSB-83-250, Statement on the Engineering Mission of the National Science Foundation over the Next Decade, 18–19 August 1983.

2. McNinch, "ERCs: How They Happened," 4.

3. Syl McNinch, *Engineering: An Expanded and More Active Role for NSF* (Washington, D.C.: NSF, 1985), 3–4. McNinch, "ERCs: How They Happened," 5. Roland W. Schmitt, "Engineering Research and International Competitiveness," in NRC, *The New Engineering Research Centers: Purposes, Goals, and Expectations* (Washington, D.C.: National Academy Press, 1986), 26.

4. Lewis Branscomb, interview with author, 21 March 1991. Knapp to White, 13 December 1983, NSB Subject Files: Engineering—General. Ralph Devries, "White House Office of Science and Technology Policy Views on the Engineering Research Centers," paper delivered to the NSF Informational Meeting on ERCs, Washington, D.C., 15 June 1984, 1, NSB Subject Files: Engineering—Research Centers. Carl Hall, interview with author, 27 March 1990.

5. Knapp to White, 13 December 1983. NSF Budget to Congress, FY 1985, ENG-1. NSB-83-127, *Issues and Analysis,* vol. 1 of *Discussion Issues, 1983: The*

Engineering Mission of NSF over the Next Decade (Washington, D.C.: NSF, June 1983), 3–4. Branscomb interview.

6. NAE, *Guidelines for Engineering Research Centers: A Report for the National Science Foundation by the National Academy of Engineering* (Washington, D.C.: NAE, 1984), letter of transmittal, vi–vii.

7. NAE, *Guidelines for ERCs,* 1–5, 8–9.

8. NSB-84-61, Memorandum, Knapp to NSB, Fiscal Year 1985 Program Announcement for Engineering Research Centers. Memorandum, Bloch to Organization, Change in the Role of the Office of Interdisciplinary Research, 30 November 1984; AD/Eng Bulletin No. 86-3, Memorandum, Suh to Organization, Change in the Organizational Status of the Office of Cross-Disciplinary Research (CDR), Directorate for Engineering, 29 January 1986, both in NSB Subject Files: Engineering—ECD. McNinch, *Engineering: An Expanded Role,* 4. "NSF Gears Up for Big New Engineering Program," *Science and Government Report* 14 (1 November 1984): 1. Eric Walker, "The Criteria Used in Selecting the First Centers," in *The New ERCs,* 44–48. Eric Walker, interview with author, 1 October 1990.

9. Staff paper, List of Recommended ERC Awards, attached to NSB-85-66, Memorandum, Bloch to NSB, Engineering Research Centers, n.d. [around February 1985]. Walker, "The Criteria Used," 49.

10. NSB-85-99, appendix B, Provisional Record of Discussion, 260th Meeting, 22 March 1985, CS:3-85:5-8. Erich Bloch, "Responsibilities in Science and Technology," 4, in NSB Book, 258th Meeting, 17–18 January 1985. NSF-PR85-22, NSF Announces Establishment of Six Major Engineering Research Centers, 3 April 1985.

11. NSB-85-66, and attached descriptions of selected ERCs (NSB-85-67–NSB-85-72). The Six Engineering Research Centers: Summary Descriptions, March 1985, NSB Subject Files: Engineering—Research Centers. Walker interview. Branscomb interview.

12. McNinch, "ERCs: How They Happened," 8–10. In fact, to keep industry involved, it was often necessary to continue minimal NSF funding beyond five years as reassurance of the Foundation's "Good Housekeeping Seal of Approval." Lynn Preston, telephone interview with author, 24 May 1991.

13. "NSF Gears Up," 3. John Adam, "Response to NSF Research Centers May Test Relationship of Science and Engineering Communities," *The [IEEE] Institute* (January 1985): 6. McNinch, *Engineering: An Expanded Role,* 4.

14. Russell Drew, as quoted in Adam, "Response to NSF Research Centers," 6.

15. "NSF Gears Up," 1, 3. Adam, "Response to NSF Research Centers," 6.

16. NRC, *The New ERCs,* vii, 1–7, 18, 19–27. "New Engineering Centers Expected to Bridge Science and Technology," *C&EN* 63 (13 May 1985): 15–16.

17. Nam Suh, "A New Experience: Lessons Learned by the NSF," in NRC, *The Engineering Research Centers: Leaders in Change* (Washington, D.C.: National Academy Press, 1987), 21, 26–27. NRC, *The New ERCs,* 37.

18. Suh, "A New Experience," 21–27.

19. Lynn Preston, interview with author, 27 March 1991. Branscomb interview. Nam Suh, interview with author, 17 April 1991. See also NSB-89-37, Memorandum, Bloch to NSB, Engineering Research Centers Third-Year Evaluation and Renewal Review for FY 1986 ERCs, 6 March 1989, and attached individual ERC evaluations, NSB-89-38–NSB-89-42. The universities losing their ERC funding were UC, Santa Barbara (Robotics Systems in Microelectronics) and Delaware/Rutgers (Composites Manufacturing) and, later, UCLA (Hazardous Substance Control).

20. Brigham Young University/University of Utah, Advanced Combustion Engineering Research Center, promotional brochure. L. Douglas Smoot, "Advanced Combustion Engineering Research Center," in *ERCs: Leaders in Change,* 191–202. NSF, Engineering Research Centers: A Partnership for Competitiveness, ACERC, information packet.

21. BYU/UU, ACERC promotional brochure. ERCs: A Partnership for Competitiveness. NSB-89-59, appendix B, Provisional Record of Discussion, 286th Meeting, 17 March 1989, CS:3-89:6.

22. "Turbulent Particle Dispersion," *Burning Issues* (August 1987): 3.

23. "New Theory Relates Coal Structure to Reaction Rates," *Burning Issues* (March 1988): 2.

24. "Studies of Hazardous Waste Combustion (Thrust Area 3)," *Burning Issues* (December 1988): 3.

25. David Pershing, telephone interview with Joan M. Zenzen, 3 April 1991.

26. David Hoeppner, telephone interview with Joan M. Zenzen, 21 March 1991.

27. ACERC telephone interviews, anonymous by request, March and April 1991.

28. Bruce Woodson, telephone interview with author, 21 May 1991.

29. Daniel I. C. Wang, "Biotechnology Process Engineering Center," in *ERCs: Leaders in Change,* 94–106. ERCs: A Partnership for Competitiveness, Biotechnology Process Engineering Center. Woodson interview.

30. Suh interview. Walker interview.

31. Memorandum, Hazen and Pollack to Schmitt, Proposed Study of Engineering Research and Priorities, 15 September 1983, NSB Subject Files: Engineering—General. NAE, *Assessment of the National Science Foundation's Engineering Research Centers Program* (Washington, D.C.: NAE, 1989), appendix A (Bloch to White, 20 September 1988), letter of transmittal, vii, 2–8, 11–12, 14.

32. NAE, *Assessment of NSF's ERCs Program,* letter of transmittal, 1–3, 12–20.

33. Ibid., 8–10, 12–14.

34. Preston telephone interview.

35. NAE, *Assessment of NSF's ERCs Program,* 19–20.

36. Preston, Walker, Woodson, Branscomb, and Suh interviews.

37.   Richard G. Cunningham, Statement to NAE Ad Hoc Panel on NSF Engineering Program on Behalf of the National Society of Professional Engineers, 22 June 1983, 4, NSB Subject Files: Engineering—General.

38.   Kathryn W. Hickerson, "Revised NSF Charter Needed to Meet Long-Term Goals for U.S. Engineering Support," *Engineering Times* (January 1984): 1–2.

39.   NSF Office of Legislative and Public Affairs, Sectional Analysis of H.R. 4822, with attached bill, 16 February 1984, NSB Subject Files: Engineering—General. S. Rept. 98-495, Senate Committee on Labor and Human Resources, *Authorizing Appropriations for the National Science Foundation, Fiscal Year 1985,* 25 May 1984, 6.

40.   Hickerson, "Revised NSF Charter Needed," 1–2. Kim McDonald, "National Science Foundation Starts to Broaden Support of Engineering Research," *Chronicle of Higher Education* (18 January 1984). The verb "starts" was revealing.

41.   Russell Drew, Statement, House Subcommittee on Science, Research and Technology, Committee on Science and Technology, 28 February 1984, 1, 6–7, NSB Subject Files: Engineering—General.

42.   Branscomb to Good, Gilkeson, Miller, Ragone, Rasmussen, Schmitt, and Knapp, National Science and Engineering Foundation, 10 February 1984, NSB Subject Files: Engineering—General (2), 1984–85.

43.   White to Walgren, 19 March 1984, NSB Subject Files: Engineering, National Academy. Knapp to Walgren, 20 March 1984; and Knapp to Hatch, 1 May 1984, both in EAK Chron 1984. Joel Widder, interview with author, 9 May 1991.

44.   Branscomb to Walgren, 20 March 1984; and Memorandum, Hall to Schmitt, Resolution of Advisory Committee for Engineering, 5 July 1984, and attached resolution, all in NSB Subject Files: Engineering—General.

45.   Robert Rosenzweig, Statement, 3; J. Thomas Ratchford, Statement, 2–3; Donald Glower, Statement, 2, 4–5; Myron Tribus, Statement, 3, all made to House Committee on Science and Technology on 21 March 1984, in NSB Subject Files: Engineering—General.

46.   White to Fuqua, and White to Walgren, 19 March 1984, both in NSB Subject Files: Engineering, National Academy.

47.   Frank Press, Statement, House Committee on Science and Technology, 21 March 1984, 2, 4, 7, NSB Subject Files: Engineering—General. Frank Press, "Amending the National Science Foundation Act," *Science* 224 (13 April 1984): 12.

48.   Lewis Branscomb, "Engineering and the National Science Foundation," *Science* 224 (27 April 1984). See also "Press, Branscomb Debate Merits of NSF Charter Changes," *Engineering Education News* 10 (June 1984): 1.

49.   "Engineering Is No Threat to the Funding of Science," *Electronics* (19 April 1984): 24.

50.   Walgren quoted by John Horgan, "US Government Speakers at IEEE Policy Conference Get Lukewarm Reception," *The [IEEE] Institute* 8 (April 1984): 1, 7.

51.   John McKelvey, "Science and Technology: The Driven and the Driver," *Technology Review* (January 1985): 38–47, esp. 41, 47. Arthur L. Donovan, "Engineering in an Increasingly Complex Society," in NRC, *Engineering in Society: Engineering Education and Practice in the United States* (Washington, D.C.: National Academy Press, 1985), 96–97.

52.   Press Release, Office of the White House Press Secretary, Ronald Reagan, Memorandum of Disapproval of H.R. 5172, 30 October 1984, NSB Subject Files: Engineering—General. Widder interview.

53.   NSF Controller's Office/Budget Division, NSF Chronological History of FY 1986 Budget Actions, June 1986, 1.

54.   Wil Lepkowski, "National Science Foundation Emphasizes Engineering Research," *Chemistry and Industry* (2 July 1984). Wil Lepkowski, "New Leadership at NSF Emphasizes Role of Technology," *C&EN* (16 July 1984): 22–24.

55.   Erich Bloch, interview with author, 23 August 1990.

56.   Suh interview.

57.   Suh interview. William Butcher, interview with Joan M. Zenzen, 14 May 1991.

58.   Bloch interview. Erich Bloch, "Responsibilities in Science and Technology," 9.

59.   William Butcher, interviews with author and Zenzen, 9 and 14 May 1991, respectively.

60.   Suh interview.

61.   Bloch to Boland, 14 December 1984, NSB Subject Files: Engineering—General. NSF-PR85-8, NSF Announces Major Reorganization of Its Directorate for Engineering, [January 1985].

62.   Butcher interviews. NSF-PR85-74, NSF Starts Program to Encourage Creative Engineering Research, 12 December 1985. Erich Bloch, Testimony, House Subcommittee on Science, Research and Technology, Committee on Science, Space and Technology, 6 March 1990, 10, NSF Office of Legislative and Public Affairs.

63.   NAE, *New Directions for Engineering in the National Science Foundation* (Washington, D.C.: NAE, 1985), 1–3, 10–11, 38–39.

64.   NAE, *Federal Actions for Improving Engineering Research and Education* (Washington, D.C.: NAE, 1986), 1–3, 7–8.

65.   NSF-86-74, Strategic Plan, Directorate for Engineering, June 1986. NSB-85-142, Strategic Plan, Directorate for Engineering. Butcher interview with Zenzen. Memorandum, Fenstermacher to Bloch, Update of the NSF Organization Chart, 30 June 1986, NSF Organizational Charts. AD/Eng Bulletin No. 89-2, Organizational Changes in the Directorate for Engineering, 27 February 1989, NSB Subject Files: Engineering—General.

66.   See Reynolds, "The Engineer in 20th-Century America," 189; Donovan, "Engineering in an Increasingly Complex Society," 97.

CHAPTER 9: NSF ENGINEERING OVER TIME

1. "Whatever Happened to . . . Project Mohole?" *IEEE Spectrum* 27 (September 1990): 58, 60. Milton Lomask, *A Minor Miracle: An Informal History of the National Science Foundation* (Washington, D.C.: NSF, 1976), 192–95. NSF Annual Reports, FY 1961, 39–41; FY 1966, 62. MPE Annual Report, FY 1961, 8, NSF History Office Files. NSB-77-227, appendix A, Provisional Record of Discussion, 189th Meeting, 22 April 1977, CS:189:13.

2. NSF, Engineering Research Centers: A Partnership for Competitiveness, Offshore Technology Research Center (OTRC), information packet.

3. James Kirtley, Jr., "Supercool Generation," *IEEE Spectrum* 20 (April 1983): 28–35. "The Superconductivity of Certain Materials," reprinted from *The Engineer,* 10 May 1929, in *Mechanical Engineering* 51 (August 1929): 588. NSF Annual Report, FY 1961, 38. See also Ruth R. Harris, "Research and Development of Superconducting Transmission Lines, 1911–1983," unpublished manuscript, History Associates Incorporated, 5 February 1985, 1–17, 33–35.

4. NSF Annual Report, FY 1987, 3–5. William Booth, "Superconductor Research Faulted for Focus on Short-Term Goals," *Washington Post,* 4 March 1990, A21.

5. See, for example, V. E. Gunlock, "Chicago Builds $230,000,000 Superhighway System," *Civil Engineering* 21 (February 1951): 38–39; and H. V. Pittman, "Bull Shoals Dam Now Completed," *Civil Engineering* 21 (November 1951): 26–32. See also American Public Works Association, *History of Public Works in the United States, 1776–1976,* ed. Ellis L. Armstrong (Chicago, Ill.: American Public Works Association, 1976).

6. NSF Annual Report, FY 1963, 12. Myle J. Holley Jr., "Prestressed Concrete Makes Rapid Progress in the United States," *Civil Engineering* 23 (January 1953): 25–29. Emory Kemp, project consultation, 17 September 1991.

7. NSB Book, 48th Meeting, 6 September 1957, 27. Ned H. Burns, Development of Continuity between Precast-Prestressed Concrete Members, February 1969, WNRC, RG 307, Acc. #307-73-003, Box 59, File: Burns, University of Texas. Kemp consultation.

8. NAE, COPEP, "Priorities in Applied Research: A Preliminary Report," 15 December 1969, attached to Memorandum, Falk to Members of Planning Council, NAE Preliminary Report on "Priorities in Applied Research," 17 December 1969, WNRC, RG 307, Acc. #307-75-052, Box 3, File: Committee on Public Engineering.

9. See National Council on Public Works Improvement, *Fragile Foundations: A Report on America's Public Works* (Washington, D.C.: GPO, 1988), 1–10, passim. Lester H. Poggemeyer, "New Frontiers in Civil Engineering: Infrastructure," *Journal of Professional Issues in Engineering* 115 (October 1989): 393–94. Kent Jessen, "The Infrastructure and the Civil Engineer," *Journal of Professional Issues in*

Engineering 110 (October 1984): 151. Dale W. Avery, Nicholas T. Zilka, and Barry W. Klein, "Bridge to the Future: Rebuilding America's Infrastructure," *Minerals Today* (January 1990): 12.

10. NRC, Building Research Board, *Annual Report 1989* (Washington, D.C.: NRC, 1989), 1. *Engineering News* quoted in *Perspectives on Urban Infrastructure,* ed. Royce Hanson (Washington, D.C.: National Academy Press, 1984), 52, chap. 1, passim. Poggemeyer, "New Frontiers," 393–94. U.S. Congress, Joint Economic Committee, Subcommittee on Economic Goals and Intergovernmental Policy, Hearings, *Infrastructure: A National Challenge,* 98th Cong., 2d sess., 29 February 1984, 1–11, 120–32. The document includes *Hard Choices.*

11. William S. Butcher, Statement, House Subcommittees on Investigations and Oversight and Transportation, Aviation and Materials, Committee on Science and Technology, 9 June 1983, 5, 8, NSB Subject Files: Engineering—General. Sanderson to Dixon, 23 August 1983, EAK Chron 1983. AD/Eng Bulletin No. 84-01, Construction Engineering and Building Research Program Established in the Division of Civil and Environmental Engineering, 28 February 1984.

12. John Scalzi, oral presentation, Advisory Committee for Engineering meeting, NSF, 29 March 1990.

13. Roland Schmitt, as quoted in NRC, Engineering Research Board, *Directions in Engineering Research: An Assessment of Opportunities and Needs* (Washington, D.C.: National Academy Press, 1987), 20.

14. H. Guyford Stever, *International Engineering in the National Science Foundation—Issues and Ideas: A Letter Report of the NAE Foreign Secretary to the Committee on International Science of the National Science Foundation* (Washington, D.C.: NAE, 1984), 4, 2–4.

15. Stever, *International Engineering—Issues and Ideas.* NAE and NRC Office of International Affairs, *Strengthening U.S. Engineering through International Cooperation: Some Recommendations for Action* (Washington, D.C.: NAE, 1987), 1–11.

16. NSF, Directorate for Scientific, Technological, and International Affairs, Division of Policy Research and Analysis, *The State of Academic Science and Engineering,* January 1990, IC 2-IC 3.

17. Stephen J. Mraz and Melissa S. Kennedy, "Engineering: 60 Years of Progress," *Machine Design* (7 September 1989): 123–24.

18. S. Levin and R. J. Buegler, "Large-Scale Systems Engineering for Airline Reservations," *Electrical Engineering* 81 (August 1962): 604–8.

19. NSF 88-3, Thomas N. Cooley and Deh-I Hsiung, *Report on Funding Trends and Balance of Activities: National Science Foundation, 1951–1988* (Washington, D.C.: NSF, December 1987), 1, 9.

20. John A. White, "Current and Emerging Issues Facing American Engineering Education," paper based on speech at International Forum on Engineering

Education, Washington, D.C., 30 November 1988, 11, NSB Subject Files: Engineering—General (3), 1986– .

21. NRC, *Directions in Engineering Research,* 1, 3–4.

22. Margaret [Windus] to Rich [Knapp assistant], 13 April 1984, with attached data, NSB Subject Files: Engineering—General (2), 1984–85.

23. NRC, *Directions in Engineering Research,* 6, 21, 43, 47. NSF Budget Summary [Request], FY 1991, 8, 17. Engineering's estimate for FY 1991 was $226 million, a 12.9 percent increase over the previous year. NSF Budget Estimate to the Congress, FY 1988, 33.

24. Memorandum and attachments, Margaret L. Windus to Dick Morrison, Basic and Applied Research, 16 November 1984, NSB Subject Files: Applied Research.

25. See, for example, Philip M. Smith, "The National Science Foundation as History," paper delivered at NSF's Fortieth Anniversary Symposium, Washington, D.C., 11 May 1990, 10, Butcher Office Files.

26. Lewis Branscomb working paper, 1979, attached to Memorandum, Windus to Morrison, 16 November 1984. Branscomb elaborated on his theme of the fallaciousness of the term "applied research" in "Physics—Used and Unused," *Physics Today,* March 1981, 9–11.

27. H. Rept. 96-61, Committee on Science and Technology, *Authorizing Appropriations to the National Science Foundation,* 96th Cong., 1st sess., 21 March 1979, 10.

28. Arthur L. Donovan, "Engineering in an Increasingly Complex Society," in NRC, *Engineering in Society: Engineering Education and Practice in the United States* (Washington, D.C.: National Academy Press, 1985), 97.

29. Erich Bloch, "Science and Engineering: A Continuum," in NRC, *The New Engineering Research Centers: Purposes, Goals, and Expectations* (Washington, D.C.: National Academy Press, 1986), 28–30. See also John P. McKelvey, "Science and Technology: The Driven and the Driver," *Technology Review* (January 1985): 38–47.

30. Lewis Branscomb, "The Changing Face of Science and Engineering," keynote address before the 150th Anniversary Symposium on Technology and Society, University of Alabama College of Engineering, 3 March 1988, 12–13. NSB-84-153, Lewis M. Branscomb, Report to the National Science Board, 10 May 1984, 12.

31. Robert White, *Taking Technological Stock: Report of the NAE President at the NAE Twenty-second Annual Meeting* (Washington, D.C.: NAE, 1986), 5–6. Mraz and Kennedy, "Engineering: 60 Years of Progress," 124–25. Thomas P. Hughes, *American Genesis: A Century of Invention and Technological Enthusiasm, 1870–1970* (New York: Viking, 1989), 462–70.

NOTES ON SOURCES

This study depends fundamentally on primary documentary sources and human resources of the National Science Foundation. From my detailed working bibliography of more than 120 pages, now deposited in the NSF history office, the following particularly useful sources are summarized. In addition to the locations specifically noted, the small NSF library has a fine collection of varied and pertinent materials.

Primary Sources: National Science Foundation

The most helpful initial source was the forty-year run of the National Science Foundation's annual reports. While these have become pictorial glossies in recent years, the earlier volumes are filled with policy and program detail—all the more important because of the relative paucity of other primary documentation from the early period, especially on the generally undernourished area of engineering. An opening statement by the National Science Board (NSB) chair (after the first few years by the NSF director—itself a significant statement of evolving relative influence) summarizes the year's research priorities, political issues, and accomplishments within the context of contemporary needs. Besides focusing on internal Foundation concerns, these reports are informative on noteworthy NSF-sponsored technical achievements and the people and institutions responsible for them.

Although the influence and direct involvement in operations of the presidentially appointed National Science Board has fluctuated, generally diminishing over time, major policy decisions have always come before the board. Thus, NSB files, with their detailed, staff-generated background papers on issues, programs, and special board-requested studies, were invaluable. Particularly so were the members' board meeting agenda books where such papers were attached or summarized. They also included the board's minutes for open and executive sessions. The NSB office has for twenty-five years or more used a chronological numbering system for all NSB documents, which makes them readily retrievable. The board meetings are also numbered as well as dated. The National Science Board office generously made its active subject files relevant to engineering available for research on this project. NSB-84-300, NSB staff member Margaret L. Windus's compilation,

National Science Board Policy Activity over the Past 10 Years, October 1984, for example, provided a compendium of policy information but also revealed evolving attitudes ("the Board has been dealing seriously with engineering since the late 1970's"). NSB clipping files were rich, although they did not always include page numbers.

A key source for NSF policy and program material bearing on engineering was papers of the Office of the Director. Those for the Alan T. Waterman years are held in Record Group 307 at the National Archives and Records Administration in Washington. The director's correspondence and subject files were particularly useful. The analogous files for later years, also in the NSF-dedicated Record Group 307, are housed at the Washington National Records Center in Suitland, Maryland.

The NSF historian's office keeps other salient historical documents from the Director's Office. These include directors' chronological files, "Diary Notes" (typed summaries of telephone conversations and informal meetings), and other pertinent documents not found elsewhere. The historian's office also houses a sizable collection of transcribed oral-history interviews, including an extensive series with leaders of the path-breaking and controversial RANN (Research Applied to National Needs) program of the 1970s. These vary in quality and depth but overall provide the color and insight of personal perspective. Annual reports of the Mathematical, Physical, and Engineering Sciences Division and the later Engineering Division contributed greatly to understanding both developments and attitudes, but unfortunately, these were later discontinued. Minutes of meetings and reports of the statutory MPE Divisional Committee and later engineering advisory committees also proved their worth. Indeed, these engineering documents were not only significant in their own right but sometimes the only internal source expressing the engineers' point of view. Some of these are also held at Suitland.

Miscellaneous files available in the NSF historian's office or the Office of Legislative and Public Affairs include office records of various NSF individuals, such as the blunt analyses of administrator Ed Todd (1968–75), other engineering subject files, and runs of press releases and other public relations materials, which confirmed facts and dates and provided the "official spin" on new directions. NSF's periodically appearing *Organizational Development* facilitated the tracking of administrative changes.

The NSF library and history office maintain copies of leaders' speeches, statements, and other writings. "McElroy Asks Expanded NSF Role," *Science* 166 (5 December 1969): 1252, excerpted from congressional testimony, revealed that director's activist, social-relevance leanings. Speeches of the articulate H. Guyford Stever were informative, especially when directed toward engineering audiences or the subject of RANN. Some were reprinted in *Science*. Erich Bloch's insightful and outspoken commentary is abundant in the professional and popular press. For the early period, the Library of Congress Manuscript Division holds numerous illustrative speeches of Vannevar Bush in its collection of his papers, while Waterman's are in the National Archives.

Valuable speeches of other prominent NSF figures of the later years included those of National Science Board chair Lewis Branscomb. His "The Changing Face of Science and Engineering," a 1988 keynote address at the 150th Anniversary Symposium on Technology and Society at the University of Alabama, disputed the familiar "linear model," which traced a path from science to engineering, and urged progress for economic competitiveness. Assistant director for Engineering John A. White's "Current and Emerging Issues Facing American Engineering Education," a paper delivered before the International Forum on Engineering Education on 30 November 1988, persuasively argued for a more inclusive profession in order to ensure the personnel strength and quality to solve current and future technological problems.

Together these sources documented decision making over time, the evolution of policy, and the personal and philosophical flavor of individual leaders' respective contributions. Correspondence, sometimes informal, with program officers, politicians, and members of the public often gave insight and perspective lost in formal reports.

A particularly valuable collection of records, focusing on the RANN program and the subsequent reorganizations that led to the formation in 1981 of the Engineering Directorate, came from the Office of the Assistant Director for Engineering. These documents include memoranda, drafts of reports, working papers, and correspondence along with published sources. Engineering planning and resources officer Paul Herer, who was with RANN from the beginning, has retained and made available his copious files of RANN documents.

NSF maintains extensive document collections relating to the legislative history of the Foundation, especially at turning points, including successive bills, congressional hearings and testimony, debate, and the statutes as finally passed. While these materials are available at the Library of Congress, their assemblage in one place greatly facilitated the research. Slightly differing bound legislative document collections on NSF's founding are in the NSF library and budget office. The history office also has a detailed file relating to the Daddario Amendments of 1968, which had a profound impact on engineering at NSF by expanding the Foundation's mission to include applied research.

The Daddario investigations began in 1965, with studies by the Library of Congress and the House science subcommittee accompanied by extensive hearings: U.S. Library of Congress, Legislative Reference Service, *The National Science Foundation: A General Review of Its First 15 Years,* Report to the Subcommittee on Science, Research and Development of the Committee on Science and Astronautics, U.S. House of Representatives, 89th Cong., 1st sess., 1965; U.S. Congress, House, Subcommittee on Science, Research and Development, *The National Science Foundation: Its Present and Future,* 89th Cong., 1st sess., 1966; and U.S. Congress, House, Subcommittee on Science, Research and Development, *Government and Science: Review of the National Science Foundation,* Hearings before Subcommittee of Committee on Science and Astronautics, 89th Cong., 1st sess., June–August 1965.

Papers documenting and summarizing recent and current NSF relations with Capitol Hill, such as those on the Manpower Act and the inclusion of engineering in the wording of the NSF Act, both from the 1980s, reside in NSF's Office of Legislative and Public Affairs. The CIS Index at the Law Library of the Library of Congress led to the legislative history of various other mandates. Most of NSF's congressional interactions involve first the House Committee on Science and Astronautics (later Science and Technology) and its Subcommittee on Science, Research and Development. On the Earthquake Hazards Reduction Act of 1977, for example, see *Earthquake Hearings,* 94th Congress, 2d sess., 22–24 June 1976; and House Rept. No. 95-286, Part II, *Reducing the Hazards of Earthquakes, and for Other Purposes,* 95th Cong., 1st sess., 9 June 1977. Other hearings of interest included U.S. Congress, House, Subcommittee on the Environment and the Atmosphere, Committee on Science and Technology, *The Costs and Effects of Chronic Exposure to Low-level Pollutants in the Environment,* 94th Cong., 1st sess., November 1975, during the RANN era; U.S. Congress, Joint Economic Committee, *Infrastructure: A National Challenge,* 98th Cong., 2d sess., 29 February 1984; and a related hearing before the House Subcommittee on Economic Development, Committee on Public Works and Transportation, *To Discuss the Final Report of the National Council on Public Works Improvement,* 100th Cong., 2d sess., 18 May 1988. The series *Public Papers of the Presidents of the United States* remains the exhaustive source for administration views.

The story of an agency whose primary function is distributing money for research to institutions and individuals must include numerical accounting as well as verbal. Bound annual NSF budget documents, held in the Foundation's Office of Budget and Control, give the figures and narrative justifications of spending proposals by year. The narratives reveal the evolution of funding priorities as the forces of outside events and the promise of technological breakthroughs are assessed. Because the budget process involves multiple stages of negotiation—from the initial request to the Office of Management and Budget through the subsequent submittal to Congress, congressional authorization, and finally appropriation—using budget numbers in the text was tricky. Individuals who spoke or wrote about budget figures were not always clear about which point in the process their numbers reflected. For the purposes of this study, figures were cited as the principals introduced them. However, in graphs and charts that delineate spending facts or trends, the numbers were obtained from the monies actually spent during the previous fiscal year, which figures might vary some, though usually not much, from the appropriation level. Budget testimony was an informative adjunct to the budget documents.

Human resources were central to the telling of NSF's engineering history. Key players in several aspects of engineering's unfolding (see my acknowledgments for list of individuals) agreed to taped interviews, which probed their respective agendas, accomplishments, and perspectives. The tapes and synopses of these interviews are available in the NSF history office. Many other former and current NSF engineering leaders also participated helpfully in untaped interviews detailing their respective roles. These plus the existing oral-history transcripts all provided rich insights available no other way, augmenting—

though of course not supplanting—the written record. No two individuals interpreted any given event or circumstances quite the same way, which both enlivened and enriched the account.

One disappointment was the NSF engineering grant files, of which 474 feet (now more) are stored at the Washington National Records Center. These are a potential treasure trove of information on technical emphases and their evolution over time as well as on individuals and institutions making significant progress in a particular area. They include individual research proposals, staff and peer evaluations of proposals, correspondence between investigators and NSF staff, and reports of progress and final outcomes. Unfortunately, these files are not catalogued systematically or in any detail. Organized chronologically by institution, they are virtually impossible to penetrate by subject in any practical manner. NSF has computerized databases of its grants, which can be searched in a number of ways, but the information given in the database and on the records center data sheets (Standard Form 135) does not overlap sufficiently to provide direct or efficient access. So while the grant files used were illuminating, the search cannot pretend to have uncovered the most illustrative or significant projects or personnel. References cite what was found on some topics in the time available.

Finally, NSF's own publications and internal reports on key programs were varyingly helpful, including detailed periodic in-house program reports for various disciplines. The RANN period generated an unusually large quantity of significant documents. Syl McNinch's 1984 summary, *The Rise and Fall of RANN (Research Applied to National Needs): A National Science Foundation Program,* edited by David Gould and Louise McIntire, gave the facts his own slant in unvarnished language, but it is a succinct, reliable paper. William Consolazio's *The Foundation's Role in the Solution of Problems of Social Relevancy: A Report to the Director* (January 1970), written on the eve of the new venture, indicated the potential of the new approach but also revealed the fears of NSF traditionalists who preferred not to stray into areas that were not strictly basic research. *A Study of the Research Applied to National Needs Program* (1975), by Harvey Averch, was an NSF self-evaluation mandated by the Office of Management and Budget, one of many critical reviews of RANN that ultimately led to its undoing, while *RANN 2: The Second Symposium on Research Applied to National Needs* (Washington, November 1976) continued to tout RANN's contributions to the solutions of societal problems. The final straw was NSF's so-called Whinnery Committee's *Report of the Science Applications Task Force to the Director of the National Science Foundation* (July 1977), which called for strengthening engineering and applied science but did not endorse the RANN concept.

Besides RANN, engineering's organizational equality and expanded mission, and the Engineering Research Centers effort have garnered wide attention and internal documentation. Syl McNinch contributed his typically crisp "Engineering Research Centers (ERC): How They Happened, Their Purpose, and Comments on Related Programs" (1984), and *Engineering: An Expanded and More Active Role for NSF* (1985). *Recent Research Reports* (NSF, 1981) gives abstracts of grants in the fields of NSF-sponsored research, including all the engineering divisions. This document indicates the scope of research being conducted at the time as well as the kinds of research receiving the most attention—in this case, heavy emphasis on finite element analysis and other computer work. Unfortunately, this report was not one of a series. In January 1990 NSF's Directorate for Scientific, Technological, and International Affairs produced a massive document called *The State of Academic Science and Engineering,* some of whose statistical tables and graphs have been borrowed for this history. It also included a section on historical perspective.

National Science Board offerings included its fourth annual report to Congress—mandated on some subject by the Daddario Amendments—*The Role of Engineers and Scientists in a National Policy for Technology* (Washington: GPO, 1972). Directed to the broad technical community, the focus of this report was an important recognition of engineering's expanding presence in NSF. The two-volume *Engineering Mission of NSF over the Next Decade* (NSB-83-127, 1983) provided factual and statistical information, helping the board define and promote what that mission should be. *The State of U.S. Science and Engineering: A View from the National Science Board* (1989) was likewise informative on engineering's achievements and "place."

Primary Sources: Other Institutions

Sources from other agencies that highlight NSF engineering included *Opportunities for Improved Management of the Research Applied to*

National Needs (RANN) Program (MWD-75-84, November 1975) prepared by the Office of the Comptroller General of the United States; and U.S. Congress, House, Committee on Science and Technology, Subcommittee on Special Studies, Investigations and Oversight and Subcommittee on Science, Research and Technology, *The National Science Foundation Program on Research Applied to National Needs (RANN): Goals and Objectives, an Investigative Report*, 94th Cong., 2d sess., March 1977. These reviews illustrate the intensifying criticism of RANN, even though both are relatively positive. A more recent outside report on the Foundation is from the United States General Accounting Office, *U.S. Science and Engineering Base: A Synthesis of Concerns about Budget and Policy Development* (GAO/RCED-87-65, 1987). It gives a GAO perspective on engineering support, including some historical material on how funding fared under successive administrations and how various agencies supported research and interacted with one another.

The National Academy of Engineering and, especially before the NAE was founded in 1964, the National Academy of Sciences paid close attention to developments at the National Science Foundation. NSF, in turn, consulted these institutions regularly to help set priorities, envision new directions, and evaluate ongoing programs. The resulting NAE and NAS reports and publications provide key records of Foundation ventures and accomplishments. Because of the august reputations of academy members, their words carried great weight with Foundation planners and lent credence and validity to NSF decisions based on them. Much of the NAE and/or NAS research was conducted by the National Research Council, which represented both honorary organizations and whose members participated by coveted invitation. The academies are headquartered in the same Washington building, and their library yielded many significant publications besides those found at NSF.

The National Research Council, *Federal Support of Basic Research in Institutions of Higher Learning* (Washington: NAS, 1964), has useful chapters on NSF's formative years and the post-Sputnik era. When the House Committee on Science and Astronautics asked the National Academy of Sciences for views on the proper federal role in support of research and development, the NAS requested fifteen prominent scientists to write essays. The resulting NAS publication, *Ba-*

sic Research and National Goals (1965), was an important guide for the thinking that produced the Daddario Amendments and eventually RANN. The NAS report, *Science, Government, and the Universities* (Seattle: University of Washington Press, 1966), presents the viewpoints of such leaders of the day as NAS president Frederick Seitz, presidential science adviser Donald Hornig, NSF director Leland Haworth, future NSB chair Philip Handler, and even President Lyndon Johnson, who again touted the value of more democratic distribution of federal funding for research. The National Academy of Engineering produced NSF-requested studies vital to the formation and early thrusts of the RANN program: *Federal Support of Applied Research* and *Priorities in Applied Research, An Initial Appraisal,* both in 1970; and *Priorities for Research Applicable to National Needs* in 1973. The NRC's *Social and Behavioral Science Programs in the National Science Foundation: Final Report* (Washington: NAS, 1976), the so-called Simon Report, was crucial because its faint praise of RANN's social research implicated the entire RANN effort and sealed its demise.

NAE President Robert M. White, asked to consider a new mission for NSF engineering, responded with *Strengthening Engineering in the National Science Foundation: A View from the President of the National Academy of Engineering* (July 1983). A special NAE committee evaluated the NSF Engineering Directorate's programs, both on past patterns and potential technological and human resources, in *New Directions for Engineering in the National Science Foundation* (1985). Sometimes the National Academy of Engineering did more general studies that NSF leaders consulted with interest. Its two-part probe of *Public Attitudes toward Engineering and Technology* (1986) is a good example. The National Research Council's *Engineering in Society: Engineering Education and Practice in the United States: Foundations of Our Techno-Economic Future* (Washington: National Academy Press, 1985), written from a technical, not historical, perspective, contains historical background on the development of various engineering disciplines and characteristics of contemporary engineering. It also includes comparisons of science and engineering and notes the influence of societal factors on engineering. It was followed by *Federal Actions for Improving Engineering Research and Education* in 1986. The NAE's *Directions in Engineering Research: An Assessment of Opportunities and Needs* (Washington: National Academy Press,

1987) is an excellent anticipation of new directions generally and specific engineering challenges.

Meanwhile, the National Academy of Engineering, at NSF's request, produced *Guidelines for Engineering Research Centers* (Washington: NAE, 1984). The NRC sponsored a symposium to evaluate the ERC program after its first year and published its proceedings as *The New Engineering Research Centers: Purposes, Goals, and Expectations* (Washington: National Academy Press, 1986). These discussions provided both factual information and interpretations and anticipations of key players. The NRC took another look at ERCs in 1987 in *The Engineering Research Centers: Leaders in Change* (Washington: National Academy Press, 1987), noting the accomplishments of the first six; and still another, by the NAE, *Assessment of the National Science Foundation's Engineering Research Centers Program,* appeared in 1989.

Published Sources

The classic work associated with the creation of the National Science Foundation is, of course, Vannevar Bush's *Science—The Endless Frontier: A Report to the President on a Program for Postwar Scientific Research* (1945; reprints, Washington: NSF, 1980, 1990), although it discusses engineering but little and mostly to warn against the dangers of applied research driving out the pure. (Some disputed that Bush, an engineer himself, held such an attitude. In another, unpublished, study for NSF the author explored this subject further.) Offered somewhat in rebuttal to *Science—The Endless Frontier,* a study by the President's Scientific Research Board (John Steelman, chair), *Science and Public Policy: A Report to the President* (Washington: GPO, 1947), advocated political control over science and was more in sync with Truman's views. Lyman Chalkley's "Prologue to the National Science Foundation," a manuscript in the NSF history office subject files and available through Ann Arbor, Mich.: University Microfilms International, 1979, is a corroborating source. Daniel J. Kevles's *Principles and Politics in Federal R & D Policy, 1945–1990, an Appreciation of the Bush Report* reviews NSF issues and the lasting impact of Bush's contributions as a preface to a fortieth-anniversary NSF reprint of *Science—The Endless Frontier.* Earlier, Kevles perceptively traced the politics of NSF's formation in "The National Science Foundation and the Debate over Postwar Research Policy, 1942–1945: A Political Interpretation of *Science—The Endless Frontier,*" *Isis* 68 (March 1977): 5–26. All of these confirm by silence that engineering was a minor player in the early Foundation.

Specific to NSF are two readable, reliable agency histories. J. Merton England's *A Patron for Pure Science: The National Science Foundation's Formative Years, 1945–57* (Washington: NSF, 1982) is a thorough, indeed definitive account of the early Foundation, but it covers a period when it was not necessary to say much about engineering—and it does not. It does make clear NSF's historic commitment to basic research. Milton Lomask's *A Minor Miracle: An Informal History of the National Science Foundation* (Washington: NSF, 1976) is briefer and breezier. It ends in the mid-1970s but includes trenchant chapters on Project Mohole and RANN. Mary Ellen Mogee wrote a detailed analysis of the RANN program for her master's thesis, "Public Policy and Organizational Change: The Creation of the RANN Program in the National Science Foundation" (George Washington University, 1973). Later, for the Congressional Research Service, she examined RANN after its demise in *Reorganization of the Research Applications Directorate in the National Science Foundation: The Directorate for Applied Science and Research Applications* (1978).

Broad historical studies of American engineering include Elting E. Morison, *From Know-how to Nowhere: The Development of American Technology* (New York: Basic Books, 1974), which insightfully discusses the nature of engineering as well as its historical development, including a fine chapter on John Jervis. James Kip Finch's *The Story of Engineering* (Garden City, N.Y.: Doubleday, 1960) presents specific engineers and engineering achievements and comments on the engineering/science relationship. Technology can precede science, Finch says. W. H. G. Armytage's *A Social History of Engineering* (London: Faber and Faber, 1961, 1976) addresses a British audience but offers interesting observations on American engineering and its popular reception from the early twentieth century to 1960. He calls his last chapter "The Endless Frontier." Terry S. Reynolds, ed., *The Engineer in America: A Historical Anthology from "Technology and Culture"* (Chicago: University of Chicago Press, 1991), is an excellent source on the development, concerns, and achievements of the profession—both its introductory essays on American engineers in the nineteenth and twentieth centuries and the selection of articles. *Technology and Culture,* the quarterly of the Society for the History of Technology, is the premier periodical on the history of engineering and the engineering/science relationship.

Shedding light on particular periods are Silvio A. Bedini's *Thinkers and Tinkers: Early American Men of Science* (New York: Charles Scribner's Sons, 1975), which considers developments pertinent to engineering, however informally defined, in the eighteenth century; and Brooke Hindle and Steven Lubar's *Engines of Change: The American Industrial Revolution, 1790–1860* (Washington: Smithsonian Institution Press, 1986), which discusses the nineteenth. Raymond Merritt, *Engineering in American Society, 1850–1875* (Lexington: University Press of Kentucky, 1969), shows the development of a professional consciousness among engineers during a period in the nineteenth century when the country was rapidly becoming industrialized and more urban. Nathan Reingold, ed., *The Sciences in the American Context: New Perspectives,* contains informative contributions on institutions that lent NSF modes of organization or operation: Kendall Birr on industrial research laboratories, Stanley Coben on private foundations, and Harvey M. Sapolsky on military agencies. Stuart W. Leslie, *The Cold War and American Science: The Military-Industrial-Academic Complex at MIT and Stanford* (Columbia University Press, 1993) discusses the changed climate of opportunity for postwar science at two major research institutions, particularly with Defense Department funding.

Biographies not noted elsewhere that enhanced this study include Talbot Hamlin, *Benjamin Henry Latrobe* (New York: Oxford University Press, 1955), chap. 23, "Latrobe as Engineer"; and David F. Larkin, *John B. Jervis, an American Engineering Pioneer* (Ames: Iowa State University Press, 1990). Together, these books reveal comparisons and contrasts between two differently trained early American engineers and include valuable accounts of their technical achievements and the political milieu in which they came to fruition. George Wise's *Willis R. Whitney, General Electric, and the Origin of U.S. Industrial Research* (New York: Columbia University Press, 1985) traces the beginnings of the great industrial research laboratories. Engineering advocate Eric A. Walker's memoir, *Now It's My Turn: Engineering My Way* (New York: Vantage Press, 1989), offers forthright and lively commentary on engineers and engineering and on his NSF and other leadership experiences.

Various engineering disciplines have their own histories, many of them commemorating their longevity at significant anniversaries—attesting to a burgeoning professionalism in the late nineteenth and early twentieth centuries. Consulted for specific points, these include: Daniel Hovey Calhoun, *The American Civil Engineer: Origins and Conflict* (Cambridge: MIT Press, 1960); Monte A. Calvert, *The Mechanical Engineer in America, 1830–1910: Professional Cultures in Conflict* (Baltimore: Johns Hopkins University Press, 1967); A. Michal McMahon, *The Making of a Profession: A Century of Electrical Engineering in America* (New York: Institute of Electrical and Electronics Engineers, 1984); Terry S. Reynolds, *75 Years of Progress: A History of the American Institute of Chemical Engineers* (New York: AIChE, 1983); and Bruce Sinclair, *A Centennial History of the American Society of Mechanical Engineers, 1880–1980* (Toronto: University of Toronto Press, 1980).

Some shorter treatments on these fields were also helpful. Terry S. Reynolds, "Defining Professional Boundaries: Chemical Engineering in the Early 20th Century," *Technology and Culture* (October 1986): 694–716; and Robert Arthur Rosenberg, "American Physics and the Origins of Electrical Engineering," *Physics Today* (October 1983): 48–54, detail the effects of new fields emerging from existing ones—chemical manufacture and academic physics, respectively. Stephen J. Mraz and Melissa S. Kennedy, "Engineering: 60 Years of Progress"; and Susan Gibson, "Celebrating 60 Years of Design Engineering," appear in *Machine Design* (7 September 1989): 112–27.

Defining terms precisely enough to capture the essence of engineering and to differentiate between science and engineering (and engineering's product, technology) has obsessed historians of these subjects for years, as has the mushier and more misused concept of applied research, as compared with basic (or fundamental or "pure") research. Innumerable articles tackle these meanings in one way or another in *Technology and Culture*. See, for example, James K. Feibleman in the Fall 1961 issue, C. David Gruender in July 1971, and Robert Friedel in October 1986. An excellent recent study, Ronald Kline's "Construing 'Technology' as 'Applied Science': Public Rhetoric of Scientists and Engineers in the United States, 1880–1945," *Isis* 86

(June 1995): 194–221, focuses on four historic periods, including the time of Bush's *Endless Frontier,* and concludes that despite criticisms of that model, many prominent engineers called their work applied science to enjoy by reflection the higher status of science. They and scientists have also "adjusted [the] boundaries" around their disciplines "to their rhetorical purposes."

A host of scholars have pursued the relationship of science and engineering more usefully by functional analysis than semantic argument. Eugene Ferguson explores the nature of engineering in *Engineering and the Mind's Eye* (Cambridge, Mass.: MIT Press, 1992), where he asserts that engineering involves a largely nonverbal, intuitive, and open-ended approach, with no one right answer. He charges that trying to force engineering into a deductive, exact science was a detriment to both engineering education and practice. Bringing a career in aeronautical engineering to bear on his historical investigations, Walter Vincenti analyzes engineering as a unique body of knowledge in *What Engineers Know and How They Know It: Analytical Studies from Aeronautical History* (Baltimore: Johns Hopkins University Press, 1990) and other studies. In "Control-Volume Analysis: A Difference in Thinking between Engineering and Physics," *Technology and Culture* 23 (April 1982): 145–75, Vincenti relates how engineers' approximations, which are based on boundary conditions, dissatisfy scientists as nonrigorous, but they provide practical solutions. Arthur Donovan, in "Thinking about Engineering," *Technology and Culture* 27 (October 1986): 674–79, makes the revealing statement that engineers are less clear on what engineering is than history specialists are. Vannevar Bush, in his memoir *Endless Horizons* (New York: Arno Press, 1975), discusses the qualities of the engineering profession in chapter 13, concluding that engineering is "derived jointly from the quiet cloisters of science and the turmoil and strife of aggressive business."

Intellectual excursions into the meaning of technology in society include Thomas P. Hughes's *American Genesis: A Century of Invention and Technological Enthusiasm, 1870–1970* (New York: Viking, 1989), which traces the changes from nineteenth-century celebrations of technology to a late-twentieth-century view that questions and criticizes technology as harmful to society. He develops the point of Lewis Mumford and others that values shape technology as well as the other way around. Editor Bruce Sinclair's provocative collection of essays, *New Perspectives on Technology and Culture* (Philadelphia: American Philosophical Society, 1986), includes Sinclair's own "Inventing a Genteel Tradition: MIT Crosses the River," on the nineteenth-century intellectual rivalry between Harvard and MIT; and Carroll W. Pursell's "'What the Senate Is to the American Commonwealth': A National Academy of Engineers," which traces the academy's long gestation period. *Readings in Technology and American Life* (New York: Oxford University Press, 1969), edited by Carroll W. Pursell, Jr., is a series of essays commenting on the connections between societal forces and technological change, including one by Representative Emilio Daddario in 1967.

Edwin T. Layton, Jr., stands tall among scholars investigating the science/engineering/technology relationship. Among his best contributions for this study, all published in *Technology and Culture,* are "Mirror-Image Twins: The Communities of Science and Technology in Nineteenth-Century America," 12 (October 1971): 562–80, which reviews technology's scientific revolution in the nineteenth century and how the different values of science and technology bore on the Hindsight and TRACES studies that so discomfited the Foundation in the 1960s; "American Ideologies of Science and Engineering," 17 (October 1976): 688–700, which notes that the meaning of science historically included engineering; and "Science as a Form of Action: The Role of the Engineering Sciences," 29 (January 1988): 82–97, and "Scientific Technology, 1845–1900: The Hydraulic Turbine and the Origin of American Industrial Research," 20 (January 1979): 64–89, which show that as engineers replaced craftsmen, technology became more scientific and systematic, engineering science evolved, and industrial research laboratories resulted. Layton's prizewinning *Revolt of the Engineers: Social Responsibility and the American Engineering Profession* (Cleveland: Press of Case Western Reserve University, 1972), which concentrates on the pre–World War II period, was relevant here to engineers' rise in social consciousness in the late 1960s and 1970s.

Those attempting to expand on Layton's "mirror-image twins" model of science and technology include Arie Rip, "Science and Technology as Dancing Partners," an appraisal of Derek de Solla Price's concept of that name; and Eda Kranakis, "Science and Technology as Intersecting Socio-cognitive Worlds," which sees an emerging hybridization as science and technology increasingly overlap. These papers were delivered at the conference Technological Development and Science in the 19th and 20th Centuries, University of Technology, Eindhoven, The Netherlands, 6–9 November 1991; they were published in *Technological Development and Science in the Industrial Age: New Perspectives on the Science-Technology Relationship,* Peter Kroes and Martijn Bakker, eds. (Dordrecht: Kluwer Academic Publishers, 1992).

The emergence of engineering science, that link between science and engineering that played so prominently in early NSF history, is discussed in many sources elsewhere listed but most helpfully by Layton (above) and, especially, David Channell. Channell's "The Harmony of Theory and Practice: The Engineering Science of W. J. M. Rankine," *Technology and Culture* 23 (January 1982): 39–52, lays out the nineteenth-century origin in Scotland of this approach, and his *History of Engineering Science: An Annotated Bibliography* (Garland, 1989) lists the important works on the topic.

Among others bringing a historical view to the science/engineering relationship are A. Rupert Hall in "Engineering and the Scientific Revolution," *Technology and Culture* 2 (Fall 1961): 333–40, which reports that by the seventeenth century, scientists considered mathematically innocent engineers to be "rude mechanicals."In the same issue, James Kip Finch, "Engineering and Science: A Historical Review and Appraisal," argues that only in the last century has science become useful to engineering, while engineering has long been and remains the means of turning science to social needs. Also in *Technology and Culture* are: Edward W. Constant, "Scientific Theory and Technological Testability: Science, Dynamometers, and Water Turbines in the 19th Century," 24 (April 1983): 183–98, one of many case studies in the journal; Robert Friedel, "Engineering in the 20th Century," 27 (October 1986): 669–73, an elaboration on his view that technology and engineering are not "coextensive," although "the engineer is the most prominent actor on the technological stage"; and, in the same issue, Samuel C. Florman, "An Engineer's Comment," a reminder that while historians may care about such relational arguments, most engineers devote their interest and commitment to the concrete.

As to which came first, NSF's traditional view that basic science is the wellspring from which useful application later flows has been both championed and questioned over the years. The DOD's *Project Hindsight* in the 1960s stirred a hornet's nest by challenging the importance of basic research to innovation in selected weapons systems. See Chalmers W. Sherwin and Raymond S. Isenson, *First Interim Report on Project Hindsight* (Washington: Office of the Director of Defense Research and Engineering, 1966), and Isenson's 1969 *Project Hindsight Final Report.* The Illinois Institute of Technology Research Institute prepared NSF's rebuttal, *Technology in Retrospect and Critical Events in Science (TRACES),* 2 vols. (Washington: NSF, 1968). An NSF-funded follow-up study was Columbus Laboratories, *Interactions of Science and Technology in the Innovative Process: Some Case Studies* (Columbus, Ohio: Battelle Memorial Institute, 1973). *Science* magazine followed this issue closely with numerous short articles during these years; see, for example, 18 November 1966, 23 June 1967, 26 January 1968. More recently, John P. McKelvey, in "Science and Technology: The Driven and the Driver," *Technology Review* (January 1985): 38–47, argues with numerous examples that technology contributes to scientific progress as importantly as it receives and uses knowledge from science to benefit humanity. Barry Barnes, in "The Science-Technology Relationship: A Model and a Query," *Social Studies of Science* 12 (February 1982)" 166–71, puts science and technology "on a par," their interaction "symbiotic."

Engineering education was a recurring issue for NSF and a perennial concern for the profession. An informative recent work is Lawrence P. Grayson's *The Making of an Engineer: An Illustrated History of Engineering Education in the United States and Canada* (New York: John Wiley & Sons, 1993), which chronicles educational milestones and provides biographical sketches of the major leaders through the years. See also his "Brief History of Engineering Education in the United States," *Engineering Education* (December 1977). Terry S. Reynolds and Bruce E. Seely's "Striving for Balance: A Hundred Years of the American Society for Engineering Education," *Journal of Engineering Education* (July 1993), traces the work and impact of America's oldest organization devoted to professional education. The ASEE was known from its founding in 1893 until 1946 as the Society for the Promotion of Engineering Education (SPEE).

Remarkable early efforts toward curriculum reform in engineering education are described in Samuel Rezneck, "The Engineering Profession Considers Its Educational Problems: A Forgotten Episode of the Centennial Exposition of 1876," *Association of American Colleges Bulletin* 43 (1957): 410–18. Contemporary sources include Lewis M. Haupt, "Discussion of Technical Education Developed at the

Joint Meeting in June, 1876," *American Institute of Mining Engineers Transactions* 5 (1876): 510–15; Benjamin F. Thomas, "Technical Education in Colleges and Universities," *Proceedings of the American Association for the Advancement of Science* 41 (August 1892): 67–79; and J. K. Freitag, "The World's Fair International Engineering Congress," *Engineering Magazine* 5 (1893): 485–92. Frederic T. Mavis also discusses 1876 events and earlier in "History of Engineering Education," *Journal of Engineering Education* 43 (December 1952): 214–21. In that same issue, Thorndike Saville reviews trends over the preceding fifty years in "Achievements in Engineering Education." W. Bernard Carlson, in "Academic Entrepreneurship and Engineering Education: Dugald C. Jackson and the MIT-GE Cooperative Engineering Course, 1907–1932," *Technology and Culture* 29 (July 1988): 536–44, presents the origins of industry-university cooperation, an approach acclaimed today.

The SPEE/ASEE produced a series of landmark studies of engineering education beginning with the so-called Mann Report, begun in 1907 and published finally in 1918. See Charles Riborg Mann, *A Study of Engineering Education Prepared for the Joint Committee on Engineering Education of the National Engineering Societies* (New York: Carnegie Foundation for the Advancement of Teaching, Bulletin No. 11, 1918), and his "Report of the Joint Committee on Engineering Education" in *Proceedings of the Society for the Promotion of Engineering Education* 26 (1918): 126–76. The Society for the Promotion of Engineering Education's *Report of the Investigation of Engineering Education, 1923–1929,* 2 vols. (Pittsburgh: University of Pittsburgh, 1930, 1934), was the famous Wickenden Report, which had an enormous influence and was the benchmark for almost twenty-five years. Engineering educator and contemporary Dugald C. Jackson confirmed that the Wickenden Report led faculties to examine their methods in *Present Status and Trends of Engineering Education in the United States* (New York: Engineers Council for Professional Development, 1939). Harry P. Hammond, William Wickenden's associate, conducted two additional studies: "Aim and Scope of Engineering Curricula," *Journal of Engineering Education* 30 (December 1939): 555–66, and "Engineering Education after the War," *Journal of Engineering Education* 34 (May 1944): 589–613. L. E. Grinter's "Report on the Evaluation of Engineering Education," *Journal of Engineering Education* 46

(April 1956): 25–63, urged and confirmed a growing emphasis on mathematics and science in engineering curriculum.

One of the most important studies on the subject, and surely the most controversial, was the American Society of Engineering Education's *Goals of Engineering Education: Final Report of the Goals Committee,* by E. A. Walker, J. M. Pettit, and G. A. Hawkins (Washington: ASEE, January 1968), which argued for broad, diverse engineering training in a five-year program. Eric A. Walker laid out his frustration with the earlier Wickenden Report in an address to a National Academy of Engineering Symposium, "What Should the National Academy of Engineering Do about Engineering Education?" on 13 October 1966 (NAS-NAE Library). Walker and his staff assistant Benjamin Nead defended their work in "The Goals Study: An Interpretation by the Chairman, ASEE Goals Committee," *Journal of Engineering Education* 57 (September 1966"): 13–19, and explained they were not seeking a consensus report. Vivid illustrations of the storm of protest that followed release of *The Goals of Engineering Education, The Preliminary Report* in October 1965 appeared in *Chemical Engineering Progress* 62 (February 1966): 17–49; *Civil Engineering* 36 (June 1966): 87; *Mechanical Engineering* 88 (August 1966): 32–35; and the *Journal of Engineering Education* 59 (September 1968): 21–30.

Other studies, aimed more directly at NSF, include the National Academy of Engineering's 1980 *Issues in Engineering Education: A Framework for Analysis,* which urged closer ties between academic and industrial engineers to promote the nation's economic health and productivity. NSF's Directorate of Science and Engineering Education (SEE) also did a strategic plan for education in 1985 called *Enhancing the Quality of Science Education in America: NSF's Role and Strategy, 1985–1990* (May 1985), which focused on an "energizing or catalytic" federal role. Two articles specific to electrical engineering, Frank E. Terman's "A Brief History of Electrical Engineering Education," *Proceedings IEEE* 64 (September 1976), and Robert Rosenberg's "The Origin of EE Education: A Matter of Degree," *IEEE Spectrum* (July 1984), illustrate the emergence of specialized subdisciplines as well as increasing professionalism.

Engineering research and its interwoven connections with engineering eduation are set forth in a fine recent study by Bruce Seely in "Research, Engineering, and Science in American Engineering Colleges:

1900–1960" *Technology and Culture* 33 (April 1993): 344–86. His tracing of the great changes in academic engineering over the years paints the setting for engineering in the early Foundation. Roger L. Geiger, in *To Advance Knowledge: The Growth of American Research Universities, 1900–1940* (New York: Oxford University Press, 1986), confirms the importance of land-grant colleges to engineering research, noting that in 1900 the ten largest engineering schools were land-grant institutions. He brings this valuable resource forward in *Research and Relevant Knowledge: American Research Universities since World War II* (New York: Oxford University Press, 1993). William E. Wickenden, of the education report fame, attributed the paucity of engineering research to the lack of capable researchers in "Research in the Engineering Colleges," *Mechanical Engineering* 51 (August 1929): 585–88. Confirming trends later seen clearly, Earle B. Norris, in "Research as Applied to Engineering," *Civil Engineering* 5 (May 1935), argued that engineers—supposedly applied scientists—had moved into pure science because they had "caught up" with scientists and had to do their own pioneering research for new knowledge. By 1946, Boris Bakhmeteff, in "Science and Engineering: Engineer's Scientific Approach Needed to Guide Man's Destiny," *Civil Engineering* 16 (March 1946), firmly supported basic engineering research, as a science, in the new NSF. Daniel J. Kevles, in "Foundations, Universities, and Trends in Support for the Physical and Biological Sciences, 1900–1992," *Daedalus* 121 (1992): 195–235, lays other funding background for scientific research. The Engineers Joint Council, under Eric Walker's leadership, published *The Nation's Engineering Research Needs 1965–1985,* which addresses not only research areas such as energy and mineral resources, transportation, and medicine but also engineering education in comments foreshadowing the Goals Study.

On the underrepresentation of certain groups in engineering, few studies exist until the relatively recent period when the shortages and unevenness became a concern. Some noteworthy ones are George Bugliarello, Vivian Cardwell, Olive Salembier, and Winifred White, eds., *Women in Engineering: Bridging the Gap between Society and Technology,* Proceedings of an Engineering Foundation Conference (Henniker, N.H.: New England College, July 1971), which details a profession in transition; NSF's own *Women and Minorities in Science and Engineering* (NSF-77-304, 1977), a statistical account timid about declaring causes but honestly concerned to change directions; and, also delineating the difficulties faced, Violet B. Haas and Carolyn C. Perrucci, eds., *Women in Scientific and Engineering Professions* (Ann Arbor: University of Michigan Press, 1984). The May 1988 issue of the American Society for Engineering Education's journal, *Engineering Education,* devoted itself to problems of human resources, both the overall "pipeline" problem and how to reach out especially to women and minorities.

The classic general work on the critical and growing role of the federal government is *Science in the Federal Government: A History of Policies and Activities to 1940* (New York: Harper & Row, 1957) by A. Hunter Dupree. It is an excellent source for background information but, unfortunately, entirely predates the National Science Foundation. Jeffrey K. Stine's *A History of Science Policy in the United States, 1940–1985,* Background Report No. 1 for the Task Force on Science Policy, House Committee on Science and Technology, 99th Cong., 2d sess., September 1986 (Washington: GPO, 1986), scans the pre–World War II period and carries the theme forward, its focus on "policy for basic and applied research," though explicitly not including "policy for engineering development or technology." Curiously ignoring the Daddario Amendments of 1968, this study is nevertheless a valuable overview. Science journalist Daniel S. Greenberg offers lively and penetrating insights on science and politics during NSF's first decade and a half in *The Politics of Pure Science* (New York: New American Library, 1967). This book was subsequently published in Britain as *The Politics of American Science* (Penguin Books) in 1969. Michael D. Reagan's *Science and the Federal Patron* (New York: Oxford University Press, 1969) contains a thoughtful chapter on the National Science Foundation in which the author notes the diminishing national role of the National Science Board and anticipates change on the heels of the passage of the Daddario Amendments. Ken Hechler's *Toward the Endless Frontier: History of the [House of Representatives] Committee on Science and Technology, 1959–79* (GPO, 1980) deals peripherally but corroboratively with the basic-versus-applied-research argument, as well as Congress's and OMB's influence over the National Science Foundation. Periodicals offering perceptive, reliable, up-to-date information and interpretation of science and/or engineering policy and NSF developments were *Science,* the journal of the American Association for the Advancement of Science, and Daniel Greenberg's *Science and Government Report.* NSF's magazine *Mosaic* contained relevant articles as well.

Roger L. Williams, *The Origins of Federal Support for Higher Education: George W. Atherton and the Land-Grant College Movement* (University Park: The Pennsylvania State University Press, 1991); Daniel J. Kevles, "Federal Legislation for Engineering Experiment Stations," *Technology and Culture* 12 (April 1971); and Carroll Pursell, "A Preface to Government Support of Research and Development: Research Legislation and NBS, 1935–41," *Technology and Culture* 9 (April 1968): 145–46, offered background on earlier federal forays pertinent to NSF. NASA SP-4103, Alex Roland's *Model Research: The National Advisory Committee for Aeronautics, 1915–1958,* vol. 1 (Washington: National Aeronautics and Space Administration, 1985), which asserts that Vannevar Bush's NSF concept was consciously modeled after NACA, ably describes and analyzes that model, including insights on the science/engineering interface. A study of Roosevelt's U.S. National Resources Committee, *Research—A National Resource* (New York: Arno Press, 1980, reprint of 1938 edition), argued for government funding of research noting the precedent for federal support since the nation's birth.

Todd Shallat's *Structures in the Stream: Water, Science, and the Rise of the U.S. Army Corps of Engineers* (Austin: University of Texas Press, 1994) presents an emphasis, hydraulic construction, of the Corps of Engineers, the government's oldest sustained engineering effort; its edifying introduction uses Benjamin Henry Latrobe as the embodiment of French and British antecedents of American engineering. Leland Johnson and Daniel Schaffer, *Oak Ridge National Laboratory: The First Fifty Years* (Knoxville: University of Tennessee Press, 1994), illustrates another federal involvement in engineering and science as well as the changing missions of the national laboratories. NSF funding of environmental studies at ORNL in the late 1960s, predating RANN, was the beginning of the laboratory's current energy-related environmental focus. American Public Works Association, *History of Public Works in the United States, 1776–1976,* ed. Ellis L. Armstrong (Chicago: APWA, 1976), offers a good overview and perspective on the nation's infrastructure, and the National Council on Public Works Improvement, *Fragile Foundations: A Report on America's Public*

Works, Final Report to the President and the Congress (Washington: GPO, February 1988), provides an infrastructure "report card" critical of America's "long pattern of disinvestment" and the high but necessary costs of improvement. More general, philosophical works on government and science include Sanford A. Lakoff, ed., *Knowledge and Power: Essays on Science and Government* (New York: The Free Press, 1966), which contains familiar views by Alan Waterman, NSB member Harvey Brooks, and others. Daniel Greenberg's contribution is on "Mohole: The Project That Went Awry."

Periodicals of the respective engineering disciplines gave a sense of priority areas of research over the period of NSF's history, their degree of involvement with or financial dependency on the National Science Foundation, and their reactions to NSF policy changes. Chief of these were *Chemical & Engineering News, Chemical Engineering Progress, AIChE Journal* [of the American Institute of Chemical Engineers], *Civil Engineering, Proceedings of the American Society of Civil Engineers, Minerals Today, Mechanical Engineering, Journal of Metals, Electrical Engineering,* and *IEEE Spectrum* [of the Institute of Electrical and Electronics Engineers]. John Walsh has ably covered National Science Foundation developments for many years for *Science.* Wil Lepkowski could be counted on for trenchant commentary on NSF in *C&EN.* John Adam has similarly informed *IEEE Spectrum* readers on NSF affairs in recent years.

An important aid for identifying major and representative engineering advances and then uncovering the most significant periodical literature on those topics was an NSF program evaluation: Report 1-87, *National Science Foundation Support for Significant Advances in Fundamental Engineering Research as Shown in Publication Acknowledgements 1965–1985,* August 1987 (revised 19 August 1987). The staff identified six areas of engineering—chemical, civil, mechanical, metallurgical, materials, and electrical and electronic (at a time when the Engineering Directorate was organized along functional, not disciplinary lines)—and asked the corresponding professional societies to identify the best papers reporting the most important progress. This study had the further purpose of ascertaining how much path-breaking research NSF, in fact, funded. It confirmed that NSF was the leading sponsor of academic engineering research overall, with 23 percent of the projects acknowledging NSF support.

Appropriate technical papers were gleaned by using the NSF program staff report and by screening the technical literature, identifying subject areas that generated new and productive avenues of research. By tying these research activities to what NSF supported through research grants and special programs, this history focused upon specific areas of research that both NSF and engineers practicing in the field were studying. Specific articles and reports are cited in the notes, but general subjects included the infrastructure, solar energy, air and water pollution, materials, superconductivity, enzyme catalysis, earthquake engineering, finite element analysis, bioengineering, manufacturing, transportation, computers, robotics, and ocean drilling.

Numerous articles in the professional journals revealed the tenor of operations of the fledgling National Science Foundation and how engineers viewed their relationship with it, many of them written by NSF personnel. They confirmed the prevailing view in the first decade that engineering research had to follow NSF's strict adherence to basic research. James F. Fairman addressed the question "Is Engineering a Profession?" in *Electrical Engineering* 69 (July 1950): 579–82. Paul E. Klopsteg, first MPE director, wrote about the possibilities for "Engineering Research in the Program of the National Science Foundation" in the *Journal of Engineering Education* 42 (January 1952): 263–69, while Andrey A. Potter, first and only engineer on the early National Science Board, revealed his commitment to basic research in "National Science Foundation Developments," *Journal of Engineering Education* 42 (October 1951): 81–85. NSF's first director, Alan T. Waterman, gave his views on "Government-Supported Research" in the same publication, 43 (November 1952): 153–54.

Once NSF funding for research got under way, more explanatory and philosophical articles appeared. *Chemical and Engineering News,* always a keen observer of NSF developments, presented "Developing a National Science Foundation" in vol. 31 (19 January 1953): 228–33, while "National Science Foundation Grants" were detailed in *Chemical Engineering Progress* 50 (March 1954): 50. Lee A. DuBridge, charter NSB member, offered "Fundamental Research and the Engineer" in *Chemical Engineering Progress* 50 (August 1954): 12, and "The Goals of University Research" in *Electrical Engineering* 73 (September 1954): 790–93. Because of their positions as NSF engineering program directors, Ralph Morgen's and Gene Nordby's contributions to the literature were important. See, for example, Ralph A. Morgen, "The National Science Foundation and Its Program," *Journal of Engineering Education* 44 (December 1953): 243–46; G. M. Nordby, "The National Science Foundation: Its Role in Electrical Engineering Research," *Electrical Engineering* 77 (September 1958): 782–85; Gene M. Nordby, "NSF's Basic-Engineering Research Program," *Mechanical Engineering* 81 (February 1959): 52–53; and Nordby and Robert N. Faiman, "NSF Promotes Basic Research in Civil Engineering," *Civil Engineering* 28 (April 1958): 64–67.

The professional press also followed engineers' growing insistence on greater recognition for the value of their work, whether basic or applied, starting with pieces like "NSF Act Needs Updating" in *Chemical and Engineering News* 37 (3 August 1959): 28–30. "NSF Broadens Research Program," in *Chemical Engineering Progress* 59 (January 1963): 18, followed the formation of the new engineering section. "Engineering and the National Science Foundation," an overview advocating change, appeared in the *Journal of Engineering Education* 53 (January 1963): 278. Charter NSB member Paul M. Gross gave a 1960s perspective in "R & D, and the Relations of Science and Government" in *Science* 142 (8 November 1963): 645–50. NSF director Leland J. Haworth pushed his view that engineering was important but engineers must submit better proposals in "In Support of Engineering Excellence," *Proceedings of the IEEE* 52 (January 1964). Herbert J. Holloman, who persistently needled the Foundation to respond to real-world needs, did so, for example, in "A National Science and Technology Policy," *IEEE Spectrum* 1 (August 1964): 89–97. "NSF May Be More Aggressive in the Future" was a hopeful commentary in *Chemical and Engineering News* 43 (5 July 1965): 26–27.

Congressman Emilio Q. Daddario set the stage for a new era for NSF in "A Revised Charter for the Science Foundation" in the widely read *Science* (1 April 1966). Other writers picking up on the new applied research theme and its implications were D. S. Greenberg in "Research Priorities: New Program at NSF Reflects Shift in Values," *Science* 170 (2 October 1970): 144–46; Robert J. Bazell in "NSF: Is Applied Research at the Take-Off Point?" *Science* 172 (25 June 1971): 1315–17; James J. Andover in "Revamping NSF," *IEEE Spectrum* 9 (October 1972): 78–79; and W. R. Marshall, Jr., in "Social Directions of Engineering," *Chemical Engineering Progress* 67 (January 1971):

11–16. Specifically and perceptively assessing the RANN program was "RANN: Growth at NSF stirs concern, but . . ." in *Science & Government Report* 2 (15 July 1972): 1–4.

During the 1980s many NSF leaders looked to engineering as the nation's best hope for confronting America's worsening international economic competitiveness posture. Press coverage was extensive and instructive on efforts at NSF to create a new, stronger mission for engineering, for example, Donald Christiansen, "Science, Technology, and the Government," *IEEE Spectrum* 18 (July 1981): 23; "Engineering Is No Threat to the Funding of Science," *Electronics* (19 April 1984); and Kathryn W. Hickerson, "Revised NSF Charter Needed to Meet Long-Term Goals for U.S. Engineering Support," *Engineering Times* (January 1984): 1–2. *C&EN* reporter Wil Lepkowski contributed "National Science Foundation Emphasizes Engineering Research" to *Chemistry & Industry* (2 July 1984) and "NSF Mulls Reorganizing Applied Science" in *Chemical & Engineering News* 58 (18 August 1980): 17–19. He probably also wrote the unattributed *C&EN* piece "NSF May Restructure Engineering Directorate" of 3 December 1984: 7–8. Kim McDonald wrote a series of thoughtful, pertinent articles for *The Chronicle of Higher Education:* "Engineering Research to Get More Emphasis in National Science Foundation's Programs" (20 September 1983); "National Science Foundation Starts to Broaden Support of Engineering Research" (18 January 1984); and "Engineering Deans Ask Congress to Give Their Field Equal Standing with Science in Mission of NSF" (29 February 1984). Among John Walsh's pieces in *Science* were "NSF under Challenge from Congress, Engineers" 209 (26 September 1980): 1499; "NSF Boosts Engineering, Applied Research," 210 (5 December 1980): 1105–6; "NSF Seeks Expanded Role in Engineering," 222 (9 December 1983): 1101–2; and "Writing Engineering's Ticket at NSF," 224 (6 April 1984).

As Congress debated the specific insertion of engineering into the NSF Act, government-oriented publications took notice, for example, "NSF Engineering Activities to Receive New Direction" in *Legislative Reports* 5 (30 September 1983) and "NSF Gears Up for Big New Engineering Program" in *Science & Government Report* 14 (1 November 1984): 1–5. NSF board chair Lewis Branscomb and National Academy of Sciences president Frank Press, disagreeing on the wisdom of tampering with the statutory wording, exchanged editorials in

Science, 13 and 27 April 1984. "Press, Branscomb Debate Merits of NSF Charter Changes," in *Engineering Education News* 10 (June 1984): 1, continued the discussion.

Finally, milestone anniversaries provided opportunities to review NSF progress, and several thoughtful critiques have appeared over the years, such as White House science advisor Donald F. Hornig's "National Science Foundation" in *Chemical and Engineering News* 43 (5 July 1965): 62–65, a fifteenth-anniversary assessment. Board member Detlev Bronk's "The National Science Foundation: Origins, Hopes, and Aspirations," in *Science* 188 (2 May 1975): 409–14, discussed the Foundation at twenty-five years—especially its evolving mission and the impact of applied research. Another former NSB member, William A. Fowler, contributed more silver-anniversary observations, "A Foundation for Research," to *Science* 188 (2 May 1975): 414–19. Highlights of NSF's fortieth-anniversary observance are captured in NSF, "Great Achievements, Great Expectations," 40th Anniversary Symposium: A Report, 11 May 1990, including NRC executive officer Philip M. Smith's "The National Science Foundation as History."

INDEX

ACERC (Advanced Combustion Engineering Research Center), as ERC program, 224–28, 230

Advanced Research Project Agency (ARPA) (DOD), Interdisciplinary Laboratories system, 166

Advanced Technology Applications . (RANN division), 98, 103

Advisory Committee for Engineering, 84–85, 88; as new name for Engineering Divisional Committee, 70; and IRPOS program. 89–90, 92–93; on NSF's interest in RANN, 102; suggested name change for NSF, 122–23

African Americans, degree holders in engineering, 205–6, 209

airplane design, 50, 174

Alaska earthquake (1964), 156

Allied Chemical Corporation, 165–66

aluminum, corrosion studies, 165

American Association for the Advancement of Science, 100

American Association of Engineering Societies (AAES), 142–43, 186, 188, 201, 236

American Association of Land-Grant Colleges and State Universities, 57; and engineering experiment stations, 29–30

American Association of State Highway Transportation Officials, 158

American Institute of Chemical Engineers (AIChE), 7, 60

American Institute of Electrical Engineers, 7

American Institute of Mining and Metallurgical Engineers, 7

American Literary, Scientific and Military Academy (later Norwich University), 9

American Motors, control by French government, 190

American Society for Engineering Education (ASEE), 30, 46, 142, 182–83, 186, 203; *Goals of Engineering Education,* 1968 study, 206; report with AAES on engineering, 187–89

American Society of Civil Engineers (ASCE), 7, 60, 123, 180; Underground Construction Research Council, 114

American Society of Mechanical Engineers (ASME), 7, 87

amorphous solids, as research area, 165

Antarctic research program, 50

antiknock compounds, in gasoline, 111

antiscience climate, effect on NSF's mission, 86–89

applied research: vs. basic research, 144–45, 147, 265–66; COPEP's studies for RANN program, 95–96; and Daddario Amendments, 78–80, 260

applied science: and Congress, 33; in land-grant schools of engineering, 16; vs. pure science, 1, 9, 79, 185; and NSF's restrictions, 61

Applied Science and Research Applications (ASRA), 130–34, 135–36

apprentice-training, 3, 4

Armenia, 1988 earthquake, 163–64

artillery, French engineers as designers, 3

ASEE/AAES reports, on engineering, 187–89

Asians, in engineering and science, 207

Astronomical, Atmospheric, Earth and Ocean Sciences, as NSF directorate, 126, 131

Atkinson, Richard, 118–19, 121, 123, 127, 143, 179; and NTF challenge, 139–40, 144; role in ASRA, 130, 132, 135; views on applied science and engineering, 133, 153

atomic bomb, 29

Atomic Energy Commission (AEC), 32, 36, 55

automation, use of computers, 170–71

automobiles: cooperative university/industry research program on, 138; in underground expressways, 115–16;

efforts, 121; status in NSF, 56, 78
societal problems, research, 256; application of electrotechnology, 99; as driving external force in NSF studies, 258, 261; NSF's creation of IRRPOS, 89–93, 97
Society for the Promotion of Engineering Education (SPEE), 13–14, 30, 181
soil liquefaction, 157, 173
solar cells, gallium arsenide, 110
solar energy, 40; RANN projects, 106–10, 126
solar homes, experimental projects, 108, 305 n.5
solid-state physics, 45
Solla Price, Derek de, 196
South Korea, Republic of, invasion of, 35, 37
Soviet Union: detonation of nuclear device, 35; output of scientists and engineers, 42, 179; research advancements, 88–89; *Sputnik* launch, 47
Spaulding, William, 184
Special Interdepartmental Committee on the Training of Scientists and Engineers (1950s), 42, 179
Sputnik launch, 47–51, 77, 79, 258
Stanford Synchrotron Radiation Laboratory, 165, 167
Stanford University, 175
State of Academic Science and Engineering, The (NSF study, 1990), 261–62
State University of New York, Buffalo (SUNY), 163
status, engineering: academic, 43–47, 185; social, 41–42, 187–88
steel, 160, 254–55
Steelman Report, 1947 study, *Science and Public Policy,* 33
Steinmetz, Charles, 11
Stevens Institute of Technology, 10–11
Stever, H. Guyford, 85–86, 123, 221, 259; and dissolution of RANN, 125, 126; role in RANN, 103, 107–8, 116, 117, 118
Stöhr, Joachim, 165

Strengthening American Science (1958 PSAC report), 49
Strengthening Engineering in the National Science Foundation (NAE report), 196–97
stress, structural: and Rankin, 12; and prestressed concrete, 63, 254–55
structures: Rankin's stress studies, 12; wearability over time, 255–56; U.S.-Japan cooperative earthquake studies, 160. *See also* buildings
Strutt, John William (Lord Rayleigh), 165
submicron components, Cornell's study of, 134, 168–69
subway systems, 116
Suh, Nam, 185, 200, 260; restructuring of NSF's engineering sector, 242–47; role in ERCs, 215, 217, 221, 222, 223–24, 230, 233
superconductivity, research challenges, 252–54
surfactants, chemical, in tunneling projects, 113
Symington, James, 160
system design and analysis, 55
systems engineering: 60, 65, 68, 107–8, 111, 116, 170; in education, 185, 197, 213, 215, 218, 219, 231; and NSF approach, 243, 245–46, 248, 262

"teaming," vs. lone investigator, 263
technology: challenges for the NSF, 250–58; national interest and Congress's concepts of research, 77–78; 19th-century America revolution, 5; role of system design and analysis, 55; on early frontier, 2; view that new flows from old, 82. *See also* Japan
Technology in Retrospect and Critical Events in Science (TRACES) (NSF 1969 study), 82
Tennessee Valley Authority, 19–20
Terman, Frederick, 143